T0211671

Springer Undergraduate Texts in Mathematics and Technology

Springer Undergraduate Texts in Mathematics and Technology (SUMAT) publishes textbooks aimed primarily at the undergraduate. Each text is designed principally for students who are considering careers either in the mathematical sciences or in technology-based areas such as engineering, finance, information technology and computer science, bioscience and medicine, optimization or industry. Texts aim to be accessible introductions to a wide range of core mathematical disciplines and their practical, real-world applications; and are fashioned both for course use and for independent study.

More information about this series at http://www.springer.com/series/7438

Fady Alajaji · Po-Ning Chen

An Introduction to Single-User Information Theory

 Springer

Fady Alajaji
Department of Mathematics
  and Statistics
Queen's University
Kingston, ON
Canada

Po-Ning Chen
Department of Electrical
  and Computer Engineering
National Chiao Tung University
Hsinchu
Taiwan, Republic of China

ISSN 1867-5506   ISSN 1867-5514 (electronic)
Springer Undergraduate Texts in Mathematics and Technology
ISBN 978-981-13-4038-3   ISBN 978-981-10-8001-2 (eBook)
https://doi.org/10.1007/978-981-10-8001-2

Mathematics Subject Classification: 94-XX, 60-XX, 68-XX, 62-XX

Printed on acid-free paper

This Springer imprint is published by the registered company Springer Nature Singapore Pte Ltd.
part of Springer Nature
The registered company address is: 152 Beach Road, #21-01/04 Gateway East, Singapore 189721,
Singapore

Preface

The reliable transmission and processing of information bearing signals over a noisy communication channel is at the heart of what we call communication. *Information theory*—founded by Claude E. Shannon in 1948 [340]—provides a mathematical framework for the theory of communication. It describes the *fundamental limits* to how efficiently one can encode information and still be able to recover it at the destination either with negligible loss or within a prescribed distortion threshold.

The purpose of this textbook is to present a concise, yet mathematically rigorous, introduction to the main pillars of information theory. It thus naturally focuses on the foundational concepts and indispensable results of the subject for single-user systems, where a single data source or message needs to be reliably processed and communicated over a noiseless or noisy point-to-point channel. The book consists of five meticulously drafted core chapters (with accompanying problems), emphasizing the key topics of information measures, lossless and lossy data compression, channel coding, and joint source–channel coding. Two appendices covering necessary and supplementary material in real analysis and in probability and stochastic processes are included and a comprehensive instructor's solutions manual is separately available. The book is well suited for a single-term first course on information theory, ranging from 10 to 15 weeks, offered to senior undergraduate and entry-level graduate students in mathematics, statistics, engineering, and computing and information sciences.

The textbook grew out of lecture notes we developed while teaching information theory over the last 20 years to students in applied mathematics, statistics, and engineering at our home universities. Over the years teaching the subject, we realized that standard textbooks, some of them undeniably outstanding, tend to cover a large amount of material (including advanced topics), which can be overwhelming and inaccessible to debutant students. They also do not always provide all the necessary technical details in the proofs of the main results. We hope that this book fills these needs by virtue of being succinct and mathematically precise and that it helps beginners acquire a profound understanding of the fundamental elements of the subject.

The textbook aims at providing a coherent introduction of the primary principles of single-user information theory. All the main Shannon coding theorems are proved in full detail (without skipping important steps) using a consistent approach based on the law of large numbers, or equivalently, the *asymptotic equipartition property* (AEP). A brief description of the topics of each chapter follows.

- Chapter 2: Information measures for discrete systems and their properties: self-information, entropy, mutual information and divergence, data processing theorem, Fano's inequality, Pinsker's inequality, simple hypothesis testing, the Neyman–Pearson lemma, the Chernoff–Stein lemma, and Rényi's information measures.
- Chapter 3: Fundamentals of lossless source coding (data compression): discrete memoryless sources, fixed-length (block) codes for asymptotically lossless compression, AEP, fixed-length source coding theorems for memoryless and stationary ergodic sources, entropy rate and redundancy, variable-length codes for lossless compression, variable-length source coding theorems for memory-less and stationary sources, prefix codes, Kraft inequality, Huffman codes, Shannon–Fano–Elias codes, and Lempel–Ziv codes.
- Chapter 4: Fundamentals of channel coding: discrete memoryless channels, block codes for data transmission, channel capacity, coding theorem for discrete memoryless channels, example of polar codes, calculation of channel capacity, channels with symmetric structures, lossless joint source–channel coding, and Shannon's separation principle.
- Chapter 5: Information measures for continuous alphabet systems and Gaussian channels: differential entropy, mutual information and divergence, AEP for continuous memoryless sources, capacity and channel coding theorem of discrete-time memoryless Gaussian channels, capacity of uncorrelated parallel Gaussian channels and the water-filling principle, capacity of correlated Gaussian channels, non-Gaussian discrete-time memoryless channels, and capacity of band-limited (continuous-time) white Gaussian channels.
- Chapter 6: Fundamentals of lossy source coding and joint source–channel coding: distortion measures, rate–distortion theorem for memoryless sources, rate–distortion theorem for stationary ergodic sources, rate–distortion function and its properties, rate–distortion function for memoryless Gaussian and Laplacian sources, lossy joint source–channel coding theorem, and Shannon limit of communication systems.
- Appendix A: Overview on suprema and limits.
- Appendix B: Overview in probability and random processes. Random variables and stochastic processes, statistical properties of random processes, Markov chains, convergence of sequences of random variables, ergodicity and laws of large numbers, central limit theorem, concavity and convexity, Jensen's inequality, Lagrange multipliers, and the Karush–Kuhn–Tucker (KKT) conditions for constrained optimization problems.

We are very much indebted to all readers, including many students, who provided valuable feedback. Special thanks are devoted to Yunghsiang S. Han, Yu-Chih Huang, Tamás Linder, Stefan M. Moser, and Vincent Y. F. Tan; their insightful and incisive comments greatly benefited the manuscript. We also thank all anonymous reviewers for their constructive and detailed criticism. Finally, we sincerely thank all our mentors and colleagues who immeasurably and positively impacted our understanding of and fondness for the field of information theory, including Lorne L. Campbell, Imre Csiszár, Lee D. Davisson, Nariman Farvardin, Thomas E. Fuja, Te Sun Han, Tamás Linder, Prakash Narayan, Adrian Papamarcou, Nam Phamdo, Mikael Skoglund, and Sergio Verdú.

Remarks to the reader: In the text, all assumptions, claims, conjectures, corollaries, definitions, examples, exercises, lemmas, observations, properties, and theorems are numbered under the same counter. For example, a lemma that immediately follows Theorem 2.1 is numbered as Lemma 2.2, instead of Lemma 2.1. Readers are welcome to submit comments to *fa@queensu.ca* or to *poning@faculty.nctu.edu.tw*.

Kingston, ON, Canada   Fady Alajaji
Hsinchu, Taiwan, Republic of China   Po-Ning Chen
2018

Acknowledgements

Thanks are given to our families for their full support during the period of writing this textbook.

Contents

List of Figures

List of Tables

Chapter 1
Introduction

1.1 Overview

Since its inception, the main role of *information theory* has been to provide the engineering and scientific communities with a mathematical framework for the theory of communication by establishing the *fundamental limits* on the performance of various communication systems. The birth of information theory was initiated with the publication of the groundbreaking works [340, 346] of Claude Elwood Shannon (1916–2001) who asserted that it is possible to send information-bearing signals at a *fixed positive rate* through a noisy communication channel with an arbitrarily small probability of error as long as the transmission rate is below a certain fixed quantity that depends on the channel statistical characteristics; he "named" this quantity *channel capacity*. He further proclaimed that random (stochastic) sources, representing data, speech or image signals, can be compressed distortion-free at a minimal rate given by the source's intrinsic amount of information, which he called *source entropy*[1] and defined in terms of the source statistics. He went on proving that if a source has an entropy that is less than the capacity of a communication channel, then the source can be reliably transmitted (with asymptotically vanishing probability of error) over the channel. He further generalized these "coding theorems" from the lossless (distortionless) to the lossy context where the source can be compressed and reproduced (possibly after channel transmission) within a tolerable distortion threshold [345].

---

[1] Shannon borrowed the term "entropy" from statistical mechanics since his quantity admits the same expression as Boltzmann's entropy [55].

© Springer Nature Singapore Pte Ltd. 2018

F. Alajaji and P.-N. Chen, *An Introduction to Single-User Information Theory*,
Springer Undergraduate Texts in Mathematics and Technology,
https://doi.org/10.1007/978-981-10-8001-2_1

Inspired and guided by the pioneering ideas of Shannon,[2] information theorists gradually expanded their interests beyond communication theory, and investigated fundamental questions in several other related fields. Among them we cite:

- statistical physics (thermodynamics, quantum information theory);
- computing and information sciences (distributed processing, compression, algorithmic complexity, resolvability);
- probability theory (large deviations, limit theorems, Markov decision processes);
- statistics (hypothesis testing, multiuser detection, Fisher information, estimation);
- stochastic control (control under communication constraints, stochastic optimization);
- economics (game theory, team decision theory, gambling theory, investment theory);
- mathematical biology (biological information theory, bioinformatics);
- information hiding, data security and privacy;
- data networks (network epidemics, self-similarity, traffic regulation theory);
- machine learning (deep neural networks, data analytics).

In this textbook, we focus our attention on the study of the basic theory of communication for single-user (point-to-point) systems for which information theory was originally conceived.

1.2   Communication System Model

A simple block diagram of a general communication system is depicted in Fig. 1.1. Let us briefly describe the role of each block in the figure.

- **Source**: The source, which usually represents data or multimedia signals, is modeled as a random process (an introduction to random processes is provided in Appendix B). It can be discrete (finite or countable alphabet) or continuous (uncountable alphabet) in value and in time.
- **Source Encoder**: Its role is to represent the source in a compact fashion by removing its unnecessary or redundant content (i.e., by compressing it).
- **Channel Encoder**: Its role is to enable the reliable reproduction of the source encoder output after its transmission through a noisy communication channel. This is achieved by adding redundancy (using usually an algebraic structure) to the source encoder output.

---

[2]See [359] for accessing most of Shannon's works, including his master's thesis [337, 338] which made a breakthrough connection between electrical switching circuits and Boolean algebra and played a catalyst role in the digital revolution, his dissertation on an algebraic framework for population genetics [339], and his seminal paper on information-theoretic cryptography [342]. Refer also to [362] for a recent (nontechnical) biography on Shannon and [146] for a broad discourse on the history of information and on the information age.

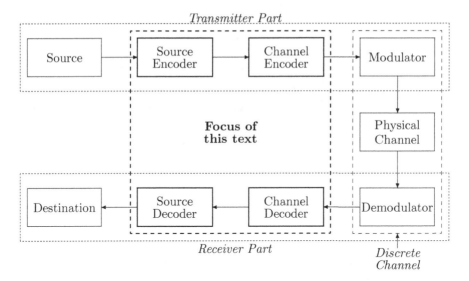

Fig. 1.1 Block diagram of a general communication system

- **Modulator**: It transforms the channel encoder output into a waveform suitable for transmission over the physical channel. This is typically accomplished by varying the parameters of a sinusoidal signal in proportion with the data provided by the channel encoder output.
- **Physical Channel**: It consists of the noisy (or unreliable) medium that the transmitted waveform traverses. It is usually modeled via a sequence of conditional (or transition) probability distributions of receiving an output given that a specific input was sent.
- **Receiver Part**: It consists of the *demodulator*, the *channel decoder*, and the *source decoder* where the reverse operations are performed. The *destination* represents the sink where the source estimate provided by the source decoder is reproduced.

In this text, we will model the concatenation of the modulator, physical channel, and demodulator via a discrete-time[3] channel with a given sequence of conditional probability distributions. Given a source and a discrete channel, our objectives will include determining the fundamental limits of how well we can construct a (source/channel) coding scheme so that:

- the smallest number of source encoder symbols can represent each source symbol distortion-free or within a prescribed distortion level D, where $D > 0$ and the channel is noiseless;

---

[3] Except for a brief interlude with the continuous-time (waveform) Gaussian channel in Chap. 5, we will consider discrete-time communication systems throughout the text.

- the largest rate of information can be transmitted over a noisy channel between the channel encoder input and the channel decoder output with an arbitrarily small probability of decoding error;
- we can guarantee that the source is transmitted over a noisy channel and reproduced at the destination within distortion D, where $D > 0$.

We refer the reader to Appendix A for the necessary background on suprema and limits; in particular, Observation A.5 (resp. Observation A.11) provides a pertinent connection between the supremum (resp., infimum) of a set and the proof of a typical channel coding (resp., source coding) theorem in information theory. Finally, Appendix B provides an overview of basic concepts from probability theory and the theory of random processes that are used in the text. The appendix also contains a brief discussion of convexity, Jensen's inequality and the Lagrange multipliers constrained optimization technique.

Chapter 2
Information Measures for Discrete Systems

In this chapter, we define Shannon's information measures[1] for discrete-time discrete-alphabet[2] systems from a probabilistic standpoint and develop their properties. Elucidating the operational significance of probabilistically defined information measures vis-a-vis the fundamental limits of coding constitutes a main objective of this book; this will be seen in the subsequent chapters.

2.1 Entropy, Joint Entropy, and Conditional Entropy

2.1.1 Self-information

Let E be an event belonging to a given event space and having probability $p_E :=$ $\Pr(E)$, where $0 \leq p_E \leq 1$. Let $\mathcal{I}(E)$ – called the *self-information* of event E [114, 135] – represent *the amount of information one gains when learning that E has occurred (or equivalently, the amount of uncertainty one had about E prior to learning that it has happened)*. A natural question to ask is "what properties should $\mathcal{I}(E)$ have?" Although the answer to this question may vary from person to person, here are some properties that $\mathcal{I}(E)$ is reasonably expected to have.

1. $\mathcal{I}(E)$ should be a decreasing function of p_E.
   In other words, this property states that $\mathcal{I}(E) = I(p_E)$, where $I(\cdot)$ is a real-valued function defined over $[0, 1]$. Furthermore, one would expect that the less likely

---

[1]More specifically, Shannon introduced the entropy, conditional entropy, and mutual information measures [340], while divergence is due to Kullback and Leibler [236, 237].

[2]By discrete alphabets, one usually means finite or countably infinite alphabets. We however focus mostly on finite-alphabet systems, although the presented information measures allow for countable alphabets (when they exist).

© Springer Nature Singapore Pte Ltd. 2018
F. Alajaji and P.-N. Chen, *An Introduction to Single-User Information Theory*,
Springer Undergraduate Texts in Mathematics and Technology,
https://doi.org/10.1007/978-981-10-8001-2_2

event E is, the more information is gained when one learns it has occurred. In other words, $I(p_E)$ is a decreasing function of p_E.

2. $I(p_E)$ should be continuous in p_E.

Intuitively, one should expect that a small change in p_E corresponds to a small change in the amount of information carried by E.

3. If E_1 and E_2 are independent events, then $\mathcal{I}(E_1 \cap E_2) = \mathcal{I}(E_1) + \mathcal{I}(E_2)$, or equivalently, $I(p_{E_1} \times p_{E_2}) = I(p_{E_1}) + I(p_{E_2})$.

This property declares that when events E_1 and E_2 are independent of each other (i.e., when they do not affect each other probabilistically), the amount of information one gains by learning that both events have jointly occurred should be equal to the sum of the amounts of information of each individual event.

Next, we show that the only function that satisfies Properties 1–3 above is the logarithmic function.

Theorem 2.1 *The* only *function defined over $p \in [0, 1]$ and satisfying*

1. *$I(p)$ is monotonically decreasing in p;*
2. *$I(p)$ is a continuous function of p for $0 \le p \le 1$;*
3. *$I(p_1 \times p_2) = I(p_1) + I(p_2)$;*

is $I(p) = -c \cdot \log_b(p)$, where c is a positive constant and the base b of the logarithm is any number larger than one.

Proof   **Step 1: Claim.** For $n = 1, 2, 3, \ldots$,

$$I\left(\frac{1}{n}\right) = -c \cdot \log_b\left(\frac{1}{n}\right),$$

where $c > 0$ is a constant.

Proof: First note that for $n = 1$, Condition 3 directly shows the claim, since it yields that $I(1) = I(1) + I(1)$. Thus $I(1) = 0 = -c \log_b(1)$.

Now let n be a fixed positive integer greater than 1. Conditions 1 and 3 respectively imply

$$n < m \implies I\left(\frac{1}{n}\right) < I\left(\frac{1}{m}\right) \tag{2.1.1}$$

and

$$I\left(\frac{1}{mn}\right) = I\left(\frac{1}{m}\right) + I\left(\frac{1}{n}\right), \tag{2.1.2}$$

where $n, m = 1, 2, 3, \ldots$. Now using (2.1.2), we can show by induction (on k) that

$$I\left(\frac{1}{n^k}\right) = k \cdot I\left(\frac{1}{n}\right) \tag{2.1.3}$$

for all nonnegative integers k.

Now for any positive integer r, there exists a nonnegative integer k such that

$$n^k \leq 2^r < n^{k+1}.$$

By (2.1.1), we obtain

$$I\left(\frac{1}{n^k}\right) \leq I\left(\frac{1}{2^r}\right) < I\left(\frac{1}{n^{k+1}}\right),$$

which together with (2.1.3), yields

$$k \cdot I\left(\frac{1}{n}\right) \leq r \cdot I\left(\frac{1}{2}\right) < (k+1) \cdot I\left(\frac{1}{n}\right).$$

Hence, since $I(1/n) > I(1) = 0$,

$$\frac{k}{r} \leq \frac{I(1/2)}{I(1/n)} \leq \frac{k+1}{r}.$$

On the other hand, by the monotonicity of the logarithm, we obtain

$$\log_b n^k \leq \log_b 2^r \leq \log_b n^{k+1} \iff \frac{k}{r} \leq \frac{\log_b(2)}{\log_b(n)} \leq \frac{k+1}{r}.$$

Therefore,

$$\left| \frac{\log_b(2)}{\log_b(n)} - \frac{I(1/2)}{I(1/n)} \right| < \frac{1}{r}.$$

Since n is fixed, and r can be made arbitrarily large, we can let $r \to \infty$ to get

$$I\left(\frac{1}{n}\right) = c \cdot \log_b(n),$$

where $c = I(1/2)/\log_b(2) > 0$. This completes the proof of the claim.
Step 2: Claim. $I(p) = -c \cdot \log_b(p)$ for positive rational number p, where $c > 0$ is a constant.

Proof: A positive rational number p can be represented by a ratio of two integers, i.e., $p = r/s$, where r and s are both positive integers. Then condition 3 yields that

$$I\left(\frac{1}{s}\right) = I\left(\frac{r}{s}\frac{1}{r}\right) = I\left(\frac{r}{s}\right) + I\left(\frac{1}{r}\right),$$

which, from Step 1, implies that

$$I(p) = I\left(\frac{r}{s}\right) = I\left(\frac{1}{s}\right) - I\left(\frac{1}{r}\right) = c \cdot \log_b s - c \cdot \log_b r = -c \cdot \log_b p.$$

Step 3: For any $p \in [0, 1]$, it follows by continuity and the density of the rationals in the reals that

$$I(p) = \lim_{a\uparrow p,\ a\ \text{rational}} I(a) = \lim_{b\downarrow p,\ b\ \text{rational}} I(b) = -c \cdot \log_b(p).$$

$\square$

The constant c above is by convention normalized to $c = 1$. Furthermore, the base b of the logarithm determines the type of units used in measuring information. When $b = 2$, the amount of information is expressed in *bits* (i.e., *binary digits*). When $b = e$ – i.e., the natural logarithm (ln) is used – information is measured in *nats* (i.e., *natural units or digits*). For example, if the event E concerns a Heads outcome from the toss of a fair coin, then its self-information is $\mathcal{I}(E) = -\log_2(1/2) = 1$ bit or $-\ln(1/2) = 0.693$ nats.

More generally, under base $b > 1$, information is in *b-ary units or digits*. For the sake of simplicity, we will use the base-2 logarithm throughout unless otherwise specified. Note that one can easily convert information units from bits to b-ary units by dividing the former by $\log_2(b)$.

2.1.2   Entropy

Let X be a discrete random variable taking values in a finite alphabet $\mathcal{X}$ under a probability distribution or probability mass function (pmf) $P_X(x) := P[X = x]$ for all $x \in \mathcal{X}$. Note that X generically represents a memoryless source, which is a discrete-time random process $\{X_n\}_{n=1}^{\infty}$ with independent and identically distributed (i.i.d.) random variables (cf. Appendix B).[3]

Definition 2.2 (*Entropy*) The entropy of a discrete random variable X with pmf $P_X(\cdot)$ is denoted by $H(X)$ or $H(P_X)$ and defined by

$$H(X) := -\sum_{x \in \mathcal{X}} P_X(x) \cdot \log_2 P_X(x) \quad \text{(bits)}.$$

Thus, $H(X)$ represents the *statistical average (mean)* amount of information one gains when learning that one of its $|\mathcal{X}|$ outcomes has occurred, where $|\mathcal{X}|$ denotes the size of alphabet $\mathcal{X}$. Indeed, we directly note from the definition that

---

[3]We will interchangeably use the notations $\{X_n\}_{n=1}^{\infty}$ and $\{X_n\}$ to denote discrete-time random processes.

$$H(X) = E[-\log_2 P_X(X)] = E[\mathcal{I}(X)],$$

where $\mathcal{I}(x) := -\log_2 P_X(x)$ is the self-information of the elementary event $\{X = x\}$. When computing the entropy, we adopt the convention

$$0 \cdot \log_2 0 = 0,$$

which can be justified by a continuity argument since $x \log_2 x \to 0$ as $x \to 0$. Also note that $H(X)$ only depends on the probability distribution of X and is not affected by the symbols that represent the outcomes. For example when tossing a fair coin, we can denote Heads by 2 (instead of 1) and Tail by 100 (instead of 0), and the entropy of the random variable representing the outcome would remain equal to $\log_2(2) = 1$ bit.

Example 2.3 Let X be a binary (valued) random variable with alphabet $\mathcal{X} = \{0, 1\}$ and pmf given by $P_X(1) = p$ and $P_X(0) = 1 - p$, where $0 \le p \le 1$ is fixed. Then $H(X) = -p \cdot \log_2 p - (1 - p) \cdot \log_2(1 - p)$. This entropy is conveniently called the *binary entropy function* and is usually denoted by $h_b(p)$: it is illustrated in Fig. 2.1. As shown in the figure, $h_b(p)$ is maximized for a uniform distribution (i.e., $p = 1/2$).

The units for $H(X)$ above are in bits as base-2 logarithm is used. Setting

$$H_D(X) := -\sum_{x \in \mathcal{X}} P_X(x) \cdot \log_D P_X(x)$$

yields the entropy in D-ary units, where $D > 1$. Note that we abbreviate $H_2(X)$ as $H(X)$ throughout the book since bits are common measure units for a coding system, and hence

$$H_D(X) = \frac{H(X)}{\log_2 D}.$$

Fig. 2.1 Binary entropy function $h_b(p)$

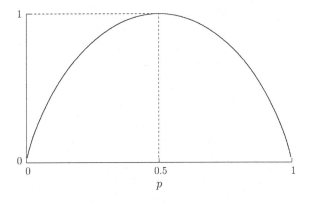

Thus

$$H_e(X) = \frac{H(X)}{\log_2(e)} = (\ln 2) \cdot H(X)$$

gives the entropy in nats, where e is the base of the natural logarithm.

2.1.3 Properties of Entropy

When developing or proving the basic properties of entropy (and other information measures), we will often use the following fundamental inequality for the logarithm(its proof is left as an exercise).

Lemma 2.4 (Fundamental inequality (FI)) *For any $x > 0$ and $D > 1$, we have that*

$$\log_D(x) \le \log_D(e) \cdot (x - 1),$$

with equality if and only if (iff) $x = 1$.

Setting $y = 1/x$ and using FI above directly yield that for any $y > 0$, we also have that

$$\log_D(y) \ge \log_D(e) \left(1 - \frac{1}{y}\right),$$

also with equality iff $y = 1$. In the above the base-D logarithm was used. Specifically, for a logarithm with base-2, the above inequalities become

$$\log_2(e) \left(1 - \frac{1}{x}\right) \le \log_2(x) \le \log_2(e) \cdot (x - 1),$$

with equality iff $x = 1$.

Lemma 2.5 (Nonnegativity) *We have $H(X) \ge 0$. Equality holds iff X is deterministic (when X is deterministic, the uncertainty of X is obviously zero).*

Proof $0 \le P_X(x) \le 1$ implies that $\log_2[1/P_X(x)] \ge 0$ for every $x \in \mathcal{X}$. Hence,

$$H(X) = \sum_{x \in \mathcal{X}} P_X(x) \log_2 \frac{1}{P_X(x)} \ge 0,$$

with equality holding iff $P_X(x) = 1$ for some $x \in \mathcal{X}$. $\square$

Lemma 2.6 (Upper bound on entropy) *If a random variable X takes values from a finite set $\mathcal{X}$, then*

$$H(X) \leq \log_2 |\mathcal{X}|,$$

where[4] $|\mathcal{X}|$ denotes the size of the set $\mathcal{X}$. Equality holds iff X is equiprobable or uniformly distributed over $\mathcal{X}$ (i.e., $P_X(x) = \frac{1}{|\mathcal{X}|}$ for all $x \in \mathcal{X}$).

Proof

$$\log_2 |\mathcal{X}| - H(X) = \log_2 |\mathcal{X}| \cdot \left[\sum_{x \in \mathcal{X}} P_X(x) \right] - \left[-\sum_{x \in \mathcal{X}} P_X(x) \log_2 P_X(x) \right]$$

$$= \sum_{x \in \mathcal{X}} P_X(x) \cdot \log_2 |\mathcal{X}| + \sum_{x \in \mathcal{X}} P_X(x) \log_2 P_X(x)$$

$$= \sum_{x \in \mathcal{X}} P_X(x) \log_2 [|\mathcal{X}| \cdot P_X(x)]$$

$$\geq \sum_{x \in \mathcal{X}} P_X(x) \cdot \log_2(e) \left(1 - \frac{1}{|\mathcal{X}| \cdot P_X(x)} \right)$$

$$= \log_2(e) \sum_{x \in \mathcal{X}} \left(P_X(x) - \frac{1}{|\mathcal{X}|} \right)$$

$$= \log_2(e) \cdot (1 - 1) = 0,$$

where the inequality follows from the FI Lemma, with equality iff ($\forall x \in \mathcal{X}$), $|\mathcal{X}| \cdot P_X(x) = 1$, which means $P_X(\cdot)$ is a uniform distribution on $\mathcal{X}$. □

Intuitively, $H(X)$ tells us how random X is. Indeed, X is deterministic (not random at all) iff $H(X) = 0$. If X is uniform (equiprobable), $H(X)$ is maximized and is equal to $\log_2 |\mathcal{X}|$.

Lemma 2.7 (Log-sum inequality) *For nonnegative numbers, $a_1, a_2, \ldots, a_n$ and $b_1, b_2, \ldots, b_n$,*

$$\sum_{i=1}^{n} \left(a_i \log_D \frac{a_i}{b_i} \right) \geq \left(\sum_{i=1}^{n} a_i \right) \log_D \frac{\sum_{i=1}^{n} a_i}{\sum_{i=1}^{n} b_i}, \tag{2.1.4}$$

with equality holding iff for all $i = 1, \ldots, n$,

$$\frac{a_i}{b_i} = \frac{\sum_{j=1}^{n} a_j}{\sum_{j=1}^{n} b_j},$$

which is a constant that does not depend on i. (By convention, $0 \cdot \log_D(0) = 0$, $0 \cdot \log_D(0/0) = 0$ and $a \cdot \log_D(a/0) = \infty$ if $a > 0$. Again, this can be justified by "continuity.")

---

[4]Note that $\log |\mathcal{X}|$ is also known as Hartley's function or entropy; Hartley was the first to suggest measuring information regardless of its content [180].

Proof Let $a := \sum_{i=1}^{n} a_i$ and $b := \sum_{i=1}^{n} b_i$. Then

$$\sum_{i=1}^{n} a_i \log_D \frac{a_i}{b_i} - a \log_D \frac{a}{b} = a \left[\sum_{i=1}^{n} \frac{a_i}{a} \log_D \frac{a_i}{b_i} - \underbrace{\left(\sum_{i=1}^{n} \frac{a_i}{a} \right)}_{=1} \log_D \frac{a}{b} \right]$$

$$= a \sum_{i=1}^{n} \frac{a_i}{a} \log_D \left[\frac{a_i}{b_i} \frac{b}{a} \right]$$

$$\geq a \log_D(e) \sum_{i=1}^{n} \frac{a_i}{a} \left[1 - \frac{b_i}{a_i} \frac{a}{b} \right]$$

$$= a \log_D(e) \left(\sum_{i=1}^{n} \frac{a_i}{a} - \sum_{i=1}^{n} \frac{b_i}{b} \right)$$

$$= a \log_D(e) (1 - 1) = 0,$$

where the inequality follows from the FI Lemma, with equality holding iff $\frac{a_i}{b_i} \frac{b}{a} = 1$ for all i; i.e., $\frac{a_i}{b_i} = \frac{a}{b} \ \forall i$.

We also provide another proof using Jensen's inequality (cf. Theorem B.18 in Appendix B). Without loss of generality, assume that $a_i > 0$ and $b_i > 0$ for every i. Jensen's inequality states that

$$\sum_{i=1}^{n} \alpha_i f(t_i) \geq f\left(\sum_{i=1}^{n} \alpha_i t_i \right)$$

for any strictly convex function $f(\cdot)$, $\alpha_i \geq 0$, and $\sum_{i=1}^{n} \alpha_i = 1$; equality holds iff t_i is a constant for all i. Hence by setting $\alpha_i = b_i / \sum_{j=1}^{n} b_j$, $t_i = a_i / b_i$, and $f(t) = t \cdot \log_D(t)$, we obtain the desired result. □

2.1.4 Joint Entropy and Conditional Entropy

Given a pair of random variables (X, Y) with a joint pmf $P_{X,Y}(\cdot, \cdot)$ defined[5] on $\mathcal{X} \times \mathcal{Y}$, the self-information of the (two-dimensional) elementary event $\{X = x, Y = y\}$ is defined by

$$\mathcal{I}(x, y) := -\log_2 P_{X,Y}(x, y).$$

This leads us to the definition of joint entropy.

---

[5]Note that $P_{XY}(\cdot, \cdot)$ is another common notation for the joint distribution $P_{X,Y}(\cdot, \cdot)$.

Definition 2.8 (*Joint entropy*) The joint entropy $H(X, Y)$ of random variables (X, Y) is defined by

$$H(X, Y) := - \sum_{(x,y)\in\mathcal{X}\times\mathcal{Y}} P_{X,Y}(x, y) \cdot \log_2 P_{X,Y}(x, y)$$
$$= E[-\log_2 P_{X,Y}(X, Y)].$$

The conditional entropy can be similarly defined as follows.

Definition 2.9 (*Conditional entropy*) Given two jointly distributed random variables X and Y, the conditional entropy $H(Y|X)$ of Y given X is defined by

$$H(Y|X) := \sum_{x\in\mathcal{X}} P_X(x) \left(-\sum_{y\in\mathcal{Y}} P_{Y|X}(y|x) \cdot \log_2 P_{Y|X}(y|x) \right), \qquad (2.1.5)$$

where $P_{Y|X}(\cdot|\cdot)$ is the conditional pmf of Y given X.

Equation (2.1.5) can be written into three different but equivalent forms:

$$H(Y|X) = - \sum_{(x,y)\in\mathcal{X}\times\mathcal{Y}} P_{X,Y}(x, y) \cdot \log_2 P_{Y|X}(y|x)$$
$$= E[-\log_2 P_{Y|X}(Y|X)]$$
$$= \sum_{x\in\mathcal{X}} P_X(x) \cdot H(Y|X = x),$$

where $H(Y|X = x) := -\sum_{y\in\mathcal{Y}} P_{Y|X}(y|x) \log_2 P_{Y|X}(y|x)$.

The relationship between joint entropy and conditional entropy is exhibited by the fact that the entropy of a pair of random variables is the entropy of one plus the conditional entropy of the other.

Theorem 2.10 (Chain rule for entropy)

$$H(X, Y) = H(X) + H(Y|X). \qquad (2.1.6)$$

Proof Since

$$P_{X,Y}(x, y) = P_X(x)P_{Y|X}(y|x),$$

we directly obtain that

$$H(X, Y) = E[-\log_2 P_{X,Y}(X, Y)]$$
$$= E[-\log_2 P_X(X)] + E[-\log_2 P_{Y|X}(Y|X)]$$
$$= H(X) + H(Y|X).$$

$\square$

By its definition, joint entropy is commutative; i.e., $H(X,Y) = H(Y,X)$. Hence,

$$H(X,Y) = H(X) + H(Y|X) = H(Y) + H(X|Y) = H(Y,X),$$

which implies that
$$H(X) - H(X|Y) = H(Y) - H(Y|X). \tag{2.1.7}$$

The above quantity is exactly equal to the mutual information which will be introduced in the next section.

The conditional entropy can be thought of in terms of a channel whose input is the random variable X and whose output is the random variable Y. $H(X|Y)$ is then called the *equivocation*[6] and corresponds to the uncertainty in the channel input from the *receiver's* point of view. For example, suppose that the set of possible outcomes of random vector (X,Y) is $\{(0,0), (0,1), (1,0), (1,1)\}$, where none of the elements has zero probability mass. When the receiver Y receives 1, he still cannot determine exactly what the sender X observes (it could be either 1 or 0); therefore, the uncertainty, from the receiver's view point, depends on the probabilities $P_{X|Y}(0|1)$ and $P_{X|Y}(1|1)$.

Similarly, $H(Y|X)$, which is called *prevarication*,[7] is the uncertainty in the channel output from the *transmitter's* point of view. In other words, the sender knows exactly what he sends, but is uncertain on what the receiver will finally obtain.

A case that is of specific interest is when $H(X|Y) = 0$. By its definition, $H(X|Y) = 0$ if X becomes deterministic after observing Y. In such a case, the uncertainty of X after giving Y is completely zero.

The next corollary can be proved similarly to Theorem 2.10.

Corollary 2.11 (Chain rule for conditional entropy)

$$H(X,Y|Z) = H(X|Z) + H(Y|X,Z).$$

2.1.5 Properties of Joint Entropy and Conditional Entropy

Lemma 2.12 (Conditioning never increases entropy) *Side information Y decreases the uncertainty about X:*
$$H(X|Y) \leq H(X),$$

with equality holding iff X and Y are independent. In other words, "conditioning" reduces entropy.

---

[6]Equivocation is an ambiguous statement one uses deliberately in order to deceive or avoid speaking the truth.

[7]Prevarication is the deliberate act of deviating from the truth (it is a synonym of "equivocation").

Proof

$$H(X) - H(X|Y) = \sum_{(x,y)\in\mathcal{X}\times\mathcal{Y}} P_{X,Y}(x,y) \cdot \log_2 \frac{P_{X|Y}(x|y)}{P_X(x)}$$

$$= \sum_{(x,y)\in\mathcal{X}\times\mathcal{Y}} P_{X,Y}(x,y) \cdot \log_2 \frac{P_{X|Y}(x|y)P_Y(y)}{P_X(x)P_Y(y)}$$

$$= \sum_{(x,y)\in\mathcal{X}\times\mathcal{Y}} P_{X,Y}(x,y) \cdot \log_2 \frac{P_{X,Y}(x,y)}{P_X(x)P_Y(y)}$$

$$\geq \left(\sum_{(x,y)\in\mathcal{X}\times\mathcal{Y}} P_{X,Y}(x,y) \right) \log_2 \frac{\sum_{(x,y)\in\mathcal{X}\times\mathcal{Y}} P_{X,Y}(x,y)}{\sum_{(x,y)\in\mathcal{X}\times\mathcal{Y}} P_X(x)P_Y(y)}$$

$$= 0,$$

where the inequality follows from the log-sum inequality, with equality holding iff

$$\frac{P_{X,Y}(x,y)}{P_X(x)P_Y(y)} = \text{constant} \quad \forall \, (x,y) \in \mathcal{X} \times \mathcal{Y}.$$

Since probability must sum to 1, the above constant equals 1, which is exactly the case of X being independent of Y. $\qquad \square$

Lemma 2.13 *Entropy is additive for independent random variables; i.e.,*

$$H(X,Y) = H(X) + H(Y) \quad \text{for independent } X \text{ and } Y.$$

Proof By the previous lemma, independence of X and Y implies $H(Y|X) = H(Y)$. Hence

$$H(X,Y) = H(X) + H(Y|X) = H(X) + H(Y).$$

$\qquad \square$

Since conditioning never increases entropy, it follows that

$$H(X,Y) = H(X) + H(Y|X) \leq H(X) + H(Y). \qquad (2.1.8)$$

The above lemma tells us that equality holds for (2.1.8) only when X is independent of Y.

A result similar to (2.1.8) also applies to the conditional entropy.

Lemma 2.14 *Conditional entropy is lower additive; i.e.,*

$$H(X_1, X_2|Y_1, Y_2) \leq H(X_1|Y_1) + H(X_2|Y_2).$$

Equality holds iff

$$P_{X_1,X_2|Y_1,Y_2}(x_1, x_2|y_1, y_2) = P_{X_1|Y_1}(x_1|y_1)P_{X_2|Y_2}(x_2|y_2)$$

for all x_1, x_2, y_1 and y_2.

Proof Using the chain rule for conditional entropy and the fact that conditioning reduces entropy, we can write

$$\begin{aligned}
H(X_1, X_2|Y_1, Y_2) &= H(X_1|Y_1, Y_2) + H(X_2|X_1, Y_1, Y_2) \\
&\leq H(X_1|Y_1, Y_2) + H(X_2|Y_1, Y_2), \quad\quad (2.1.9) \\
&\leq H(X_1|Y_1) + H(X_2|Y_2). \quad\quad\quad\quad (2.1.10)
\end{aligned}$$

For (2.1.9), equality holds iff X_1 and X_2 are conditionally independent given (Y_1, Y_2): $P_{X_1,X_2|Y_1,Y_2}(x_1, x_2|y_1, y_2) = P_{X_1|Y_1,Y_2}(x_1|y_1, y_2)P_{X_2|Y_1,Y_2}(x_2|y_1, y_2)$. For (2.1.10), equality holds iff X_1 is conditionally independent of Y_2 given Y_1 (i.e., $P_{X_1|Y_1,Y_2}(x_1|y_1, y_2) = P_{X_1|Y_1}(x_1|y_1)$), and X_2 is conditionally independent of Y_1 given Y_2 (i.e., $P_{X_2|Y_1,Y_2}(x_2|y_1, y_2) = P_{X_2|Y_2}(x_2|y_2)$). Hence, the desired equality condition of the lemma is obtained.   □

2.2 Mutual Information

For two random variables X and Y, the *mutual information* between X and Y is the reduction in the *uncertainty* of Y due to the knowledge of X (or vice versa). A dual definition of mutual information states that it is the average amount of *information* that Y has (or contains) about X or X has (or contains) about Y.

We can think of the mutual information between X and Y in terms of a channel whose input is X and whose output is Y. Thereby the reduction of the uncertainty is by definition the total uncertainty of X (i.e., $H(X)$) minus the uncertainty of X after observing Y (i.e., $H(X|Y)$). Mathematically, it is

$$\text{mutual information} = I(X; Y) := H(X) - H(X|Y). \quad\quad (2.2.1)$$

It can be easily verified from (2.1.7) that mutual information is symmetric; i.e., $I(X; Y) = I(Y; X)$.

2.2.1 Properties of Mutual Information

Lemma 2.15   1. $I(X; Y) = \displaystyle\sum_{x \in \mathcal{X}} \sum_{y \in \mathcal{Y}} P_{X,Y}(x, y) \log_2 \frac{P_{X,Y}(x, y)}{P_X(x)P_Y(y)}$.

2. $I(X; Y) = I(Y; X) = H(Y) - H(Y|X)$.
3. $I(X; Y) = H(X) + H(Y) - H(X, Y)$.

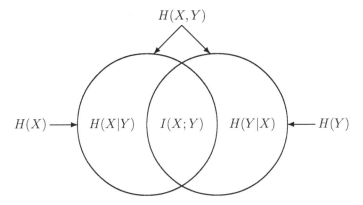

Fig. 2.2 Relation between entropy and mutual information

4. $I(X;Y) \le H(X)$ with equality holding iff X is a function of Y (i.e., $X = f(Y)$ for some function $f(\cdot)$).
5. $I(X;Y) \ge 0$ with equality holding iff X and Y are independent.
6. $I(X;Y) \le \min\{\log_2 |\mathcal{X}|, \log_2 |\mathcal{Y}|\}$.

Proof Properties 1, 2, 3, and 4 follow immediately from the definition. Property 5 is a direct consequence of Lemma 2.12. Property 6 holds iff $I(X;Y) \le \log_2 |\mathcal{X}|$ and $I(X;Y) \le \log_2 |\mathcal{Y}|$. To show the first inequality, we write $I(X;Y) = H(X) - H(X|Y)$, use the fact that $H(X|Y)$ is nonnegative and apply Lemma 2.6. A similar proof can be used to show that $I(X;Y) \le \log_2 |\mathcal{Y}|$. $\qquad\square$

The relationships between $H(X), H(Y), H(X, Y), H(X|Y), H(Y|X)$, and $I(X;Y)$ can be illustrated by the Venn diagram in Fig. 2.2.

2.2.2 Conditional Mutual Information

The conditional mutual information, denoted by $I(X;Y|Z)$, is defined as the common uncertainty between X and Y under the knowledge of Z:

$$I(X;Y|Z) := H(X|Z) - H(X|Y, Z). \qquad (2.2.2)$$

Lemma 2.16 (Chain rule for mutual information) *Defining the joint mutual information between X and the pair (Y, Z) as in (2.2.1) by*

$$I(X;Y, Z) := H(X) - H(X|Y, Z),$$

we have

$$I(X;Y, Z) = I(X;Y) + I(X;Z|Y) = I(X;Z) + I(X;Y|Z).$$

Proof Without loss of generality, we only prove the first equality:

$$
\begin{aligned}
I(X;Y,Z) &= H(X) - H(X|Y,Z) \\
&= H(X) - H(X|Y) + H(X|Y) - H(X|Y,Z) \\
&= I(X;Y) + I(X;Z|Y).
\end{aligned}
$$

$\square$

The above lemma can be read as follows: the information that (Y, Z) has about X is equal to the information that Y has about X plus the information that Z has about X when Y is already known.

2.3 Properties of Entropy and Mutual Information for Multiple Random Variables

Theorem 2.17 (Chain rule for entropy) *Let X_1, X_2, ..., X_n be drawn according to $P_{X^n}(x^n) := P_{X_1,\ldots,X_n}(x_1,\ldots,x_n)$, where we use the common superscript notation to denote an n-tuple: $X^n := (X_1,\ldots,X_n)$ and $x^n := (x_1,\ldots,x_n)$.*
Then

$$
H(X_1, X_2, \ldots, X_n) = \sum_{i=1}^{n} H(X_i|X_{i-1}, \ldots, X_1),
$$

where $H(X_i|X_{i-1},\ldots,X_1) := H(X_1)$ for $i = 1$. (The above chain rule can also be written as:

$$
H(X^n) = \sum_{i=1}^{n} H(X_i|X^{i-1}),
$$

where $X^i := (X_1, \ldots, X_i)$.)

Proof From (2.1.6),

$$
H(X_1, X_2, \ldots, X_n) = H(X_1, X_2, \ldots, X_{n-1}) + H(X_n|X_{n-1}, \ldots, X_1). \qquad (2.3.1)
$$

Once again, applying (2.1.6) to the first term of the right-hand side of (2.3.1), we have

$$
H(X_1, X_2, \ldots, X_{n-1}) = H(X_1, X_2, \ldots, X_{n-2}) + H(X_{n-1}|X_{n-2}, \ldots, X_1).
$$

The desired result can then be obtained by repeatedly applying (2.1.6). $\square$

Theorem 2.18 (Chain rule for conditional entropy)

$$H(X_1, X_2, \ldots, X_n | Y) = \sum_{i=1}^{n} H(X_i | X_{i-1}, \ldots, X_1, Y).$$

Proof The theorem can be proved similarly to Theorem 2.17.   □

If $X^n = (X_1, \ldots, X_n)$ and $Y^m = (Y_1, \ldots, Y_m)$ are jointly distributed random vectors (of not necessarily equal lengths), then their joint mutual information is given by

$$I(X_1, \ldots, X_n; Y_1, \ldots, Y_m) := H(X_1, \ldots, X_n) - H(X_1, \ldots, X_n | Y_1, \ldots, Y_m).$$

Theorem 2.19 (Chain rule for mutual information)

$$I(X_1, X_2, \ldots, X_n; Y) = \sum_{i=1}^{n} I(X_i; Y | X_{i-1}, \ldots, X_1),$$

where $I(X_i; Y | X_{i-1}, \ldots, X_1) := I(X_1; Y)$ *for* $i = 1$.

Proof This can be proved by first expressing mutual information in terms of entropy and conditional entropy, and then applying the chain rules for entropy and conditional entropy.   □

Theorem 2.20 (Independence bound on entropy)

$$H(X_1, X_2, \ldots, X_n) \leq \sum_{i=1}^{n} H(X_i).$$

Equality holds iff all the X_i's are independent of each other.[8]

Proof By applying the chain rule for entropy,

$$H(X_1, X_2, \ldots, X_n) = \sum_{i=1}^{n} H(X_i | X_{i-1}, \ldots, X_1)$$

$$\leq \sum_{i=1}^{n} H(X_i).$$

Equality holds iff each conditional entropy is equal to its associated entropy, that iff X_i is independent of $(X_{i-1}, \ldots, X_1)$ for all i.   □

---

[8] This condition is equivalent to requiring that X_i be independent of $(X_{i-1}, \ldots, X_1)$ for all i. The equivalence can be directly proved using the chain rule for joint probabilities, i.e., $P_{X^n}(x^n) = \prod_{i=1}^{n} P_{X_i | X_1^{i-1}}(x_i | x_1^{i-1})$; it is left as an exercise.

Theorem 2.21 (Bound on mutual information) *If $\{(X_i, Y_i)\}_{i=1}^n$ is a process satisfying the conditional independence assumption $P_{Y^n|X^n} = \prod_{i=1}^n P_{Y_i|X_i}$, then*

$$I(X_1, \ldots, X_n; Y_1, \ldots, Y_n) \le \sum_{i=1}^n I(X_i; Y_i),$$

with equality holding iff $\{X_i\}_{i=1}^n$ are independent.

Proof From the independence bound on entropy, we have

$$H(Y_1, \ldots, Y_n) \le \sum_{i=1}^n H(Y_i).$$

By the conditional independence assumption, we have

$$
\begin{aligned}
H(Y_1, \ldots, Y_n | X_1, \ldots, X_n) &= E\left[-\log_2 P_{Y^n|X^n}(Y^n|X^n)\right] \\
&= E\left[-\sum_{i=1}^n \log_2 P_{Y_i|X_i}(Y_i|X_i)\right] \\
&= \sum_{i=1}^n H(Y_i|X_i).
\end{aligned}
$$

Hence,

$$
\begin{aligned}
I(X^n; Y^n) &= H(Y^n) - H(Y^n|X^n) \\
&\le \sum_{i=1}^n H(Y_i) - \sum_{i=1}^n H(Y_i|X_i) \\
&= \sum_{i=1}^n I(X_i; Y_i),
\end{aligned}
$$

with equality holding iff $\{Y_i\}_{i=1}^n$ are independent, which holds iff $\{X_i\}_{i=1}^n$ are independent. $\square$

2.4 Data Processing Inequality

Recalling that the Markov chain relationship $X \to Y \to Z$ means that X and Z are conditional independent given Y (cf. Appendix B), we have the following result.

Lemma 2.22 (Data processing inequality) (This is also called the *data processing lemma*.) *If* $X \to Y \to Z$, *then*

$$I(X;Y) \geq I(X;Z).$$

Proof Since $X \to Y \to Z$, we directly have that $I(X;Z|Y) = 0$. By the chain rule for mutual information,

$$
\begin{aligned}
I(X;Z) + I(X;Y|Z) &= I(X;Y,Z) & (2.4.1) \\
&= I(X;Y) + I(X;Z|Y) \\
&= I(X;Y). & (2.4.2)
\end{aligned}
$$

Since $I(X;Y|Z) \geq 0$, we obtain that $I(X;Y) \geq I(X;Z)$ with equality holding iff $I(X;Y|Z) = 0$. $\qquad\square$

The data processing inequality means that the mutual information will not increase after processing. This result is somewhat counterintuitive since given two random variables X and Y, we might believe that applying a well-designed processing scheme to Y, which can be generally represented by a mapping $g(Y)$, could possibly increase the mutual information. However, for any $g(\cdot)$, $X \to Y \to g(Y)$ forms a Markov chain which implies that data processing cannot increase mutual information. A communication context for the data processing lemma is depicted in Fig. 2.3, and summarized in the next corollary.

Corollary 2.23 *For jointly distributed random variables X and Y and any function $g(\cdot)$, we have $X \to Y \to g(Y)$ and*

$$I(X;Y) \geq I(X;g(Y)).$$

We also note that if Z obtains all the information about X through Y, then knowing Z will not help increase the mutual information between X and Y; this is formalized in the following.

Corollary 2.24 *If $X \to Y \to Z$, then*

$$I(X;Y|Z) \leq I(X;Y).$$

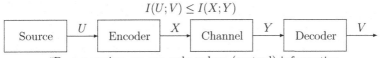

"By processing, we can only reduce (mutual) information, but the processed information may be in a more *useful* form!"

Fig. 2.3 Communication context of the data processing lemma

Proof The proof directly follows from (2.4.1) and (2.4.2). $\square$

It is worth pointing out that it is possible that $I(X; Y|Z) > I(X; Y)$ when X, Y and Z do *not* form a Markov chain. For example, let X and Y be independent equiprobable binary zero-one random variables, and let $Z = X + Y$. Then,

$$\begin{aligned} I(X; Y|Z) &= H(X|Z) - H(X|Y, Z) \\ &= H(X|Z) \\ &= P_Z(0)H(X|z = 0) + P_Z(1)H(X|z = 1) + P_Z(2)H(X|z = 2) \\ &= 0 + 0.5 + 0 \\ &= 0.5 \text{ bits}, \end{aligned}$$

which is clearly larger than $I(X; Y) = 0$.

Finally, we observe that we can extend the data processing inequality for a sequence of random variables forming a Markov chain (see (B.3.5) in Appendix B for the definition) as follows.

Corollary 2.25 *If* $X_1 \to X_2 \to \cdots \to X_n$, *then for any* $i, j, k.l$ *such that* $1 \le i \le j \le k \le l \le n$, *we have that*

$$I(X_i; X_l) \le I(X_j; X_k).$$

2.5 Fano's Inequality

Fano's inequality [113, 114] is a useful tool widely employed in information theory to prove converse results for coding theorems (as we will see in the following chapters).

Lemma 2.26 (Fano's inequality) *Let* X *and* Y *be two random variables, correlated in general, with alphabets* $\mathcal{X}$ *and* $\mathcal{Y}$, *respectively, where* $\mathcal{X}$ *is finite but* $\mathcal{Y}$ *can be countably infinite. Let* $\hat{X} := g(Y)$ *be an estimate of* X *from observing* Y, *where* $g : \mathcal{Y} \to \mathcal{X}$ *is a given estimation function. Define the probability of error as*

$$P_e := \Pr[\hat{X} \ne X].$$

Then the following inequality holds

$$H(X|Y) \le h_b(P_e) + P_e \cdot \log_2(|\mathcal{X}| - 1), \tag{2.5.1}$$

where $h_b(x) := -x \log_2 x - (1 - x) \log_2(1 - x)$ *for* $0 \le x \le 1$ *is the binary entropy function (see Example 2.3).*

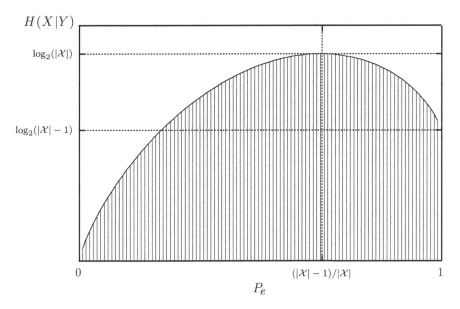

Fig. 2.4 Permissible $(P_e, H(X|Y))$ region due to Fano's inequality

Observation 2.27

- Note that when $P_e = 0$, we obtain that $H(X|Y) = 0$ (see (2.5.1)) as intuition suggests, since if $P_e = 0$, then $\hat{X} = g(Y) = X$ (with probability 1) and thus $H(X|Y) = H(g(Y)|Y) = 0$.
- Fano's inequality yields upper and lower bounds on P_e in terms of $H(X|Y)$. This is illustrated in Fig. 2.4, where we plot the region for the pairs $(P_e, H(X|Y))$ that are permissible under Fano's inequality. In the figure, the boundary of the permissible (dashed) region is given by the function

$$f(P_e) := h_b(P_e) + P_e \cdot \log_2(|\mathcal{X}| - 1),$$

the right-hand side of (2.5.1). We obtain that when

$$\log_2(|\mathcal{X}| - 1) < H(X|Y) \le \log_2(|\mathcal{X}|),$$

P_e can be upper and lower bounded as follows:

$$0 < \inf\{a : f(a) \ge H(X|Y)\} \le P_e \le \sup\{a : f(a) \ge H(X|Y)\} < 1.$$

Furthermore, when
$$0 < H(X|Y) \le \log_2(|\mathcal{X}| - 1),$$

only the lower bound holds:

$$P_e \geq \inf\{a : f(a) \geq H(X|Y)\} > 0.$$

Thus for all nonzero values of $H(X|Y)$, we obtain a lower bound (of the same form above) on P_e; the bound implies that if $H(X|Y)$ is bounded away from zero, P_e is also bounded away from zero.

- A weaker but simpler version of Fano's inequality can be directly obtained from (2.5.1) by noting that $h_b(P_e) \leq 1$:

$$H(X|Y) \leq 1 + P_e \log_2(|\mathcal{X}| - 1), \tag{2.5.2}$$

which in turn yields that

$$P_e \geq \frac{H(X|Y) - 1}{\log_2(|\mathcal{X}| - 1)} \quad \text{(for } |\mathcal{X}| > 2)$$

which is weaker than the above lower bound on P_e.

Proof of Lemma 2.26 Define a new random variable,

$$E := \begin{cases} 1, & \text{if } g(Y) \neq X \\ 0, & \text{if } g(Y) = X \end{cases}.$$

Then using the chain rule for conditional entropy, we obtain

$$\begin{aligned} H(E, X|Y) &= H(X|Y) + H(E|X, Y) \\ &= H(E|Y) + H(X|E, Y). \end{aligned}$$

Observe that E is a function of X and Y; hence, $H(E|X, Y) = 0$. Since conditioning never increases entropy, $H(E|Y) \leq H(E) = h_b(P_e)$. The remaining term, $H(X|E, Y)$, can be bounded as follows:

$$\begin{aligned} H(X|E, Y) &= \Pr[E = 0]H(X|Y, E = 0) + \Pr[E = 1]H(X|Y, E = 1) \\ &\leq (1 - P_e) \cdot 0 + P_e \cdot \log_2(|\mathcal{X}| - 1), \end{aligned}$$

since $X = g(Y)$ for $E = 0$, and given $E = 1$, we can upper bound the conditional entropy by the logarithm of the number of remaining outcomes, i.e., $(|\mathcal{X}| - 1)$. Combining these results completes the proof. □

Fano's inequality cannot be improved in the sense that the lower bound, $H(X|Y)$, can be achieved for some specific cases. Any bound that can be achieved in some cases is often referred to as *sharp*.[9] From the proof of the above lemma, we can observe

---

[9]**Definition**. A bound is said to be *sharp* if the bound is achievable for *some specific* cases. A bound is said to be *tight* if the bound is achievable for *all* cases.

that equality holds in Fano's inequality, if $H(E|Y) = H(E)$ and $H(X|Y, E = 1) = \log_2(|\mathcal{X}| - 1)$. The former is equivalent to E being independent of Y, and the latter holds iff $P_{X|Y}(\cdot|y)$ is uniformly distributed over the set $\mathcal{X} \setminus \{g(y)\}$. We can therefore create an example in which equality holds in Fano's inequality.

Example 2.28 Suppose that X and Y are two independent random variables which are both uniformly distributed on the alphabet $\{0, 1, 2\}$. Let the estimating function be given by $g(y) = y$. Then

$$P_e = \Pr[g(Y) \neq X] = \Pr[Y \neq X] = 1 - \sum_{x=0}^{2} P_X(x)P_Y(x) = \frac{2}{3}.$$

In this case, equality is achieved in Fano's inequality, i.e.,

$$h_b\left(\frac{2}{3}\right) + \frac{2}{3} \cdot \log_2(3 - 1) = H(X|Y) = H(X) = \log_2 3.$$

To conclude this section, we present an alternative proof for Fano's inequality to illustrate the use of the data processing inequality and the FI Lemma.

Alternative Proof of Fano's inequality: Noting that $X \to Y \to \hat{X}$ form a Markov chain, we directly obtain via the data processing inequality that

$$I(X; Y) \geq I(X; \hat{X}),$$

which implies that

$$H(X|Y) \leq H(X|\hat{X}).$$

Thus, if we show that $H(X|\hat{X})$ is no larger than the right-hand side of (2.5.1), the proof of (2.5.1) is complete.
Noting that

$$P_e = \sum_{x \in \mathcal{X}} \sum_{\hat{x} \in \mathcal{X}: \hat{x} \neq x} P_{X,\hat{X}}(x, \hat{x})$$

and

$$1 - P_e = \sum_{x \in \mathcal{X}} \sum_{\hat{x} \in \mathcal{X}: \hat{x} = x} P_{X,\hat{X}}(x, \hat{x}) = \sum_{x \in \mathcal{X}} P_{X,\hat{X}}(x, x),$$

we obtain that

$$H(X|\hat{X}) - h_b(P_e) - P_e \log_2(|\mathcal{X}| - 1)$$

$$= \sum_{x \in \mathcal{X}} \sum_{\hat{x} \in \mathcal{X}: \hat{x} \neq x} P_{X,\hat{X}}(x, \hat{x}) \log_2 \frac{1}{P_{X|\hat{X}}(x|\hat{x})} + \sum_{x \in \mathcal{X}} P_{X,\hat{X}}(x, x) \log_2 \frac{1}{P_{X|\hat{X}}(x|x)}$$

$$- \left[\sum_{x \in \mathcal{X}} \sum_{\hat{x} \in \mathcal{X}: \hat{x} \neq x} P_{X,\hat{X}}(x, \hat{x}) \right] \log_2 \frac{(|\mathcal{X}| - 1)}{P_e} + \left[\sum_{x \in \mathcal{X}} P_{X,\hat{X}}(x, x) \right] \log_2(1 - P_e)$$

$$= \sum_{x \in \mathcal{X}} \sum_{\hat{x} \in \mathcal{X}: \hat{x} \neq x} P_{X,\hat{X}}(x, \hat{x}) \log_2 \frac{P_e}{P_{X|\hat{X}}(x|\hat{x})(|\mathcal{X}| - 1)}$$

$$+ \sum_{x \in \mathcal{X}} P_{X,\hat{X}}(x, x) \log_2 \frac{1 - P_e}{P_{X|\hat{X}}(x|x)} \tag{2.5.3}$$

$$\leq \log_2(e) \sum_{x \in \mathcal{X}} \sum_{\hat{x} \in \mathcal{X}: \hat{x} \neq x} P_{X,\hat{X}}(x, \hat{x}) \left[\frac{P_e}{P_{X|\hat{X}}(x|\hat{x})(|\mathcal{X}| - 1)} - 1 \right]$$

$$+ \log_2(e) \sum_{x \in \mathcal{X}} P_{X,\hat{X}}(x, x) \left[\frac{1 - P_e}{P_{X|\hat{X}}(x|x)} - 1 \right]$$

$$= \log_2(e) \left[\frac{P_e}{(|\mathcal{X}| - 1)} \sum_{x \in \mathcal{X}} \sum_{\hat{x} \in \mathcal{X}: \hat{x} \neq x} P_{\hat{X}}(\hat{x}) - \sum_{x \in \mathcal{X}} \sum_{\hat{x} \in \mathcal{X}: \hat{x} \neq x} P_{X,\hat{X}}(x, \hat{x}) \right]$$

$$+ \log_2(e) \left[(1 - P_e) \sum_{x \in \mathcal{X}} P_{\hat{X}}(x) - \sum_{x \in \mathcal{X}} P_{X,\hat{X}}(x, x) \right]$$

$$= \log_2(e) \left[\frac{P_e}{(|\mathcal{X}| - 1)}(|\mathcal{X}| - 1) - P_e \right] + \log_2(e) [(1 - P_e) - (1 - P_e)]$$

$$= 0,$$

where the inequality follows by applying the FI Lemma to each logarithm term in (2.5.3). □

2.6 Divergence and Variational Distance

In addition to the probabilistically defined *entropy* and *mutual information*, another measure that is frequently considered in information theory is *divergence*. In this section, we define this measure and study its statistical properties.

Definition 2.29 (*Divergence*) Given two discrete random variables X and $\hat{X}$ defined over a common alphabet $\mathcal{X}$, the divergence or the *Kullback–Leibler divergence or dis-*

tance[10] (other names are *relative entropy* and *discrimination*) is denoted by $D(X \| \hat{X})$ or $D(P_X \| P_{\hat{X}})$ and defined by[11]

$$D(X \| \hat{X}) = D(P_X \| P_{\hat{X}}) := E_X \left[\log_2 \frac{P_X(X)}{P_{\hat{X}}(X)} \right] = \sum_{x \in \mathcal{X}} P_X(x) \log_2 \frac{P_X(x)}{P_{\hat{X}}(x)}.$$

In other words, the divergence $D(P_X \| P_{\hat{X}})$ is the expectation (with respect to P_X) of the *log-likelihood ratio* $\log_2[P_X / P_{\hat{X}}]$ of distribution P_X against distribution $P_{\hat{X}}$. $D(X \| \hat{X})$ can be viewed as a measure of "distance" or "dissimilarity" between distributions P_X and $P_{\hat{X}}$. $D(X \| \hat{X})$ is also called *relative entropy* since it can be regarded as a measure of the inefficiency of mistakenly assuming that the distribution of a source is $P_{\hat{X}}$ when the true distribution is P_X. For example, if we know the true distribution P_X of a source, then we can construct a lossless data compression code with average codeword length achieving entropy $H(X)$ (this will be studied in the next chapter). If, however, we mistakenly thought that the "true" distribution is $P_{\hat{X}}$ and employ the "best" code corresponding to $P_{\hat{X}}$, then the resultant average codeword length becomes

$$\sum_{x \in \mathcal{X}} [-P_X(x) \cdot \log_2 P_{\hat{X}}(x)].$$

As a result, the *relative* difference between the resultant average codeword length and $H(X)$ is the *relative entropy* $D(X \| \hat{X})$. Hence, divergence is a measure of the system cost (e.g., storage consumed) paid due to mis-classifying the system statistics.

Note that when computing divergence, we follow the convention that

$$0 \cdot \log_2 \frac{0}{p} = 0 \quad \text{and} \quad p \cdot \log_2 \frac{p}{0} = \infty \quad \text{for } p > 0.$$

We next present some properties of the divergence and discuss its relation with entropy and mutual information.

Lemma 2.30 (Nonnegativity of divergence)

$$D(X \| \hat{X}) \geq 0,$$

with equality iff $P_X(x) = P_{\hat{X}}(x)$ for all $x \in \mathcal{X}$ (i.e., the two distributions are equal).

---

[10]As noted in Footnote 1, this measure was originally introduced by Kullback and Leibler [236, 237].

[11]In order to be consistent with the units (in bits) adopted for entropy and mutual information, we will also use the base-2 logarithm for divergence unless otherwise specified.

Proof

$$D(X \,\|\, \hat{X}) = \sum_{x \in \mathcal{X}} P_X(x) \log_2 \frac{P_X(x)}{P_{\hat{X}}(x)}$$

$$\geq \left(\sum_{x \in \mathcal{X}} P_X(x) \right) \log_2 \frac{\sum_{x \in \mathcal{X}} P_X(x)}{\sum_{x \in \mathcal{X}} P_{\hat{X}}(x)}$$

$$= 0,$$

where the second step follows from the log-sum inequality with equality holding iff for every $x \in \mathcal{X}$,

$$\frac{P_X(x)}{P_{\hat{X}}(x)} = \frac{\sum_{a \in \mathcal{X}} P_X(a)}{\sum_{b \in \mathcal{X}} P_{\hat{X}}(b)} = 1,$$

or equivalently $P_X(x) = P_{\hat{X}}(x)$ for all $x \in \mathcal{X}$.   □

Lemma 2.31 (Mutual information and divergence)

$$I(X; Y) = D(P_{X,Y} \| P_X \times P_Y),$$

where $P_{X,Y}(\cdot, \cdot)$ is the joint distribution of the random variables X and Y and $P_X(\cdot)$ and $P_Y(\cdot)$ are the respective marginals.

Proof The observation follows directly from the definitions of divergence and mutual information.   □

Definition 2.32 (*Refinement of distribution*) Given the distribution P_X on $\mathcal{X}$, divide $\mathcal{X}$ into k mutually disjoint sets, $\mathcal{U}_1, \mathcal{U}_2, \ldots, \mathcal{U}_k$, satisfying

$$\mathcal{X} = \bigcup_{i=1}^{k} \mathcal{U}_i.$$

Define a new distribution P_U on $\mathcal{U} = \{1, 2, \ldots, k\}$ as

$$P_U(i) = \sum_{x \in \mathcal{U}_i} P_X(x).$$

Then P_X is called a *refinement* (or more specifically, a *k-refinement*) of P_U.

Let us briefly discuss the relation between the *processing* of information and its *refinement*. Processing of information can be modeled as a (many-to-one) mapping, and refinement is actually the reverse operation. Recall that the *data processing lemma* shows that mutual information can never increase due to *processing*. Hence, if one wishes to increase mutual information, he should "anti-process" (or refine) the involved statistics.

From Lemma 2.31, the mutual information can be viewed as the divergence of a joint distribution against the product distribution of the marginals. It is therefore reasonable to expect that a similar effect due to *processing* (or a reverse effect due to *refinement*) should also apply to *divergence*. This is shown in the next lemma.

Lemma 2.33 (Refinement cannot decrease divergence) *Let P_X and $P_{\hat{X}}$ be the refinements (k-refinements) of P_U and $P_{\hat{U}}$ respectively. Then*

$$D(P_X \| P_{\hat{X}}) \geq D(P_U \| P_{\hat{U}}).$$

Proof By the log-sum inequality, we obtain that for any $i \in \{1, 2, \ldots, k\}$

$$\sum_{x \in \mathcal{U}_i} P_X(x) \log_2 \frac{P_X(x)}{P_{\hat{X}}(x)} \geq \left(\sum_{x \in \mathcal{U}_i} P_X(x) \right) \log_2 \frac{\sum_{x \in \mathcal{U}_i} P_X(x)}{\sum_{x \in \mathcal{U}_i} P_{\hat{X}}(x)}$$

$$= P_U(i) \log_2 \frac{P_U(i)}{P_{\hat{U}}(i)}, \tag{2.6.1}$$

with equality iff

$$\frac{P_X(x)}{P_{\hat{X}}(x)} = \frac{P_U(i)}{P_{\hat{U}}(i)}$$

for all $x \in \mathcal{U}$. Hence,

$$D(P_X \| P_{\hat{X}}) = \sum_{i=1}^{k} \sum_{x \in \mathcal{U}_i} P_X(x) \log_2 \frac{P_X(x)}{P_{\hat{X}}(x)}$$

$$\geq \sum_{i=1}^{k} P_U(i) \log_2 \frac{P_U(i)}{P_{\hat{U}}(i)}$$

$$= D(P_U \| P_{\hat{U}}),$$

with equality iff

$$\frac{P_X(x)}{P_{\hat{X}}(x)} = \frac{P_U(i)}{P_{\hat{U}}(i)}$$

for all i and $x \in \mathcal{U}_i$. □

Observation 2.34 One drawback of adopting the divergence as a measure between two distributions is that it does not meet the symmetry requirement of a true distance,[12] since interchanging its two arguments may yield different quantities. In

---

[12]Given a non-empty set A, the function $d : A \times A \to [0, \infty)$ is called a *distance* or *metric* if it satisfies the following properties.

1. Nonnegativity: $d(a, b) \geq 0$ for every $a, b \in A$ with equality holding iff $a = b$.
2. Symmetry: $d(a, b) = d(b, a)$ for every $a, b \in A$.

other words, $D(P_X \| P_{\hat{X}}) \neq D(P_{\hat{X}} \| P_X)$ in general. (It also does not satisfy the triangle inequality.) Thus, divergence is not a true distance or metric. Another measure which is a true distance, called *variational distance*, is sometimes used instead.

Definition 2.35 (*Variational distance*) The *variational distance* (also known as the $\mathcal{L}_1$-distance) between two distributions P_X and $P_{\hat{X}}$ with common alphabet $\mathcal{X}$ is defined by

$$\|P_X - P_{\hat{X}}\| := \sum_{x \in \mathcal{X}} \left| P_X(x) - P_{\hat{X}}(x) \right|.$$

Lemma 2.36 *The variational distance satisfies*

$$\|P_X - P_{\hat{X}}\| = 2 \cdot \sup_{E \subset \mathcal{X}} \left| P_X(E) - P_{\hat{X}}(E) \right| = 2 \cdot \sum_{x \in \mathcal{X} : P_X(x) > P_{\hat{X}}(x)} \left(P_X(x) - P_{\hat{X}}(x) \right).$$

Proof We first show that $\|P_X - P_{\hat{X}}\| = 2 \cdot \sum_{x \in \mathcal{X} : P_X(x) > P_{\hat{X}}(x)} \left(P_X(x) - P_{\hat{X}}(x) \right)$. Setting $\mathcal{A} := \{x \in \mathcal{X} : P_X(x) > P_{\hat{X}}(x)\}$, we have

$$\begin{aligned}
\|P_X - P_{\hat{X}}\| &= \sum_{x \in \mathcal{X}} \left| P_X(x) - P_{\hat{X}}(x) \right| \\
&= \sum_{x \in \mathcal{A}} \left| P_X(x) - P_{\hat{X}}(x) \right| + \sum_{x \in \mathcal{A}^c} \left| P_X(x) - P_{\hat{X}}(x) \right| \\
&= \sum_{x \in \mathcal{A}} \left(P_X(x) - P_{\hat{X}}(x) \right) + \sum_{x \in \mathcal{A}^c} \left(P_{\hat{X}}(x) - P_X(x) \right) \\
&= \sum_{x \in \mathcal{A}} \left(P_X(x) - P_{\hat{X}}(x) \right) + P_{\hat{X}}\left(\mathcal{A}^c\right) - P_X\left(\mathcal{A}^c\right) \\
&= \sum_{x \in \mathcal{A}} \left(P_X(x) - P_{\hat{X}}(x) \right) + P_X(\mathcal{A}) - P_{\hat{X}}(\mathcal{A}) \\
&= \sum_{x \in \mathcal{A}} \left(P_X(x) - P_{\hat{X}}(x) \right) + \sum_{x \in \mathcal{A}} \left(P_X(x) - P_{\hat{X}}(x) \right) \\
&= 2 \cdot \sum_{x \in \mathcal{A}} \left(P_X(x) - P_{\hat{X}}(x) \right),
\end{aligned}$$

where $\mathcal{A}^c$ denotes the complement set of $\mathcal{A}$.

We next prove that $\|P_X - P_{\hat{X}}\| = 2 \cdot \sup_{E \subset \mathcal{X}} \left| P_X(E) - P_{\hat{X}}(E) \right|$ by showing that each quantity is greater than or equal to the other. For any set $E \subset \mathcal{X}$, we can write

---

3. Triangle inequality: $d(a, b) + d(b, c) \geq d(a, c)$ for every $a, b, c \in A$.

$$\|P_X - P_{\hat{X}}\| = \sum_{x \in \mathcal{X}} |P_X(x) - P_{\hat{X}}(x)|$$

$$= \sum_{x \in E} |P_X(x) - P_{\hat{X}}(x)| + \sum_{x \in E^c} |P_X(x) - P_{\hat{X}}(x)|$$

$$\geq \left| \sum_{x \in E} [P_X(x) - P_{\hat{X}}(x)] \right| + \left| \sum_{x \in E^c} [P_X(x) - P_{\hat{X}}(x)] \right|$$

$$= |P_X(E) - P_{\hat{X}}(E)| + |P_X(E^c) - P_{\hat{X}}(E^c)|$$

$$= |P_X(E) - P_{\hat{X}}(E)| + |P_{\hat{X}}(E) - P_X(E)|$$

$$= 2 \cdot |P_X(E) - P_{\hat{X}}(E)|.$$

Thus $\|P_X - P_{\hat{X}}\| \geq 2 \cdot \sup_{E \subset \mathcal{X}} |P_X(E) - P_{\hat{X}}(E)|$. Conversely, we have that

$$2 \cdot \sup_{E \subset \mathcal{X}} |P_X(E) - P_{\hat{X}}(E)| \geq 2 \cdot |P_X(\mathcal{A}) - P_{\hat{X}}(\mathcal{A})|$$

$$= |P_X(\mathcal{A}) - P_{\hat{X}}(\mathcal{A})| + |P_{\hat{X}}(\mathcal{A}^c) - P_X(\mathcal{A}^c)|$$

$$= \left| \sum_{x \in \mathcal{A}} [P_X(x) - P_{\hat{X}}(x)] \right| + \left| \sum_{x \in \mathcal{A}^c} [P_{\hat{X}}(x) - P_X(x)] \right|$$

$$= \sum_{x \in \mathcal{A}} |P_X(x) - P_{\hat{X}}(x)| + \sum_{x \in \mathcal{A}^c} |P_X(x) - P_{\hat{X}}(x)|$$

$$= \|P_X - P_{\hat{X}}\|.$$

Therefore, $\|P_X - P_{\hat{X}}\| = 2 \cdot \sup_{E \subset \mathcal{X}} |P_X(E) - P_{\hat{X}}(E)|$. $\square$

Lemma 2.37 (Variational distance vs divergence: Pinsker's inequality)

$$D(X \| \hat{X}) \geq \frac{\log_2(e)}{2} \cdot \|P_X - P_{\hat{X}}\|^2.$$

This result is referred to as Pinsker's inequality.

Proof

1. With $\mathcal{A} := \{x \in \mathcal{X} : P_X(x) > P_{\hat{X}}(x)\}$, we have from the previous lemma that

$$\|P_X - P_{\hat{X}}\| = 2[P_X(\mathcal{A}) - P_{\hat{X}}(\mathcal{A})].$$

2. Define two random variables U and $\hat{U}$ as

$$U = \begin{cases} 1, & \text{if } X \in \mathcal{A}, \\ 0, & \text{if } X \in \mathcal{A}^c, \end{cases}$$

and

$$\hat{U} = \begin{cases} 1, & \text{if } \hat{X} \in \mathcal{A}, \\ 0, & \text{if } \hat{X} \in \mathcal{A}^c. \end{cases}$$

Then P_X and $P_{\hat{X}}$ are refinements (2-refinements) of P_U and $P_{\hat{U}}$, respectively. From Lemma 2.33, we obtain that

$$D(P_X \| P_{\hat{X}}) \geq D(P_U \| P_{\hat{U}}).$$

3. The proof is complete if we show that

$$
\begin{aligned}
D(P_U \| P_{\hat{U}}) &\geq 2 \log_2(e)[P_X(\mathcal{A}) - P_{\hat{X}}(\mathcal{A})]^2 \\
&= 2 \log_2(e)[P_U(1) - P_{\hat{U}}(1)]^2.
\end{aligned}
$$

For ease of notations, let $p = P_U(1)$ and $q = P_{\hat{U}}(1)$. Then proving the above inequality is equivalent to showing that

$$p \cdot \ln \frac{p}{q} + (1-p) \cdot \ln \frac{1-p}{1-q} \geq 2(p-q)^2.$$

Define

$$f(p,q) := p \cdot \ln \frac{p}{q} + (1-p) \cdot \ln \frac{1-p}{1-q} - 2(p-q)^2,$$

and observe that

$$\frac{df(p,q)}{dq} = (p-q)\left(4 - \frac{1}{q(1-q)}\right) \leq 0 \quad \text{for } q \leq p.$$

Thus, $f(p,q)$ is non-increasing in q for $q \leq p$. Also note that $f(p,q) = 0$ for $q = p$. Therefore,

$$f(p,q) \geq 0 \quad \text{for } q \leq p.$$

The proof is completed by noting that

$$f(p,q) \geq 0 \quad \text{for } q \geq p,$$

since $f(1-p, 1-q) = f(p,q)$.

□

Observation 2.38 The above lemma tells us that for a sequence of distributions $\{(P_{X_n}, P_{\hat{X}_n})\}_{n \geq 1}$, when $D(P_{X_n} \| P_{\hat{X}_n})$ goes to zero as n goes to infinity, $\|P_{X_n} - P_{\hat{X}_n}\|$ goes to zero as well. But the converse does not necessarily hold. For a quick counterexample, let

$$P_{X_n}(0) = 1 - P_{X_n}(1) = \frac{1}{n} > 0$$

and

$$P_{\hat{X}_n}(0) = 1 - P_{\hat{X}_n}(1) = 0.$$

In this case,

$$D(P_{X_n} \| P_{\hat{X}_n}) \to \infty$$

since by convention, $(1/n) \cdot \log_2((1/n)/0) = \infty$. However,

$$\|P_X - P_{\hat{X}}\| = 2 \left[P_X \left(\{x : P_X(x) > P_{\hat{X}}(x)\} \right) - P_{\hat{X}} \left(\{x : P_X(x) > P_{\hat{X}}(x)\} \right) \right]$$

$$= \frac{2}{n} \to 0.$$

We however can upper bound $D(P_X \| P_{\hat{X}})$ by the variational distance between P_X and $P_{\hat{X}}$ when $D(P_X \| P_{\hat{X}}) < \infty$.

Lemma 2.39 *If $D(P_X \| P_{\hat{X}}) < \infty$, then*

$$D(P_X \| P_{\hat{X}}) \leq \frac{\log_2(e)}{\min_{\{x : P_X(x) > 0\}} \min\{P_X(x), P_{\hat{X}}(x)\}} \cdot \|P_X - P_{\hat{X}}\|.$$

Proof Without loss of generality, we assume that $P_X(x) > 0$ for all $x \in \mathcal{X}$. Since $D(P_X \| P_{\hat{X}}) < \infty$, we have that $P_X(x) > 0$ implies that $P_{\hat{X}}(x) > 0$. Let

$$t := \min_{\{x \in \mathcal{X} : P_X(x) > 0\}} \min\{P_X(x), P_{\hat{X}}(x)\}.$$

Then for all $x \in \mathcal{X}$,

$$\ln \frac{P_X(x)}{P_{\hat{X}}(x)} \leq \left| \ln \frac{P_X(x)}{P_{\hat{X}}(x)} \right|$$

$$\leq \left| \max_{\min\{P_X(x), P_{\hat{X}}(x)\} \leq s \leq \max\{P_X(x), P_{\hat{X}}(x)\}} \frac{d \ln(s)}{ds} \right| \cdot |P_X(x) - P_{\hat{X}}(x)|$$

$$= \frac{1}{\min\{P_X(x), P_{\hat{X}}(x)\}} \cdot |P_X(x) - P_{\hat{X}}(x)|$$

$$\leq \frac{1}{t} \cdot |P_X(x) - P_{\hat{X}}(x)|.$$

Hence,

$$
\begin{aligned}
D(P_X \| P_{\hat{X}}) &= \log_2(e) \sum_{x \in \mathcal{X}} P_X(x) \cdot \ln \frac{P_X(x)}{P_{\hat{X}}(x)} \\
&\leq \frac{\log_2(e)}{t} \sum_{x \in \mathcal{X}} P_X(x) \cdot |P_X(x) - P_{\hat{X}}(x)| \\
&\leq \frac{\log_2(e)}{t} \sum_{x \in \mathcal{X}} |P_X(x) - P_{\hat{X}}(x)| \\
&= \frac{\log_2(e)}{t} \cdot \|P_X - P_{\hat{X}}\|.
\end{aligned}
$$

$\square$

The next lemma discusses the effect of side information on divergence. As stated in Lemma 2.12, side information usually reduces entropy; it, however, increases divergence. One interpretation of these results is that side information is *useful*. Regarding entropy, side information provides us more information, so uncertainty decreases. As for divergence, it is the measure or index of how easy one can differentiate the source from two candidate distributions. The larger the divergence, the easier one can tell these two distributions apart and make the right guess. At an extreme case, when divergence is zero, one can never tell which distribution is the right one, since both produce the same source. So, when we obtain more information (side information), we should be able to make a better decision on the source statistics, which implies that the divergence should be larger.

Definition 2.40 (*Conditional divergence*) Given three discrete random variables, X, $\hat{X}$, and Z, where X and $\hat{X}$ have a common alphabet $\mathcal{X}$, we define the conditional divergence between X and $\hat{X}$ given Z by

$$
\begin{aligned}
D(X \| \hat{X} | Z) = D(P_{X|Z} \| P_{\hat{X}|Z} | P_Z) &:= \sum_{z \in \mathcal{Z}} P_Z(z) \sum_{x \in \mathcal{X}} P_{X|Z}(x|z) \log \frac{P_{X|Z}(x|z)}{P_{\hat{X}|Z}(x|z)} \\
&= \sum_{z \in \mathcal{Z}} \sum_{x \in \mathcal{X}} P_{X,Z}(x, z) \log \frac{P_{X|Z}(x|z)}{P_{\hat{X}|Z}(x|z)}.
\end{aligned}
$$

In other words, it is the conditional divergence between $P_{X|Z}$ and $P_{\hat{X}|Z}$ given P_Z and it is nothing but the expected value with respect to $P_{X,Z}$ of the log-likelihood ratio $\log \frac{P_{X|Z}}{P_{\hat{X}|Z}}$.

Similarly, the conditional divergence between $P_{X|Z}$ and $P_{\hat{X}}$ given P_Z is defined as

$$
D(P_{X|Z} \| P_{\hat{X}} | P_Z) := \sum_{z \in \mathcal{Z}} P_Z(z) \sum_{x \in \mathcal{X}} P_{X|Z}(x|z) \log \frac{P_{X|Z}(x|z)}{P_{\hat{X}}(z)}.
$$

Lemma 2.41 (Conditional mutual information and conditional divergence) *Given three discrete random variables X, Y and Z with alphabets $\mathcal{X}$, $\mathcal{Y}$ and $\mathcal{Z}$, respectively, and joint distribution $P_{X,Y,Z}$, we have*

$$I(X;Y|Z) = D(P_{X,Y|Z} \| P_{X|Z} P_{Y|Z} | P_Z)$$

$$= \sum_{x \in \mathcal{X}} \sum_{y \in \mathcal{Y}} \sum_{z \in \mathcal{Z}} P_{X,Y,Z}(x,y,z) \log_2 \frac{P_{X,Y|Z}(x,y|z)}{P_{X|Z}(x|z) P_{Y|Z}(y|z)},$$

where $P_{X,Y|Z}$ is the conditional joint distribution of X and Y given Z, and $P_{X|Z}$ and $P_{Y|Z}$ are the conditional distributions of X and Y, respectively, given Z.

Proof The proof follows directly from the definition of conditional mutual information (2.2.2) and the above definition of conditional divergence. □

Lemma 2.42 (Chain rule for divergence)
Let P_{X^n} and Q_{X^n} be two joint distributions on $\mathcal{X}^n$. We have that

$$D(P_{X_1,X_2} \| Q_{X_1,X_2}) = D(P_{X_1} \| Q_{X_1}) + D(P_{X_2|X_1} \| Q_{X_2|X_1} | P_{X_1}),$$

and more generally,

$$D(P_{X^n} \| Q_{X^n}) = \sum_{i=1}^{n} D(P_{X_i|X^{i-1}} \| Q_{X_i|X^{i-1}} | P_{X^{i-1}}),$$

where $D(P_{X_i|X^{i-1}} \| Q_{X_i|X^{i-1}} | P_{X^{i-1}}) := D(P_{X_1} \| Q_{X_1})$ for $i = 1$.

Proof The proof readily follows from the above divergence definitions. □

Lemma 2.43 (Conditioning never decreases divergence) *For three discrete random variables, X, $\hat{X}$, and Z, where X and $\hat{X}$ have a common alphabet $\mathcal{X}$, we have that*

$$D(P_{X|Z} \| P_{\hat{X}|Z} | P_Z) \geq D(P_X \| P_{\hat{X}}).$$

Proof

$$D(P_{X|Z} \| P_{\hat{X}|Z} | P_Z) - D(P_X \| P_{\hat{X}})$$

$$= \sum_{z \in \mathcal{Z}} \sum_{x \in \mathcal{X}} P_{X,Z}(x,z) \cdot \log_2 \frac{P_{X|Z}(x|z)}{P_{\hat{X}|Z}(x|z)} - \sum_{x \in \mathcal{X}} P_X(x) \cdot \log_2 \frac{P_X(x)}{P_{\hat{X}}(x)}$$

$$= \sum_{z \in \mathcal{Z}} \sum_{x \in \mathcal{X}} P_{X,Z}(x,z) \cdot \log_2 \frac{P_{X|Z}(x|z)}{P_{\hat{X}|Z}(x|z)} - \sum_{x \in \mathcal{X}} \left(\sum_{z \in \mathcal{Z}} P_{X,Z}(x,z) \right) \cdot \log_2 \frac{P_X(x)}{P_{\hat{X}}(x)}$$

$$= \sum_{z \in \mathcal{Z}} \sum_{x \in \mathcal{X}} P_{X,Z}(x,z) \cdot \log_2 \frac{P_{X|Z}(x|z) P_{\hat{X}}(x)}{P_{\hat{X}|Z}(x|z) P_X(x)}$$

$$\geq \sum_{z \in \mathcal{Z}} \sum_{x \in \mathcal{X}} P_{X,Z}(x, z) \cdot \log_2(e) \left(1 - \frac{P_{\hat{X}|Z}(x|z) P_X(x)}{P_{X|Z}(x|z) P_{\hat{X}}(x)} \right) \quad \text{(by the FI Lemma)}$$

$$= \log_2(e) \left(1 - \sum_{x \in \mathcal{X}} \frac{P_X(x)}{P_{\hat{X}}(x)} \sum_{z \in \mathcal{Z}} P_Z(z) P_{\hat{X}|Z}(x|z) \right)$$

$$= \log_2(e) \left(1 - \sum_{x \in \mathcal{X}} \frac{P_X(x)}{P_{\hat{X}}(x)} P_{\hat{X}}(x) \right)$$

$$= \log_2(e) \left(1 - \sum_{x \in \mathcal{X}} P_X(x) \right) = 0,$$

with equality holding iff for all x and z,

$$\frac{P_X(x)}{P_{\hat{X}}(x)} = \frac{P_{X|Z}(x|z)}{P_{\hat{X}|Z}(x|z)}.$$

$\square$

Note that it is not necessary that

$$D(P_{X|Z} \| P_{\hat{X}|\hat{Z}} | P_Z) \geq D(P_X \| P_{\hat{X}}),$$

where Z and $\hat{Z}$ also have a common alphabet. In other words, side information is helpful for divergence only when it provides information on the similarity or difference of the two distributions. In the above case, Z only provides information about X, and $\hat{Z}$ provides information about $\hat{X}$; so the divergence certainly cannot be expected to increase. The next lemma shows that if the pair $(Z, \hat{Z})$ is independent component-wise of the pair $(X, \hat{X})$, then the side information of $(Z, \hat{Z})$ does not help in improving the divergence of X against $\hat{X}$.

Lemma 2.44 (Independent side information does not change divergence) *If X is independent of Z and $\hat{X}$ is independent of $\hat{Z}$, where X and Z share a common alphabet with $\hat{X}$ and $\hat{Z}$, respectively, then*

$$D(P_{X|Z} \| P_{\hat{X}|\hat{Z}} | P_Z) = D(P_X \| P_{\hat{X}}).$$

Proof This can be easily justified from the divergence definitions. $\square$

Corollary 2.45 (Additivity of divergence under independence) *If X is independent of Z and $\hat{X}$ is independent of $\hat{Z}$, where X and Z share a common alphabet with $\hat{X}$ and $\hat{Z}$, respectively, then*

$$D(P_{X,Z} \| P_{\hat{X},\hat{Z}}) = D(P_X \| P_{\hat{X}}) + D(P_Z \| P_{\hat{Z}}).$$

2.7 Convexity/Concavity of Information Measures

We next address the convexity/concavity properties of information measures with respect to the distributions on which they are defined. Such properties will be useful when optimizing the information measures over distribution spaces.

Lemma 2.46

1. $H(P_X)$ *is a concave function of* P_X, *namely*

$$H(\lambda P_X + (1 - \lambda)P_{\widetilde{X}}) \geq \lambda H(P_X) + (1 - \lambda)H(P_{\widetilde{X}})$$

   for all $\lambda \in [0, 1]$.
2. *Noting that* $I(X;Y)$ *can be rewritten as* $I(P_X, P_{Y|X})$, *where*

$$I(P_X, P_{Y|X}) := \sum_{x \in \mathcal{X}} \sum_{y \in \mathcal{Y}} P_{Y|X}(y|x)P_X(x) \log_2 \frac{P_{Y|X}(y|x)}{\sum_{a \in \mathcal{X}} P_{Y|X}(y|a)P_X(a)},$$

   then $I(X;Y)$ *is a concave function of* P_X *(for fixed* $P_{Y|X}$*), and a convex function of* $P_{Y|X}$ *(for fixed* P_X*).*
3. $D(P_X \| P_{\hat{X}})$ *is convex with respect to both the first argument* P_X *and the second argument* $P_{\hat{X}}$. *It is also convex in the pair* $(P_X, P_{\hat{X}})$; *i.e., if* $(P_X, P_{\hat{X}})$ *and* $(Q_X, Q_{\hat{X}})$ *are two pairs of probability mass functions, then*

$$D(\lambda P_X + (1 - \lambda)Q_X \| \lambda P_{\hat{X}} + (1 - \lambda)Q_{\hat{X}})$$
$$\leq \lambda \cdot D(P_X \| P_{\hat{X}}) + (1 - \lambda) \cdot D(Q_X \| Q_{\hat{X}}), \tag{2.7.1}$$

for all $\lambda \in [0, 1]$.

Proof 1. The proof uses the log-sum inequality:

$$H(\lambda P_X + (1 - \lambda)P_{\widetilde{X}}) - \left[\lambda H(P_X) + (1 - \lambda)H(P_{\widetilde{X}})\right]$$

$$= \lambda \sum_{x \in \mathcal{X}} P_X(x) \log_2 \frac{P_X(x)}{\lambda P_X(x) + (1 - \lambda)P_{\widetilde{X}}(x)}$$

$$+ (1 - \lambda) \sum_{x \in \mathcal{X}} P_{\widetilde{X}}(x) \log_2 \frac{P_{\widetilde{X}}(x)}{\lambda P_X(x) + (1 - \lambda)P_{\widetilde{X}}(x)}$$

$$\geq \lambda \left(\sum_{x \in \mathcal{X}} P_X(x)\right) \log_2 \frac{\sum_{x \in \mathcal{X}} P_X(x)}{\sum_{x \in \mathcal{X}}[\lambda P_X(x) + (1 - \lambda)P_{\widetilde{X}}(x)]}$$

$$+ (1 - \lambda) \left(\sum_{x \in \mathcal{X}} P_{\widetilde{X}}(x)\right) \log_2 \frac{\sum_{x \in \mathcal{X}} P_{\widetilde{X}}(x)}{\sum_{x \in \mathcal{X}}[\lambda P_X(x) + (1 - \lambda)P_{\widetilde{X}}(x)]}$$

$$= 0,$$

with equality holding iff $P_X(x) = P_{\widetilde{X}}(x)$ for all x.

2. We first show the concavity of $I(P_X, P_{Y|X})$ with respect to P_X. Let $\bar{\lambda} = 1 - \lambda$.

$$I(\lambda P_X + \bar{\lambda} P_{\widetilde{X}}, P_{Y|X}) - \lambda I(P_X, P_{Y|X}) - \bar{\lambda} I(P_{\widetilde{X}}, P_{Y|X})$$

$$= \lambda \sum_{y \in \mathcal{Y}} \sum_{x \in \mathcal{X}} P_X(x) P_{Y|X}(y|x) \log_2 \frac{\sum_{x \in \mathcal{X}} P_X(x) P_{Y|X}(y|x)}{\sum_{x \in \mathcal{X}} [\lambda P_X(x) + \bar{\lambda} P_{\widetilde{X}}(x)] P_{Y|X}(y|x)}$$

$$+ \bar{\lambda} \sum_{y \in \mathcal{Y}} \sum_{x \in \mathcal{X}} P_{\widetilde{X}}(x) P_{Y|X}(y|x) \log_2 \frac{\sum_{x \in \mathcal{X}} P_{\widetilde{X}}(x) P_{Y|X}(y|x)}{\sum_{x \in \mathcal{X}} [\lambda P_X(x) + \bar{\lambda} P_{\widetilde{X}}(x)] P_{Y|X}(y|x)}$$

$$\geq 0, \quad \text{(by the log-sum inequality)}$$

with equality holding iff

$$\sum_{x \in \mathcal{X}} P_X(x) P_{Y|X}(y|x) = \sum_{x \in \mathcal{X}} P_{\widetilde{X}}(x) P_{Y|X}(y|x)$$

for all $y \in \mathcal{Y}$. We now turn to the convexity of $I(P_X, P_{Y|X})$ with respect to $P_{Y|X}$. For ease of notation, let $P_{Y_\lambda}(y) := \lambda P_Y(y) + \bar{\lambda} P_{\widetilde{Y}}(y)$, and $P_{Y_\lambda|X}(y|x) := \lambda P_{Y|X}(y|x) + \bar{\lambda} P_{\widetilde{Y}|X}(y|x)$. Then

$$\lambda I(P_X, P_{Y|X}) + \bar{\lambda} I(P_X, P_{\widetilde{Y}|X}) - I(P_X, \lambda P_{Y|X} + \bar{\lambda} P_{\widetilde{Y}|X})$$

$$= \lambda \sum_{x \in \mathcal{X}} \sum_{y \in \mathcal{Y}} P_X(x) P_{Y|X}(y|x) \log_2 \frac{P_{Y|X}(y|x)}{P_Y(y)}$$

$$+ \bar{\lambda} \sum_{x \in \mathcal{X}} \sum_{y \in \mathcal{Y}} P_X(x) P_{\widetilde{Y}|X}(y|x) \log_2 \frac{P_{\widetilde{Y}|X}(y|x)}{P_{\widetilde{Y}}(y)}$$

$$- \sum_{x \in \mathcal{X}} \sum_{y \in \mathcal{Y}} P_X(x) P_{Y_\lambda|X}(y|x) \log_2 \frac{P_{Y_\lambda|X}(y|x)}{P_{Y_\lambda}(y)}$$

$$= \lambda \sum_{x \in \mathcal{X}} \sum_{y \in \mathcal{Y}} P_X(x) P_{Y|X}(y|x) \log_2 \frac{P_{Y|X}(y|x) P_{Y_\lambda}(y)}{P_Y(y) P_{Y_\lambda|X}(y|x)}$$

$$+ \bar{\lambda} \sum_{x \in \mathcal{X}} \sum_{y \in \mathcal{Y}} P_X(x) P_{\widetilde{Y}|X}(y|x) \log_2 \frac{P_{\widetilde{Y}|X}(y|x) P_{Y_\lambda}(y)}{P_{\widetilde{Y}}(y) P_{Y_\lambda|X}(y|x)}$$

$$\geq \lambda \log_2(e) \sum_{x \in \mathcal{X}} \sum_{y \in \mathcal{Y}} P_X(x) P_{Y|X}(y|x) \left(1 - \frac{P_Y(y) P_{Y_\lambda|X}(y|x)}{P_{Y|X}(y|x) P_{Y_\lambda}(y)} \right)$$

$$+ \bar{\lambda} \log_2(e) \sum_{x \in \mathcal{X}} \sum_{y \in \mathcal{Y}} P_X(x) P_{\widetilde{Y}|X}(y|x) \left(1 - \frac{P_{\widetilde{Y}}(y) P_{Y_\lambda|X}(y|x)}{P_{\widetilde{Y}|X}(y|x) P_{Y_\lambda}(y)} \right)$$

$$= 0,$$

where the inequality follows from the FI Lemma, with equality holding iff

$$(\forall x \in \mathcal{X}, y \in \mathcal{Y}) \frac{P_Y(y)}{P_{Y|X}(y|x)} = \frac{P_{\widetilde{Y}}(y)}{P_{\widetilde{Y}|X}(y|x)}.$$

3. For ease of notation, let $P_{X_\lambda}(x) := \lambda P_X(x) + (1-\lambda)P_{\widetilde{X}}(x)$.

$$\lambda D(P_X \| P_{\hat{X}}) + (1-\lambda)D(P_{\widetilde{X}} \| P_{\hat{X}}) - D(P_{X_\lambda} \| P_{\hat{X}})$$

$$= \lambda \sum_{x \in \mathcal{X}} P_X(x) \log_2 \frac{P_X(x)}{P_{X_\lambda}(x)} + (1-\lambda) \sum_{x \in \mathcal{X}} P_{\widetilde{X}}(x) \log_2 \frac{P_{\widetilde{X}}(x)}{P_{X_\lambda}(x)}$$

$$= \lambda D(P_X \| P_{X_\lambda}) + (1-\lambda)D(P_{\widetilde{X}} \| P_{X_\lambda})$$

$$\geq 0$$

by the nonnegativity of the divergence, with equality holding iff $P_X(x) = P_{\widetilde{X}}(x)$ for all x. Similarly, by letting $P_{\hat{X}_\lambda}(x) := \lambda P_{\hat{X}}(x) + (1-\lambda)P_{\widetilde{X}}(x)$, we obtain

$$\lambda D(P_X \| P_{\hat{X}}) + (1-\lambda)D(P_X \| P_{\widetilde{X}}) - D(P_X \| P_{\hat{X}_\lambda})$$

$$= \lambda \sum_{x \in \mathcal{X}} P_X(x) \log_2 \frac{P_{\hat{X}_\lambda}(x)}{P_{\hat{X}}(x)} + (1-\lambda) \sum_{x \in \mathcal{X}} P_X(x) \log_2 \frac{P_{\hat{X}_\lambda}(x)}{P_{\widetilde{X}}(x)}$$

$$\geq \frac{\lambda}{\ln 2} \sum_{x \in \mathcal{X}} P_X(x) \left(1 - \frac{P_{\hat{X}}(x)}{P_{\hat{X}_\lambda}(x)}\right) + \frac{(1-\lambda)}{\ln 2} \sum_{x \in \mathcal{X}} P_X(x) \left(1 - \frac{P_{\widetilde{X}}(x)}{P_{\hat{X}_\lambda}(x)}\right)$$

$$= \log_2(e) \left(1 - \sum_{x \in \mathcal{X}} P_X(x) \frac{\lambda P_{\hat{X}}(x) + (1-\lambda)P_{\widetilde{X}}(x)}{P_{\hat{X}_\lambda}(x)}\right)$$

$$= 0,$$

where the inequality follows from the FI Lemma, with equality holding iff $P_{\widetilde{X}}(x) = P_{\hat{X}}(x)$ for all x.

Finally, by the log-sum inequality, for each $x \in \mathcal{X}$, we have

$$\left(\lambda P_X(x) + (1-\lambda)Q_X(x)\right) \log_2 \frac{\lambda P_X(x) + (1-\lambda)Q_X(x)}{\lambda P_{\hat{X}}(x) + (1-\lambda)Q_{\hat{X}}(x)}$$

$$\leq \lambda P_X(x) \log_2 \frac{\lambda P_X(x)}{\lambda P_{\hat{X}}(x)} + (1-\lambda)Q_X(x) \log_2 \frac{(1-\lambda)Q_X(x)}{(1-\lambda)Q_{\hat{X}}(x)}.$$

Summing over x, we yield (2.7.1).

Note that the last result (convexity of $D(P_X \| P_{\hat{X}})$ in the pair $(P_X, P_{\hat{X}})$) actually implies the first two results: just set $P_{\hat{X}} = Q_{\hat{X}}$ to show convexity in the first argument P_X, and set $P_X = Q_X$ to show convexity in the second argument $P_{\hat{X}}$. $\square$

Observation 2.47 (*Applications of information measures*) In addition to playing a critical role in communications and information theory, the above information measures, as well as their extensions (such as Rényi's information measures, see Sect. 2.9) and their counterparts for continuous-alphabet systems (see Chap. 5) have been applied in many domains. Recall that entropy measures statistical uncertainty, divergence measures statistical dissimilarity, and mutual information quantifies statistical dependence or information transfer in stochastic systems.

One example where entropy has been extensively employed is in the so-called *maximum entropy principle* methods. This principle, originally espoused by Jaynes [201–203] who saw a close connection between statistical mechanics and information theory, states that given past observations, the probability distribution that best characterizes current statistical behavior is the one with the largest entropy. In other words, given prior data constraints (expressed in the form of moments or averages), the best representative distribution should be, beyond satisfying the constraints, the least informative or as unbiased as possible.

Indeed, entropy together with divergence and mutual information have been used as powerful tools in a wide range of fields, including image processing, computer vision, pattern recognition and machine learning [48, 89, 111, 123, 163, 253, 384, 385], cryptography and data privacy [6, 7, 31, 41, 53, 65, 66, 94, 197, 213, 214, 260, 264, 269, 270, 327, 328, 342, 403, 413, 414], quantum information theory, quantum cryptography and computing [40, 105, 188, 408], biology and molecular communication [2, 59, 179, 278, 390], stochastic control under communication constraints [376, 377, 418], neuroscience [60, 287, 374], natural language processing and linguistics [181, 259, 297, 363], and economics [82, 353, 379].

2.8 Fundamentals of Hypothesis Testing

One of the fundamental problems in statistics is to decide between two alternative explanations for the observed data. For example, when gambling, one may wish to test whether the game is fair or not. Similarly, a sequence of observations on the market may reveal information on whether or not a new product is successful. These are examples of the simplest form of the hypothesis testing problem, which is usually named *simple hypothesis testing*.

Hypothesis testing has also close connections with information theory. For example, we will see that the divergence plays a key role in the asymptotic error analysis of Neyman–Pearson hypothesis testing (see Lemma 2.49).

The simple hypothesis testing problem can be formulated as follows:

Problem Let $X_1, \ldots, X_n$ be a sequence of observations which is drawn according to either a "null hypothesis" distribution P_{X^n} or an "alternative hypothesis" distribution $P_{\hat{X}^n}$. The hypotheses are usually denoted by

- $H_0 : P_{X^n}$
- $H_1 : P_{\hat{X}^n}$

Based on one sequence of observations x^n, one has to decide which of the hypotheses is true. This is denoted by a decision mapping $\phi(\cdot)$, where

$$\phi(x^n) = \begin{cases} 0, \text{ if distribution of } X^n \text{ is classified to be } P_{X^n}; \\ 1, \text{ if distribution of } X^n \text{ is classified to be } P_{\hat{X}^n}. \end{cases}$$

Accordingly, the possible observed sequences are divided into two groups:

$$\text{Acceptance region for } H_0 : \{x^n \in \mathcal{X}^n : \phi(x^n) = 0\}$$
$$\text{Acceptance region for } H_1 : \{x^n \in \mathcal{X}^n : \phi(x^n) = 1\}.$$

Hence, depending on the true distribution, there are two types of error probabilities:

$$\text{Type I error} : \alpha_n = \alpha_n(\phi) := P_{X^n}\left(\{x^n \in \mathcal{X}^n : \phi(x^n) = 1\}\right)$$
$$\text{Type II error} : \beta_n = \beta_n(\phi) := P_{\hat{X}^n}\left(\{x^n \in \mathcal{X}^n : \phi(x^n) = 0\}\right).$$

The choice of the decision mapping is dependent on the optimization criterion. Two of the most frequently used ones in information theory are

1. Bayesian hypothesis testing.
Here, $\phi(\cdot)$ is chosen so that the Bayesian cost

$$\pi_0 \alpha_n + \pi_1 \beta_n$$

is minimized, where π_0 and π_1 are the prior probabilities for the null and alternative hypotheses, respectively. The mathematical expression for Bayesian testing is

$$\min_{\{\phi\}} \left[\pi_0 \alpha_n(\phi) + \pi_1 \beta_n(\phi)\right].$$

2. Neyman–Pearson hypothesis testing subject to a fixed test level.
Here, $\phi(\cdot)$ is chosen so that the type II error β_n is minimized subject to a constant bound on the type I error; i.e.,

$$\alpha_n \leq \varepsilon,$$

where $\varepsilon > 0$ is fixed. The mathematical expression for Neyman–Pearson testing is

$$\min_{\{\phi : \alpha_n(\phi) \leq \varepsilon\}} \beta_n(\phi).$$

The set $\{\phi\}$ considered in the minimization operation could have two different ranges: range over *deterministic* rules, and range over *randomization* rules. The main difference between a randomization rule and a deterministic rule is that the former allows the mapping $\phi(x^n)$ to be random on $\{0, 1\}$ for some x^n, while the latter only accepts deterministic assignments to $\{0, 1\}$ for all x^n. For example, a randomization rule for specific observations $\tilde{x}^n$ can be

$$\phi(\tilde{x}^n) = \begin{cases} 0, & \text{with probability } 0.2, \\ 1, & \text{with probability } 0.8. \end{cases}$$

The Neyman–Pearson lemma shows the well-known fact that the *likelihood ratio test* is always the optimal test [281].

Lemma 2.48 (Neyman–Pearson Lemma) *For a simple hypothesis testing problem, define an acceptance region for the null hypothesis through the* likelihood ratio *as*

$$\mathcal{A}_n(\tau) := \left\{ x^n \in \mathcal{X}^n : \frac{P_{X^n}(x^n)}{P_{\hat{X}^n}(x^n)} > \tau \right\},$$

and let

$$\alpha_n^* := P_{X^n} \left\{ \mathcal{A}_n^c(\tau) \right\}$$

and

$$\beta_n^* := P_{\hat{X}^n} \left\{ \mathcal{A}_n(\tau) \right\}.$$

Then for type I error α_n and type II error β_n associated with another choice of acceptance region for the null hypothesis, we have

$$\alpha_n \leq \alpha_n^* \implies \beta_n \geq \beta_n^*.$$

Proof Let $\mathcal{B}$ be a choice of acceptance region for the null hypothesis. Then

$$
\begin{aligned}
\alpha_n + \tau\beta_n &= \sum_{x^n \in \mathcal{B}^c} P_{X^n}(x^n) + \tau \sum_{x^n \in \mathcal{B}} P_{\hat{X}^n}(x^n) \\
&= \sum_{x^n \in \mathcal{B}^c} P_{X^n}(x^n) + \tau \left[1 - \sum_{x^n \in \mathcal{B}^c} P_{\hat{X}^n}(x^n) \right] \\
&= \tau + \sum_{x^n \in \mathcal{B}^c} \left[P_{X^n}(x^n) - \tau P_{\hat{X}^n}(x^n) \right].
\end{aligned}
\tag{2.8.1}
$$

Observe that (2.8.1) is minimized by choosing $\mathcal{B} = \mathcal{A}_n(\tau)$. Hence,

$$\alpha_n + \tau\beta_n \geq \alpha_n^* + \tau\beta_n^*,$$

which immediately implies the desired result. $\square$

The Neyman–Pearson lemma indicates that no other choices of acceptance regions can simultaneously improve both type I and type II errors of the likelihood ratio test. Indeed, from (2.8.1), it is clear that for any α_n and β_n, one can always find a likelihood ratio test that performs as good. Therefore, the likelihood ratio test is an optimal test. The statistical properties of the likelihood ratio thus become essential in hypothesis testing. Note that, when the observations are i.i.d. under both hypotheses, the divergence, which is the statistical expectation of the log-likelihood ratio, plays an important role in hypothesis testing (for non-memoryless observations, one is then concerned with the *divergence rate*, an extended notion of divergence for systems with memory which will be defined in the following chapter) as the *exponent* of the best type II error. More specifically, we have the following result, known as the Chernoff–Stein lemma [78].

Lemma 2.49 (Chernoff–Stein lemma) *For a sequence of i.i.d. observations X^n which is possibly drawn from either the null hypothesis distribution P_{X^n} or the alternative hypothesis distribution $P_{\hat{X}^n}$, the best type II error satisfies*

$$\lim_{n \to \infty} -\frac{1}{n} \log_2 \beta_n^*(\varepsilon) = D(P_X \| P_{\hat{X}}),$$

for any $\varepsilon \in (0, 1)$, where $\beta_n^*(\varepsilon) = \min_{\alpha_n \leq \varepsilon} \beta_n$, and α_n and β_n are the type I and type II errors, respectively.

Proof Forward Part: In this part, we prove that there exists an acceptance region for the null hypothesis such that

$$\liminf_{n \to \infty} -\frac{1}{n} \log_2 \beta_n(\varepsilon) \geq D(P_X \| P_{\hat{X}}).$$

Step 1: Divergence typical set. For any $\delta > 0$, define the divergence typical set as

$$\mathcal{A}_n(\delta) := \left\{ x^n \in \mathcal{X}^n : \left| \frac{1}{n} \log_2 \frac{P_{X^n}(x^n)}{P_{\hat{X}^n}(x^n)} - D(P_X \| P_{\hat{X}}) \right| < \delta \right\}.$$

Note that any sequence x^n in this set satisfies

$$P_{\hat{X}^n}(x^n) \leq P_{X^n}(x^n) 2^{-n(D(P_X \| P_{\hat{X}}) - \delta)}.$$

Step 2: Computation of type I error. The observations being i.i.d., we have by the weak law of large numbers that

$$P_{X^n}(\mathcal{A}_n(\delta)) \to 1 \qquad \text{as } n \to \infty.$$

Hence,

$$\alpha_n = P_{X^n}(\mathcal{A}_n^c(\delta)) < \varepsilon$$

for sufficiently large n.

Step 3: Computation of type II error.

$$\begin{aligned}
\beta_n(\varepsilon) &= P_{\hat{X}^n}(\mathcal{A}_n(\delta)) \\
&= \sum_{x^n \in \mathcal{A}_n(\delta)} P_{\hat{X}^n}(x^n) \\
&\leq \sum_{x^n \in \mathcal{A}_n(\delta)} P_{X^n}(x^n) 2^{-n(D(P_X \| P_{\hat{X}}) - \delta)} \\
&= 2^{-n(D(P_X \| P_{\hat{X}}) - \delta)} \sum_{x^n \in \mathcal{A}_n(\delta)} P_{X^n}(x^n) \\
&= 2^{-n(D(P_X \| P_{\hat{X}}) - \delta)}(1 - \alpha_n).
\end{aligned}$$

Hence,

$$-\frac{1}{n} \log_2 \beta_n(\varepsilon) \geq D(P_X \| P_{\hat{X}}) - \delta + \frac{1}{n} \log_2(1 - \alpha_n),$$

which implies that

$$\liminf_{n \to \infty} -\frac{1}{n} \log_2 \beta_n(\varepsilon) \geq D(P_X \| P_{\hat{X}}) - \delta.$$

The above inequality is true for any $\delta > 0$; therefore,

$$\liminf_{n \to \infty} -\frac{1}{n} \log_2 \beta_n(\varepsilon) \geq D(P_X \| P_{\hat{X}}).$$

Converse Part: We next prove that for any acceptance region $\mathcal{B}_n$ for the null hypothesis satisfying the type I error constraint, i.e.,

$$\alpha_n(\mathcal{B}_n) = P_{X^n}(\mathcal{B}_n^c) \leq \varepsilon,$$

its type II error $\beta_n(\mathcal{B}_n)$ satisfies

$$\limsup_{n \to \infty} -\frac{1}{n} \log_2 \beta_n(\mathcal{B}_n) \leq D(P_X \| P_{\hat{X}}).$$

We have

$$
\begin{aligned}
\beta_n(\mathcal{B}_n) = P_{\hat{X}^n}(\mathcal{B}_n) &\geq P_{\hat{X}^n}(\mathcal{B}_n \cap \mathcal{A}_n(\delta)) \\
&\geq \sum_{x^n \in \mathcal{B}_n \cap \mathcal{A}_n(\delta)} P_{\hat{X}^n}(x^n) \\
&\geq \sum_{x^n \in \mathcal{B}_n \cap \mathcal{A}_n(\delta)} P_{X^n}(x^n) 2^{-n(D(P_X \| P_{\hat{X}}) + \delta)} \\
&= 2^{-n(D(P_X \| P_{\hat{X}}) + \delta)} P_{X^n}(\mathcal{B}_n \cap \mathcal{A}_n(\delta)) \\
&\geq 2^{-n(D(P_X \| P_{\hat{X}}) + \delta)} \left[1 - P_{X^n}(\mathcal{B}_n^c) - P_{X^n}\left(\mathcal{A}_n^c(\delta) \right) \right] \\
&= 2^{-n(D(P_X \| P_{\hat{X}}) + \delta)} \left[1 - \alpha_n(\mathcal{B}_n) - P_{X^n}\left(\mathcal{A}_n^c(\delta) \right) \right] \\
&\geq 2^{-n(D(P_X \| P_{\hat{X}}) + \delta)} \left[1 - \varepsilon - P_{X^n}\left(\mathcal{A}_n^c(\delta) \right) \right].
\end{aligned}
$$

Hence,

$$
-\frac{1}{n} \log_2 \beta_n(\mathcal{B}_n) \leq D(P_X \| P_{\hat{X}}) + \delta + \frac{1}{n} \log_2 \left[1 - \varepsilon - P_{X^n}\left(\mathcal{A}_n^c(\delta) \right) \right],
$$

which, upon noting that $\lim_{n \to \infty} P_{X^n}\left(\mathcal{A}_n^c(\delta) \right) = 0$ (by the weak law of large numbers), implies that

$$
\limsup_{n \to \infty} -\frac{1}{n} \log_2 \beta_n(\mathcal{B}_n) \leq D(P_X \| P_{\hat{X}}) + \delta.
$$

The above inequality is true for any $\delta > 0$; therefore,

$$
\limsup_{n \to \infty} -\frac{1}{n} \log_2 \beta_n(\mathcal{B}_n) \leq D(P_X \| P_{\hat{X}}).
$$

$\square$

2.9 Rényi's Information Measures

We close this chapter by briefly introducing generalized information measures due to Rényi [317], which subsume Shannon's measures as limiting cases.

Definition 2.50 (*Rényi's entropy*) Given a parameter $\alpha > 0$ with $\alpha \neq 1$, and given a discrete random variable X with alphabet $\mathcal{X}$ and distribution P_X, its Rényi entropy of order α is given by

$$
H_\alpha(X) = \frac{1}{1 - \alpha} \log \left(\sum_{x \in \mathcal{X}} P_X(x)^\alpha \right). \tag{2.9.1}
$$

As in case of the Shannon entropy, the base of the logarithm determines the units; if the base is D, Rényi's entropy is in D-ary units. Other notations for $H_\alpha(X)$ are $H(X; \alpha)$, $H_\alpha(P_X)$, and $H(P_X; \alpha)$.

Definition 2.51 (*Rényi's divergence*) Given a parameter $0 < \alpha < 1$, and two discrete random variables X and $\hat{X}$ with common alphabet $\mathcal{X}$ and distribution P_X and $P_{\hat{X}}$, respectively, then the Rényi divergence of order α between X and $\hat{X}$ is given by

$$D_\alpha(X \| \hat{X}) = \frac{1}{\alpha - 1} \log \left(\sum_{x \in \mathcal{X}} \left[P_X^\alpha(x) P_{\hat{X}}^{1-\alpha}(x) \right] \right). \qquad (2.9.2)$$

This definition can be extended to $\alpha > 1$ if $P_{\hat{X}}(x) > 0$ for all $x \in \mathcal{X}$. Other notations for $D_\alpha(X \| \hat{X})$ are $D(X \| \hat{X}; \alpha)$, $D_\alpha(P_X \| P_{\hat{X}})$ and $D(P_X \| P_{\hat{X}}; \alpha)$.

As in the case of Shannon's information measures, the base of the logarithm indicates the units of the measure and can be changed from 2 to an arbitrary $b > 1$. In the next lemma, whose proof is left as an exercise, we note that in the limit of α tending to 1, Shannon's entropy and divergence can be recovered from Rényi's entropy and divergence, respectively.

Lemma 2.52 *When $\alpha \to 1$, we have the following:*

$$\lim_{\alpha \to 1} H_\alpha(X) = H(X) \qquad (2.9.3)$$

and

$$\lim_{\alpha \to 1} D_\alpha(X \| \hat{X}) = D(X \| \hat{X}). \qquad (2.9.4)$$

Observation 2.53 (*Operational meaning of Rényi's information measures*) Rényi's entropy has been shown to have an operational characterization for many problems, including lossless variable-length source coding under an exponential cost constraint [54, 67, 68, 310] (see also Observation 3.30 in Chap. 3), buffer overflow in source coding [206], fixed-length source coding [76, 86] and others areas [1, 20, 36, 308, 318]. Furthermore, Rényi's divergence has played a prominent role in hypothesis testing questions [17, 86, 186, 225, 279, 280].

Observation 2.54 (*α-mutual information*) While Rényi did not propose a mutual information of order α that generalizes Shannon's mutual information, there are at least three different possible definitions of such measure due to Sibson [352], Arimoto [28] and Csiszár [86], respectively. We refer the reader to [86, 395] for discussions on the properties and merits of these different measures.

Observation 2.55 (*Information measures for continuous distributions*) Note that the above information measures defined for discrete distributions can similarly be defined for continuous distributions admitting densities with the usual straightforward modifications (with densities replacing pmf's and integrals replacing summations). See Chap. 5 for a study of Shannon's differential entropy and divergence for continuous distributions (also for continuous distributions, closed-form expressions for Shannon's differential entropy and Rényi's entropy can be found in [360] and expressions for Rényi's divergence are derived in [144, 246]).

Problems

1. Prove the FI Lemma.
2. Show that the two conditions in Footnote 8 are equivalent.
3. For a finite-alphabet random variable X, show that $H(X) \leq \log_2 |\mathcal{X}|$ using the log-sum inequality.
4. Given a pair of random variables (X, Y), is $H(X|Y) = H(Y|X)$? If not, when do we have equality?
5. Given a discrete random variable X with alphabet $\mathcal{X} \subset \{1, 2, \ldots\}$, what is the relationship between $H(X)$ and $H(Y)$ when Y is defined as follows.

   (a) $Y = \log_2(X)$.
   (b) $Y = X^2$.

6. Show that the entropy of a function f of X is less than or equal to the entropy of X.

   Hint: By the chain rule for entropy,

   $$H(X, f(X)) = H(X) + H(f(X)|X) = H(f(X)) + H(X|f(X)).$$

7. Show that $H(Y|X) = 0$ iff Y is a function of X.
8. Give examples of:

   (a) $I(X; Y|Z) < I(X; Y)$.
   (b) $I(X; Y|Z) > I(X; Y)$.

   Hint: For (a), create example for $I(X; Y|Z) = 0$ and $I(X; Y) > 0$. For (b), create example for $I(X; Y) = 0$ and $I(X; Y|Z) > 0$.

9. Let the joint distribution of X and Y be:

X \ Y	0	1
0	$\frac{1}{4}$	0
1	$\frac{1}{2}$	$\frac{1}{4}$

   Draw the Venn diagram for

   $$H(X), H(Y), H(X|Y), H(Y|X), H(X, Y) \text{ and } I(X; Y),$$

   and indicate the quantities (in bits) for each area of the Venn diagram.

10. *Maximal discrete entropy.* Prove that, of all probability mass functions for a non-negative integer-valued random variable with mean μ, the geometric distribution, given by

$$P_Z(z) = \frac{1}{1+\mu}\left(\frac{\mu}{1+\mu}\right)^z, \quad \text{for } z = 0, 1, 2, \ldots,$$

has the largest entropy.

Hint: Let X be a nonnegative integer-valued random variable with mean μ. Show that $H(X) - H(Z) = -D(P_X \| P_Z) \leq 0$, with equality iff $P_X = P_Z$.

11. *Inequalities:* Which of the following inequalities are $\geq, =, \leq$? Label each with $\geq, =,$ or $\leq$ and justify your answer.

(a) $H(X|Z) - H(X|Y)$ versus $I(X; Y|Z)$.
(b) $H(X|g(Y))$ versus $H(X|Y)$, where g is a function.
(c) $H(X_2|X_1)$ versus $(1/2)H(X_1, X_2)$, where X_1 and X_2 are identically distributed on a common alphabet $\mathcal{X}$ (i.e., $P_{X_1}(a) = P_{X_2}(a)$ for all $a \in \mathcal{X}$).

12. *Entropy of invertible functions:* Given random variables X and Z with finite alphabets $\mathcal{X}$ and $\mathcal{Z}$, respectively, define a new random variable Y as $Y = f_X(Z)$ with alphabet $\mathcal{Y} := \{f_x(z) : x \in \mathcal{X} \text{ and } z \in \mathcal{Z}\}$, where for each $x \in \mathcal{X}$, the function $f_x : \mathcal{Z} \to \mathcal{Y}$ is invertible.

(a) Prove that $H(Y|X) = H(Z|X)$.
(b) Show that $H(Y) \geq H(Z)$ when X and Z are independent.
(c) Verify via a counterexample that $H(Y|Z) \neq H(X|Z)$ in general.
(d) Under what conditions does $H(Y|Z) = H(X|Z)$?

13. Let $X_1 \to X_2 \to X_3 \to \cdots \to X_n$ form a Markov chain (see (B.3.5) in Appendix B). Show that:

(a) $I(X_1; X_2, \ldots, X_n) = I(X_1; X_2)$.
(b) For any n, $H(X_1|X_{n-1}) \leq H(X_1|X_n)$.

14. *Refinement cannot decrease entropy:* Given integer $m \geq 1$ and a random variable X with distribution P_X, let $\{\mathcal{U}_1, \ldots, \mathcal{U}_m\}$ be a partition on the alphabet $\mathcal{X}$ of X; i.e., $\bigcup_{i=1}^{m} \mathcal{U}_i = \mathcal{X}$ and $\mathcal{U}_j \cap \mathcal{U}_k = \emptyset$ for all $j \neq k$. Now let U denote a random variable with alphabet $\{1, \ldots, m\}$ and distribution $P_U(i) = P_X(\mathcal{U}_i)$ for $1 \leq i \leq m$. In this case, X is called a *refinement* (or an *m-refinement*) of U. Show that

$$H(X) \geq H(U).$$

15. Provide examples for the following inequalities (see Definition 2.40 for the definition of conditional divergence).

(a) $D(P_{X|Z} \| P_{\hat{X}|\hat{Z}} | P_Z) > D(P_X \| P_{\hat{X}})$.
(b) $D(P_{X|Z} \| P_{\hat{X}|\hat{Z}} | P_Z) < D(P_X \| P_{\hat{X}})$.

16. Prove that the binary divergence defined by

$$D(p\|q) := p \log_2 \frac{p}{q} + (1-p) \log_2 \frac{1-p}{1-q}$$

satisfies

$$D(p\|q) \le \log_2(e) \frac{(p-q)^2}{q(1-q)}$$

for $0 < p < 1$ and $0 < q < 1$.
Hint: Use the FI Lemma.

17. Let $\{p_1, p_2, \ldots, p_m\}$ be a set of positive real numbers with $\sum_{i=1}^{m} p_i = 1$. If $\{q_1, q_2, \ldots, q_m\}$ is any other set of positive real numbers with $\sum_{i=1}^{m} q_i = \alpha$, where $\alpha > 0$ is a constant, show that

$$\sum_{i=1}^{m} p_i \log \frac{1}{p_i} \le \sum_{i=1}^{m} p_i \log \frac{1}{q_i} + \log \alpha.$$

Give a necessary and sufficient condition for equality.

18. *An alternative form for mutual information*: Given jointly distributed random variables X and Y with alphabets $\mathcal{X}$ and $\mathcal{Y}$, respectively, and with joint distribution $P_{X,Y} = P_{Y|X} P_X$, show that the mutual information $I(X;Y)$ can be written as

$$I(X;Y) = D(P_{Y|X} \| P_Y | P_X)$$
$$= D(P_{Y|X} \| Q_Y | P_X) - D(P_Y \| Q_Y)$$

for any distribution Q_Y on $\mathcal{Y}$, where P_Y is the marginal distribution of $P_{X,Y}$ on $\mathcal{Y}$ and

$$D(P_{Y|X} \| Q_Y | P_X) = \sum_{x \in \mathcal{X}} P_X(x) \sum_{y \in \mathcal{Y}} P_{Y|X}(y|x) \log \frac{P_{Y|X}(y|x)}{Q_Y(y)}$$

is the conditional divergence between $P_{Y|X}$ and Q_Y given P_X; see Definition 2.40.

19. Let X and Y be jointly distributed discrete random variables. Show that

$$I(X;Y) \ge I(f(X); g(Y)),$$

where $f(\cdot)$ and $g(\cdot)$ are given functions.

20. *Data processing inequality for the divergence*: Given a conditional distribution $P_{Y|X}$ defined on the alphabets $\mathcal{X}$ and $\mathcal{Y}$, let P_X and Q_X be two distributions on $\mathcal{X}$ and let P_Y and Q_Y be two corresponding distributions on $\mathcal{Y}$ defined by

$$P_Y(y) = \sum_{x \in \mathcal{X}} P_{Y|X}(y|x) P_X(x)$$

and

$$Q_Y(y) = \sum_{x \in \mathcal{X}} P_{Y|X}(y|x) Q_X(x)$$

for all $y \in \mathcal{Y}$. Show that

$$D(P_X \| Q_X) \geq D(P_Y \| Q_Y).$$

21. *An application of Jensen's inequality:* Let X and X' be two discrete independent random variables with common alphabet $\mathcal{X}$ and distributions P_X and $P_{X'}$, respectively.

   (a) Use Jensen's inequality to show that

   $$2^{E_X[\log_2 P_{X'}(X)]} \leq E_X[P_{X'}(X)].$$

   (b) Show that
   $$\Pr[X = X'] \geq 2^{-H(X) - D(X \| X')}.$$

   (c) If $\mathcal{X} = \{0, 1\}$, X' is uniformly distributed and $P_X(0) = p = 1 - P_X(1)$, evaluate the tightness of the bound in (b) as a function of p.

22. *Hölder's inequality:*

   (a) Given two probability vectors $(p_1, p_2, \ldots, p_m)$ and $(q_1, q_2, \ldots, q_m)$ on the set $\{1, 2, \ldots, m\}$, apply Jensen's inequality to show that for any $0 < \lambda < 1$,

   $$\sum_{i=1}^{m} q_i^{\lambda} p_i^{1-\lambda} \leq 1,$$

   and give a necessary and sufficient condition for equality.

   (b) Given positive real numbers a_i and b_i $i = 1, 2, \ldots, m$, show via an appropriate use of the bound in (a) that for any $0 < \lambda < 1$,

   $$\sum_{i=1}^{m} a_i b_i \leq \left[\sum_{i=1}^{m} a_i^{\frac{1}{\lambda}} \right]^{\lambda} \left[\sum_{i=1}^{m} b_i^{\frac{1}{1-\lambda}} \right]^{1-\lambda},$$

   with equality iff for some constant c,

   $$a_i^{\frac{1}{\lambda}} = c b_i^{\frac{1}{1-\lambda}}$$

   for all i. This inequality is known as *Hölder's inequality*. In the special case of $\lambda = 1/2$, the bound is referred to as the *Cauchy–Schwarz inequality*.

(c) Another form of Hölder's inequality is as follows:

$$\sum_{i=1}^{m} p_i a_i b_i \le \left[\sum_{i=1}^{m} p_i a_i^{\frac{1}{\lambda}} \right]^{\lambda} \left[\sum_{i=1}^{m} p_i b_i^{\frac{1}{1-\lambda}} \right]^{1-\lambda},$$

where $(p_1, p_2, \ldots, p_m)$ is a probability vector as in (a).

Prove this inequality using (b), and show that equality holds iff for some constant c,

$$p_i a_i^{\frac{1}{\lambda}} = c p_i b_i^{\frac{1}{1-\lambda}}$$

for all i.

Note: We refer the reader to [135, 176] for a variety of other useful inequalities.

23. *Inequality of arithmetic and geometric means*:

(a) Show that

$$\sum_{i=1}^{n} a_i \ln x_i \le \ln \left(\sum_{i=1}^{n} a_i x_i \right),$$

where $x_1, x_2, \ldots, x_n$ are arbitrary positive numbers, and $a_1, a_2, \ldots, a_n$ are positive numbers such that $\sum_{i=1}^{n} a_i = 1$.

(b) Deduce from the above the *inequality of the arithmetic and geometric means*:

$$x_1^{a_1} x_2^{a_2} \ldots x_n^{a_n} \le \sum_{i=1}^{n} a_i x_i.$$

24. Consider two distributions $P(\cdot)$ and $Q(\cdot)$ on the alphabet $\mathcal{X} = \{a_1, \ldots, a_k\}$ such that $Q(a_i) > 0$ for all $i = 1, \ldots, k$. Show that

$$\sum_{i=1}^{k} \frac{(P(a_i))^2}{Q(a_i)} \ge 1.$$

25. Let X be a discrete random variable with alphabet $\mathcal{X}$ and distribution P_X. Let $f : \mathcal{X} \to \mathbb{R}$ be a real-valued function, and let α be an arbitrary real number.

(a) Show that

$$H(X) \le \alpha E[f(X))] + \log_2 \left(\sum_{x \in \mathcal{X}} 2^{-\alpha f(x)} \right),$$

with equality iff $P_X(x) = \frac{1}{A} 2^{-\alpha f(x)}$ for $x \in \mathcal{X}$, where $A := \sum_{x \in \mathcal{X}} 2^{-\alpha f(x)}$.

(b) Show that for a positive integer-valued random variable N (such that $E[N] > 1$ without loss of generality), the following holds:

$$H(N) \le \log_2(E[N]) + \log_2 e.$$

Hint: First use part (a) with $f(N) = N$ and $\alpha = \log_2 \frac{E[N]}{E[N]-1}$.

26. *Fano's inequality for list decoding*: Let X and Y be two random variables with alphabets $\mathcal{X}$ and $\mathcal{Y}$, respectively, where $\mathcal{X}$ is finite and $\mathcal{Y}$ can be countably infinite. Given a fixed integer $m \ge 1$, define

$$\hat{X}^m := (g_1(Y), g_2(Y), \ldots, g_m(Y))$$

as the list of estimates of X obtained by observing Y, where $g_i : \mathcal{Y} \to \mathcal{X}$ is a given estimation function for $i = 1, 2, \ldots, m$. Define the probability of list decoding error as

$$P_e^{(m)} := \Pr\left[\hat{X}_1 \ne X, \hat{X}_2 \ne X, \ldots, \hat{X}_m \ne X\right].$$

(a) Show that

$$H(X|Y) \le h_b(P_e^{(m)}) + P_e^{(m)} \log_2(|\mathcal{X}| - u) + (1 - P_e^{(m)}) \log_2(u), \quad (2.10.1)$$

where

$$u := \sum_{x \in \mathcal{X}} \sum_{\hat{x}^m \in \mathcal{X}^m : \hat{x}_i = x \text{ for some } i} P_{\hat{X}^m}(\hat{x}^m).$$

Note: When $m = 1$, we obtain that $u = 1$ and the right-hand side of (2.10.1) reduces to the original Fano inequality (cf. the right-hand side of (2.5.1)).

Hint: Show that $H(X|Y) \le H(X|\hat{X}^m)$ and that $H(X|\hat{X}^m)$ is less than the right-hand side of the above inequality.

(b) Use (2.10.1) to deduce the following weaker version of Fano's inequality for list decoding (see [5], [216], [313, Appendix 3.E]):

$$H(X|Y) \le h_b(P_e^{(m)}) + P_e^{(m)} \log_2 |\mathcal{X}| + (1 - P_e^{(m)}) \log_2 m. \quad (2.10.2)$$

27. *Fano's inequality for ternary partitioning of the observation space*: In Problem 26, $P_e^{(m)}$ and u can actually be expressed as

$$P_e^{(m)} = \sum_{x \in \mathcal{X}} \sum_{\hat{x}^m \notin \mathcal{U}_x} P_{X,\hat{X}^m}(x, \hat{x}^m) = \sum_{x \in \mathcal{X}} \sum_{y \notin \mathcal{Y}_x} P_{X,Y}(x, y)$$

and

$$u = \sum_{x \in \mathcal{X}} \sum_{\hat{x}^m \in \mathcal{U}_x} P_{\hat{X}^m}(\hat{x}^m) = \sum_{x \in \mathcal{X}} \sum_{y \in \mathcal{Y}_x} P_Y(y),$$

respectively, where

$$\mathcal{U}_x := \left\{ \hat{x}^m \in \mathcal{X}^m : \hat{x}_i = x \text{ for some } i \right\}$$

and

$$\mathcal{Y}_x := \{ y \in \mathcal{Y} : g_i(y) = x \text{ for some } i \}.$$

Thus, given $x \in \mathcal{X}$, $\mathcal{Y}_x$ and $\mathcal{Y}_x^c$ form a *binary partition* on the observation space $\mathcal{Y}$.

Now consider again random variables X and Y with alphabets $\mathcal{X}$ and $\mathcal{Y}$, respectively, where $\mathcal{X}$ is finite and $\mathcal{Y}$ can be countably infinite, and assume that for each $x \in \mathcal{X}$, we are given a *ternary partition* $\{\mathcal{S}_x, \mathcal{T}_x, \mathcal{V}_x\}$ on the observation space $\mathcal{Y}$, where the sets $\mathcal{S}_x$, $\mathcal{T}_x$ and $\mathcal{V}_x$ are mutually disjoint and their union equals $\mathcal{Y}$. Define

$$p := \sum_{x \in \mathcal{X}} \sum_{y \in \mathcal{S}_x} P_{X,Y}(x, y), \quad q := \sum_{x \in \mathcal{X}} \sum_{y \in \mathcal{T}_x} P_{X,Y}(x, y), \quad r := \sum_{x \in \mathcal{X}} \sum_{y \in \mathcal{V}_x} P_{X,Y}(x, y)$$

and

$$s := \sum_{x \in \mathcal{X}} \sum_{y \in \mathcal{S}_x} P_Y(y), \quad t := \sum_{x \in \mathcal{X}} \sum_{y \in \mathcal{T}_x} P_Y(y), \quad v := \sum_{x \in \mathcal{X}} \sum_{y \in \mathcal{V}_x} P_Y(y).$$

Note that $p + q + r = 1$ and $s + t + v = |\mathcal{X}|$. Show that

$$H(X|Y) \le H(p, q, r) + p \log_2(s) + q \log_2(t) + r \log_2(v), \qquad (2.10.3)$$

where

$$H(p, q, r) = p \log_2 \frac{1}{p} + q \log_2 \frac{1}{q} + r \log_2 \frac{1}{r}.$$

Note: When $\mathcal{V}_x = \emptyset$ for all $x \in \mathcal{X}$, we obtain that $\mathcal{S}_x = \mathcal{T}_x^c$ for all x, $r = v = 0$ and $p = 1 - q$; as a result, inequality (2.10.3) acquires a similar expression as (2.10.1) with p standing for the probability of error.

28. *ϵ-Independence*: Let X and Y be two jointly distributed random variables with finite respective alphabets $\mathcal{X}$ and $\mathcal{Y}$ and joint pmf $P_{X,Y}$ defined on $\mathcal{X} \times \mathcal{Y}$. Given a fixed $\epsilon > 0$, random variable Y is said to be *ϵ-independent* from random variable X if

$$\sum_{x \in \mathcal{X}} P_X(x) \sum_{y \in \mathcal{Y}} |P_{Y|X}(y|x) - P_Y(y)| < \epsilon,$$

where P_X and P_Y are the marginal pmf's of X and Y, respectively, and $P_{Y|X}$ is the conditional pmf of Y given X. Show that

$$I(X;Y) < \frac{\log_2(e)}{2}\epsilon^2$$

is a sufficient condition for Y to be ϵ-independent from X, where $I(X;Y)$ is the mutual information (in bits) between X and Y.

29. *Rényi's entropy*: Given a fixed positive integer $n > 1$, consider an n-ary valued random variable X with alphabet $\mathcal{X} = \{1, 2, \ldots, n\}$ and distribution described by the probabilities $p_i := \Pr[X = i]$, where $p_i > 0$ for each $i = 1, \ldots, n$. Given $\alpha > 0$ and $\alpha \neq 1$, the Rényi entropy of X (see Definition 2.9.1) is given by

$$H_\alpha(X) := \frac{1}{1 - \alpha} \log_2 \left(\sum_{i=1}^{n} p_i^\alpha \right).$$

(a) Show that

$$\sum_{i=1}^{n} p_i^r > 1 \quad \text{if } r < 1,$$

and that

$$\sum_{i=1}^{n} p_i^r < 1 \quad \text{if } r > 1.$$

Hint: Show that the function $f(r) = \sum_{i=1}^{n} p_i^r$ is decreasing in r, where $r > 0$.
(b) Show that

$$0 \leq H_\alpha(X) \leq \log_2 n.$$

Hint: Use (a) for the lower bound, and use Jensen's inequality (with the convex function $f(y) = y^{\frac{1}{1-\alpha}}$, for $y > 0$) for the upper bound.

30. *Rényi's entropy and divergence*: Consider two discrete random variables X and $\hat{X}$ with common alphabet $\mathcal{X}$ and distribution P_X and $P_{\hat{X}}$, respectively.

(a) Prove Lemma 2.52.
(b) Find a distribution Q on $\mathcal{X}$ in terms of α and P_X such that the following holds:

$$H_\alpha(X) = H(X) + \frac{1}{1 - \alpha} D(P_X \| Q).$$

Chapter 3
Lossless Data Compression

3.1 Principles of Data Compression

As mentioned in Chap. 1, data compression describes methods of representing a source by a code whose average codeword length (or code rate) is acceptably small. The representation can be *lossless* (or *asymptotically lossless*) where the reconstructed source is identical (or identical with vanishing error probability) to the original source; or *lossy* where the reconstructed source is allowed to deviate from the original source, usually within an acceptable threshold. We herein focus on lossless data compression.

Since a memoryless source is modeled as a random variable, the *averaged* codeword length of a codebook is calculated based on the probability distribution of that random variable. For example, consider a ternary memoryless source X with three possible outcomes and probability distribution given by

$$P_X(x = \text{outcome}_A) = 0.5;$$
$$P_X(x = \text{outcome}_B) = 0.25;$$
$$P_X(x = \text{outcome}_C) = 0.25.$$

Suppose that a binary codebook is designed for this source, in which outcome_A, outcome_B, and outcome_C are, respectively, encoded as 0, 10, and 11. Then, the average codeword length (in bits per source outcome) is

$$\text{length}(0) \cdot P_X(\text{outcome}_A) + \text{length}(10) \cdot P_X(\text{outcome}_B)$$
$$+ \text{length}(11) \cdot P_X(\text{outcome}_C)$$
$$= 0.5 + 2 \times 0.25 + 2 \times 0.25$$
$$= 1.5 \text{ bits.}$$

© Springer Nature Singapore Pte Ltd. 2018   55
F. Alajaji and P.-N. Chen, *An Introduction to Single-User Information Theory*,
Springer Undergraduate Texts in Mathematics and Technology,
https://doi.org/10.1007/978-981-10-8001-2_3

There are usually no constraints on the basic structure of a code. In the case where the codeword length for each source outcome can be different, the code is called a *variable-length code*. When the codeword lengths of all source outcomes are equal, the code is referred to as a *fixed-length code*. It is obvious that the minimum average codeword length among all variable-length codes is no greater than that among all fixed-length codes, since the latter is a subclass of the former. We will see in this chapter that the smallest achievable average code rate for variable-length and fixed-length codes coincide for sources with *good* probabilistic characteristics, such as stationarity and ergodicity. But for more general sources with memory, the two quantities are different (e.g., see [172]).

For fixed-length codes, the sequence of adjacent codewords is concatenated together for storage or transmission purposes, and some punctuation mechanism— such as marking the beginning of each codeword or delineating internal sub-blocks for synchronization between encoder and decoder—is normally considered an implicit part of the codewords. Due to constraints on space or processing capability, the sequence of source symbols may be too long for the encoder to deal with all at once; therefore, segmentation before encoding is often necessary. For example, suppose that we need to encode using a binary code the grades of a class with 100 students. There are three grade levels: A, B, and C. By observing that there are 3^{100} possible grade combinations for 100 students, a straightforward code design requires

$$\lceil \log_2(3^{100}) \rceil = 159 \text{ bits}$$

to encode these combinations (by enumerating them). Now suppose that the encoder facility can only process 16 bits at a time. Then, the above code design becomes infeasible and segmentation is unavoidable. Under such constraint, we may encode grades of 10 students at a time, which requires

$$\lceil \log_2(3^{10}) \rceil = 16 \text{ bits}.$$

As a consequence, for a class of 100 students, the code requires 160 bits in total.

In the above example, the letters in the grade set $\{A, B, C\}$ and the letters from the code alphabet $\{0, 1\}$ are often called *source symbols* and *code symbols*, respectively. When the code alphabet is binary (as in the previous two examples), the code symbols are referred to as *code bits* or simply *bits* (as already used). A tuple (or grouped sequence) of source symbols is called a *sourceword*, and the resulting encoded tuple consisting of code symbols is called a *codeword*. (In the above example, each sourceword consists of 10 source symbols (student grades) and each codeword consists of 16 bits.)

Note that, during the encoding process, the sourceword lengths do not have to be equal. In this text, however, we only consider the case where the sourcewords have a fixed length throughout the encoding process (except for the Lempel–Ziv code briefly discussed at the end of this chapter), but we will allow the codewords to have

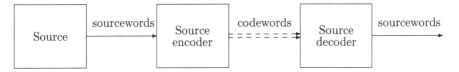

Fig. 3.1 Block diagram of a data compression system

fixed or variable lengths as defined earlier.[1] The block diagram of a source coding system is depicted in Fig. 3.1.

When adding segmentation mechanisms to fixed-length codes, the codes can be loosely divided into two groups. The first consists of *block codes* in which the encoding (or decoding) of the next segment of source symbols is independent of the previous segments. If the encoding/decoding of the next segment, somehow, retains and uses some knowledge of earlier segments, the code is called a fixed-length *tree code*. As we will not investigate such codes in this text, we can use "block codes" and "fixed-length codes" as synonyms.

In this chapter, we first consider data compression for block codes in Sect. 3.2. Data compression for variable-length codes is then addressed in Sect. 3.3.

3.2 Block Codes for Asymptotically Lossless Compression

3.2.1 Block Codes for Discrete Memoryless Sources

We first focus on the study of asymptotically lossless data compression of discrete memoryless sources via block (fixed-length) codes. Such sources were already defined in the previous chapter (see Sect. 2.1.2) and in Appendix B; but we nevertheless recall their definition.

Definition 3.1 (Discrete memoryless source) A discrete memoryless source (DMS) $\{X_n\}_{n=1}^{\infty}$ consists of a sequence of i.i.d. random variables, $X_1, X_2, X_3, \ldots$, all taking values in a common finite alphabet $\mathcal{X}$. In particular, if $P_X(\cdot)$ is the common distribution or probability mass function (pmf) of the X_i's, then

$$P_{X^n}(x_1, x_2, \ldots, x_n) = \prod_{i=1}^{n} P_X(x_i).$$

---

[1] In other words, our fixed-length codes are actually "fixed-to-fixed length codes" and our variable-length codes are "fixed-to-variable length codes" since, in both cases, a fixed number of source symbols is mapped onto codewords with fixed and variable lengths, respectively.

Definition 3.2 An (n, M) block code with blocklength n and size M (which can be a function of n in general,[2] i.e., $M = M_n$) for a discrete source $\{X_n\}_{n=1}^{\infty}$ is a set $\mathcal{C}_n = \{c_1, c_2, \ldots, c_M\} \subseteq \mathcal{X}^n$ consisting of M reproduction (or reconstruction) words, where each reproduction word is a sourceword (an n-tuple of source symbols).

  To simplify the exposition, we make an abuse of notation by writing $\mathcal{C}_n = (n, M)$ to mean that $\mathcal{C}_n$ is a block code with blocklength n and size M.

Observation 3.3 One can binary-index (or enumerate) the reproduction words in $\mathcal{C}_n = \{c_1, c_2, \ldots, c_M\}$ using $k := \lceil \log_2 M \rceil$ bits. As such k-bit words in $\{0, 1\}^k$ are usually stored for retrieval at a later date, the (n, M) block code can be represented by an encoder–decoder pair of functions (f, g), where the encoding function

$$f : \mathcal{X}^n \rightarrow \{0, 1\}^k$$

maps each sourceword x^n to a k-bit word $f(x^n)$ which we call a *codeword*. Then, the decoding function

$$g : \{0, 1\}^k \rightarrow \{c_1, c_2, \ldots, c_M\}$$

is a retrieving operation that produces the reproduction words. Since the codewords are binary-valued, such a block code is called a *binary code*. More generally, a *D-ary block code* (where $D > 1$ is an integer) would use an encoding function $f : \mathcal{X}^n \rightarrow \{0, 1, \ldots, D - 1\}^k$ where each codeword $f(x^n)$ contains k D-ary code symbols.

  Furthermore, since the behavior of block codes is investigated for sufficiently large n and M (tending to infinity), it is legitimate to replace $\lceil \log_2 M \rceil$ by $\log_2 M$ for the case of binary codes. With this convention, the *data compression rate* or *code rate* is

$$\frac{k}{n} = \frac{1}{n} \log_2 M \text{(in bits per source symbol)}.$$

Similarly, for D-ary codes, the rate is

$$\frac{k}{n} = \frac{1}{n} \log_D M \text{ (in D-ary code symbols per source symbol)}.$$

  For computational convenience, *nats* (under the natural logarithm) can be used instead of *bits* or *D-ary code symbols*; in this case, the code rate becomes

$$\frac{1}{n} \log M \text{ (in nats per source symbol)}.$$

---

[2]In the literature, both (n, M) and (M, n) have been used to denote a block code with blocklength n and size M. For example, [415, p. 149] adopts the former one, while [83, p. 193] uses the latter. We use the (n, M) notation since $M = M_n$ is a function of n in general.

The block code's operation can be symbolically represented as[3]

$$(x_1, x_2, \ldots, x_n) \rightarrow c_m \in \{c_1, c_2, \ldots, c_M\}.$$

This procedure will be repeated for each consecutive block of length n, i.e.,

$$\cdots (x_{3n}, \ldots, x_{31})(x_{2n}, \ldots, x_{21})(x_{1n}, \ldots, x_{11}) \rightarrow \cdots |c_{m_3}|c_{m_2}|c_{m_1},$$

where "|" reflects the necessity of "punctuation mechanism" or "synchronization mechanism" for consecutive source block coders.

The next theorem provides a key tool for proving Shannon's source coding theorem.

Theorem 3.4 (Shannon–McMillan–Breiman) (Asymptotic equipartition property or AEP[4]) *If $\{X_n\}_{n=1}^\infty$ is a DMS with entropy $H(X)$, then*

$$-\frac{1}{n}\log_2 P_{X^n}(X_1, \ldots, X_n) \rightarrow H(X) \quad \text{in probability.}$$

In other words, for any $\delta > 0$,

$$\lim_{n \rightarrow \infty} \Pr\left\{\left|-\frac{1}{n}\log_2 P_{X^n}(X_1, \ldots, X_n) - H(X)\right| > \delta\right\} = 0.$$

Proof This theorem follows by first observing that for an i.i.d. sequence $\{X_n\}_{n=1}^\infty$,

$$-\frac{1}{n}\log_2 P_{X^n}(X_1, \ldots, X_n) = \frac{1}{n}\sum_{i=1}^n \left[-\log_2 P_X(X_i)\right]$$

and that the sequence $\{-\log_2 P_X(X_i)\}_{i=1}^\infty$ is i.i.d., and then applying the weak law of large numbers on the latter sequence. ☐

The AEP indeed constitutes an "information theoretic" analog of the weak law of large numbers as it states that if $\{-\log_2 P_X(X_i)\}_{i=1}^\infty$ is an i.i.d. sequence, then for any $\delta > 0$,

$$\Pr\left\{\left|\frac{1}{n}\sum_{i=1}^n \left[-\log_2 P_X(X_i)\right] - H(X)\right| \leq \delta\right\} \rightarrow 1 \quad \text{as} \quad n \rightarrow \infty.$$

As a consequence of the AEP, all the probability mass will be ultimately placed on the *weakly δ-typical set*, which is defined as

---

[3] When one uses an encoder–decoder pair (f, g) to describe the block code, the code's operation can be expressed as $c_m = g(f(x^n))$.

[4] This theorem, which is also called the *entropy stability property*, is due to Shannon [340], McMillan [267], and Breiman [58].

$$\mathcal{F}_n(\delta) := \left\{ x^n \in \mathcal{X}^n : \left| -\frac{1}{n} \log_2 P_{X^n}(x^n) - H(X) \right| \le \delta \right\}$$

$$= \left\{ x^n \in \mathcal{X}^n : \left| -\frac{1}{n} \sum_{i=1}^{n} \log_2 P_X(x_i) - H(X) \right| \le \delta \right\}.$$

Note that since the source is memoryless, for any $x^n \in \mathcal{F}_n(\delta)$, $-(1/n) \log_2 P_{X^n}(x^n)$, the normalized self-information of x^n, is equal to $(1/n) \sum_{i=1}^{n} \left[-\log_2 P_X(x_i) \right]$, which is the empirical (arithmetic) average self-information or "apparent" entropy of the source. Thus, a sourceword x^n is δ-typical if it yields an apparent source entropy within δ of the "true" source entropy $H(X)$. Note that the sourcewords in $\mathcal{F}_n(\delta)$ are nearly *equiprobable* or *equally surprising* (cf. Property 1 of Theorem 3.5); this justifies naming Theorem 3.4 by AEP.

Theorem 3.5 (Consequence of the AEP) *Given a DMS $\{X_n\}_{n=1}^{\infty}$ with entropy $H(X)$ and any δ greater than zero, the weakly δ-typical set $\mathcal{F}_n(\delta)$ satisfies the following.*

1. *If $x^n \in \mathcal{F}_n(\delta)$, then*

$$2^{-n(H(X)+\delta)} \le P_{X^n}(x^n) \le 2^{-n(H(X)-\delta)}.$$

2. $P_{X^n}\left(\mathcal{F}_n^c(\delta)\right) < \delta$ *for sufficiently large n, where the superscript "c" denotes the complementary set operation.*
3. $|\mathcal{F}_n(\delta)| > (1-\delta)2^{n(H(X)-\delta)}$ *for sufficiently large n, and $|\mathcal{F}_n(\delta)| \le 2^{n(H(X)+\delta)}$ for every n, where $|\mathcal{F}_n(\delta)|$ denotes the number of elements in $\mathcal{F}_n(\delta)$.*

Note: The above theorem also holds if we define the typical set using the base-D logarithm $\log_D$ for any $D > 1$ instead of the base-2 logarithm; in this case, one just needs to appropriately change the base of the exponential terms in the above theorem (by replacing 2^x terms with D^x terms) and also substitute $H(X)$ with $H_D(X)$.

Proof Property 1 is an immediate consequence of the definition of $\mathcal{F}_n(\delta)$. Property 2 is a direct consequence of the AEP, since the AEP states that for a fixed $\delta > 0$, $\lim_{n \to \infty} P_{X^n}(\mathcal{F}_n(\delta)) = 1$; i.e., $\forall \varepsilon > 0$, there exists $n_0 = n_0(\varepsilon)$ such that for all $n \ge n_0$,

$$P_{X^n}(\mathcal{F}_n(\delta)) > 1 - \varepsilon.$$

In particular, setting $\varepsilon = \delta$ yields the result. We nevertheless provide a direct proof of Property 2 as we give an explicit expression for n_0: observe that by Chebyshev's inequality,[5]

$$P_{X^n}(\mathcal{F}_n^c(\delta)) = P_{X^n} \left(\left\{ x^n \in \mathcal{X}^n : \left| -\frac{1}{n} \log_2 P_{X^n}(x^n) - H(X) \right| > \delta \right\} \right)$$

$$\le \frac{\sigma_X^2}{n\delta^2}$$

$$< \delta,$$

---

[5]Chebyshev's inequality as well as its proof can be found on p. 287 in Appendix B.

for $n > \sigma_X^2/\delta^3$, where the variance

$$\sigma_X^2 := \mathrm{Var}[-\log_2 P_X(X)] = \sum_{x \in \mathcal{X}} P_X(x)\left[\log_2 P_X(x)\right]^2 - (H(X))^2$$

is a constant[6] independent of n.

To prove Property 3, we have from Property 1 that

$$1 \geq \sum_{x^n \in \mathcal{F}_n(\delta)} P_{X^n}(x^n) \geq \sum_{x^n \in \mathcal{F}_n(\delta)} 2^{-n(H(X)+\delta)} = |\mathcal{F}_n(\delta)|2^{-n(H(X)+\delta)},$$

and, using Properties 2 and 1, we have that

$$1 - \delta < 1 - \frac{\sigma_X^2}{n\delta^2} \leq \sum_{x^n \in \mathcal{F}_n(\delta)} P_{X^n}(x^n) \leq \sum_{x^n \in \mathcal{F}_n(\delta)} 2^{-n(H(X)-\delta)} = |\mathcal{F}_n(\delta)|2^{-n(H(X)-\delta)},$$

for $n \geq \sigma_X^2/\delta^3$. $\qquad\square$

Note that for any $n > 0$, a block code $\mathcal{C}_n = (n, M)$ is said to be *uniquely decodable* or *completely lossless* if its set of reproduction words is trivially equal to the set of all source n-tuples: $\{c_1, c_2, \ldots, c_M\} = \mathcal{X}^n$. In this case, if we are binary-indexing the reproduction words using an encoding–decoding pair (f, g), every sourceword x^n will be assigned to a distinct binary codeword $f(x^n)$ of length $k = \log_2 M$ and all the binary k-tuples are the image under f of some sourceword. In other words, f is a bijective (injective and surjective) map and hence invertible with the decoding map $g = f^{-1}$ and $M = |\mathcal{X}|^n = 2^k$. Thus, the code rate is $(1/n) \log_2 M = \log_2 |\mathcal{X}|$ bits/source symbol.

Now the question becomes: can we achieve a better (i.e., smaller) compression rate? The answer is affirmative: we can achieve a compression rate equal to the source entropy $H(X)$ (in bits), which can be significantly smaller than $\log_2 |\mathcal{X}|$ when this source is strongly nonuniformly distributed, if we give up unique decodability (for every n) and allow n to be sufficiently large to asymptotically achieve lossless reconstruction by having an arbitrarily small (but positive) probability of decoding error

$$P_e(\mathcal{C}_n) := P_{X^n}\{x^n \in \mathcal{X}^n : g(f(x^n)) \neq x^n\}.$$

---

[6]In the proof, we assume that the variance $\sigma_X^2 = \mathrm{Var}[-\log_2 P_X(X)] < \infty$. This holds since the source alphabet is finite:

$$\mathrm{Var}[-\log_2 P_X(X)] \leq E[(\log_2 P_X(X))^2] = \sum_{x \in \mathcal{X}} P_X(x)(\log_2 P_X(x))^2$$

$$\leq \sum_{x \in \mathcal{X}} \frac{4}{e^2}[\log_2(e)]^2 = \frac{4}{e^2}[\log_2(e)]^2 \times |\mathcal{X}| < \infty.$$

Thus, block codes herein can perform data compression that is *asymptotically lossless* with respect to blocklength; this contrasts with variable-length codes which can be completely lossless (uniquely decodable) for every finite blocklength.

We now can formally state and prove Shannon's asymptotically lossless source coding theorem for block codes. The theorem will be stated for general D-ary block codes, representing the source entropy $H_D(X)$ in D-ary code symbol/source symbol as the smallest (infimum) possible compression rate for asymptotically lossless D-ary block codes. Without loss of generality, the theorem will be proved for the case of $D = 2$. The idea behind the proof of the forward (achievability) part is basically to binary-index the source sequence in the weakly δ-typical set $\mathcal{F}_n(\delta)$ to a binary codeword (starting from index one with corresponding k-tuple codeword $0\cdots01$); and to encode all sourcewords outside $\mathcal{F}_n(\delta)$ to a default all-zero binary codeword, which certainly cannot be reproduced distortionless due to its many-to-one-mapping property. The resultant code rate is $(1/n)\lceil\log_2(|\mathcal{F}_n(\delta)| + 1)\rceil$ bits per source symbol. As revealed in the Shannon–McMillan–Breiman AEP theorem and its consequence, almost all the probability mass will be on $\mathcal{F}_n(\delta)$ as n is sufficiently large, and hence, the probability of non-reconstructable source sequences can be made arbitrarily small. A simple example for the above coding scheme is illustrated in Table 3.1. The converse part of the proof will establish (by expressing the probability of correct decoding in terms of the δ-typical set and also using the consequence of the AEP) that for any sequence of D-ary codes with rate strictly below the source entropy, their probability of error cannot asymptotically vanish (is bounded away from zero). Actually, a *stronger* result is proven: it is shown that their probability of error not only does not asymptotically vanish, it actually ultimately grows to 1 (this is why we call this part a "strong" converse).

Theorem 3.6 (Shannon's source coding theorem)   *Given integer $D > 1$, consider a discrete memoryless source $\{X_n\}_{n=1}^{\infty}$ with entropy $H_D(X)$. Then the following hold.*

- *Forward part (achievability): For any $0 < \varepsilon < 1$, there exists $0 < \delta < \varepsilon$ and a sequence of D-ary block codes $\{\mathcal{C}_n = (n, M_n)\}_{n=1}^{\infty}$ with*

$$\limsup_{n\to\infty} \frac{1}{n}\log_D M_n \le H_D(X) + \delta \qquad (3.2.1)$$

satisfying

$$P_e(\mathcal{C}_n) < \varepsilon \qquad (3.2.2)$$

for all sufficiently large n, where $P_e(\mathcal{C}_n)$ denotes the probability of error (or decoding error) for block code $\mathcal{C}_n$.[7]

---

[7]Note that (3.2.2) is equivalent to $\limsup_{n\to\infty} P_e(\mathcal{C}_n) \le \varepsilon$. Since ε can be made arbitrarily small, the forward part actually indicates the existence of a sequence of D-ary block codes $\{\mathcal{C}_n\}_{n=1}^{\infty}$ satisfying (3.2.1) such that $\limsup_{n\to\infty} P_e(\mathcal{C}_n) = 0$. Based on this, the converse should be that any sequence of D-ary block codes satisfying (3.2.3) satisfies $\limsup_{n\to\infty} P_e(\mathcal{C}_n) > 0$. However, the so-called *strong* converse actually gives a stronger consequence: $\limsup_{n\to\infty} P_e(\mathcal{C}_n) = 1$ (as ϵ can be made arbitrarily small).

Table 3.1 An example of the δ-typical set with $n = 2$ and $\delta = 0.4$, where $\mathcal{F}_2(0.4) = \{AB, AC,$ BA, BB, BC, CA, CB$\}$. The codeword set is $\{001(AB), 010(AC), 011(BA), 100(BB), 101(BC),$ $110(CA), 111(CB), 000(AA, AD, BD, CC, CD, DA, DB, DC, DD)\}$, where the parenthesis following each binary codeword indicates those sourcewords that are encoded to this codeword. The source distribution is $P_X(A) = 0.4$, $P_X(B) = 0.3$, $P_X(C) = 0.2$, and $P_X(D) = 0.1$

Source	$\left\| -\dfrac{1}{2} \sum_{i=1}^{2} \log_2 P_X(x_i) - H(X) \right\|$	Codeword	Reconstructed source sequence
AA	0.525 bits $\notin \mathcal{F}_2(0.4)$	000	Ambiguous
AB	0.317 bits $\in \mathcal{F}_2(0.4)$	001	AB
AC	0.025 bits $\in \mathcal{F}_2(0.4)$	010	AC
AD	0.475 bits $\notin \mathcal{F}_2(0.4)$	000	Ambiguous
BA	0.317 bits $\in \mathcal{F}_2(0.4)$	011	BA
BB	0.109 bits $\in \mathcal{F}_2(0.4)$	100	BB
BC	0.183 bits $\in \mathcal{F}_2(0.4)$	101	BC
BD	0.683 bits $\notin \mathcal{F}_2(0.4)$	000	Ambiguous
CA	0.025 bits $\in \mathcal{F}_2(0.4)$	110*	CA
CB	0.183 bits $\in \mathcal{F}_2(0.4)$	111	CB
CC	0.475 bits $\notin \mathcal{F}_2(0.4)$	000	Ambiguous
CD	0.975 bits $\notin \mathcal{F}_2(0.4)$	000	Ambiguous
DA	0.475 bits $\notin \mathcal{F}_2(0.4)$	000	Ambiguous
DB	0.683 bits $\notin \mathcal{F}_2(0.4)$	000	Ambiguous
DC	0.975 bits $\notin \mathcal{F}_2(0.4)$	000	Ambiguous
DD	1.475 bits $\notin \mathcal{F}_2(0.4)$	000	Ambiguous

- *Strong converse part:*   *For any $0 < \varepsilon < 1$, any sequence of D-ary block codes $\{\mathcal{C}_n = (n, M_n)\}_{n=1}^{\infty}$ with*

$$\limsup_{n \to \infty} \frac{1}{n} \log_D M_n < H_D(X) \qquad (3.2.3)$$

satisfies

$$P_e(\mathcal{C}_n) > 1 - \varepsilon$$

for all n sufficiently large.

Proof Forward Part: Without loss of generality, we will prove the result for the case of binary codes (i.e., $D = 2$). Also, recall that subscript D in $H_D(X)$ will be dropped (i.e., omitted) specifically when $D = 2$.

Given $0 < \varepsilon < 1$, fix δ such that $0 < \delta < \varepsilon$ and choose $n > 2/\delta$. Now construct a binary $\mathcal{C}_n$ block code by simply mapping the $\delta/2$-typical sourcewords x^n onto distinct not all-zero binary codewords of length $k := \lceil \log_2 M_n \rceil$ bits. In other words, binary-index (cf. Observation 3.3) the sourcewords in $\mathcal{F}_n(\delta/2)$ with the following encoding map:

$$x^n \rightarrow \begin{cases} \text{binary index of } x^n, & \text{if } x^n \in \mathcal{F}_n(\delta/2), \\ \text{all-zero codeword}, & \text{if } x^n \notin \mathcal{F}_n(\delta/2). \end{cases}$$

Then by the Shannon–McMillan–Breiman AEP theorem, we obtain that

$$M_n = |\mathcal{F}_n(\delta/2)| + 1 \leq 2^{n(H(X)+\delta/2)} + 1 < 2 \cdot 2^{n(H(X)+\delta/2)} < 2^{n(H(X)+\delta)}$$

for $n > 2/\delta$. Hence, a sequence of $\mathcal{C}_n = (n, M_n)$ block code satisfying (3.2.1) is established. It remains to show that the error probability for this sequence of (n, M_n) block code can be made smaller than ε for all sufficiently large n.

By the Shannon–McMillan–Breiman AEP theorem,

$$P_{X^n}(\mathcal{F}_n^c(\delta/2)) < \frac{\delta}{2} \quad \text{for all sufficiently large } n.$$

Consequently, for those n satisfying the above inequality, and being bigger than $2/\delta$,

$$P_e(\mathcal{C}_n) \leq P_{X^n}(\mathcal{F}_n^c(\delta/2)) < \delta \leq \varepsilon.$$

(For the last step, the reader can refer to Table 3.1 to confirm that only the "ambiguous" sequences outside the typical set contribute to the probability of error.)

Strong Converse Part: Fix any sequence of block codes $\{\mathcal{C}_n\}_{n=1}^{\infty}$ with

$$\limsup_{n \to \infty} \frac{1}{n} \log_2 |\mathcal{C}_n| < H(X).$$

Let $\mathcal{S}_n$ be the set of source symbols that can be correctly decoded through the $\mathcal{C}_n$-coding system. (A quick example is depicted in Fig. 3.2.) Then $|\mathcal{S}_n| = |\mathcal{C}_n|$. By choosing δ small enough with $\varepsilon/2 > \delta > 0$, and by definition of the limsup operation, we have

$$(\exists N_0)(\forall n > N_0) \quad \frac{1}{n} \log_2 |\mathcal{S}_n| = \frac{1}{n} \log_2 |\mathcal{C}_n| < H(X) - 2\delta,$$

which implies
$$|\mathcal{S}_n| < 2^{n(H(X)-2\delta)} \qquad \text{for } n > N_0.$$

Furthermore, from Property 2 of the consequence of the AEP, we obtain that

$$(\exists N_1)(\forall n > N_1) \quad P_{X^n}(\mathcal{F}_n^c(\delta)) < \delta.$$

Consequently, for $n > N := \max\{N_0, N_1, \log_2(2/\varepsilon)/\delta\}$, the probability of correctly block decoding satisfies

Source Symbols
$\mathcal{S}_n$

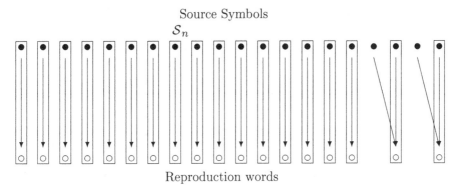

Reproduction words

Fig. 3.2 Possible code $\mathcal{C}_n$ and its corresponding $\mathcal{S}_n$. The solid box indicates the decoding mapping from $\mathcal{C}_n$ back to $\mathcal{S}_n$

$$1 - P_e(\mathcal{C}_n) = \sum_{x^n \in \mathcal{S}_n} P_{X^n}(x^n)$$

$$= \sum_{x^n \in \mathcal{S}_n \cap \mathcal{F}_n^c(\delta)} P_{X^n}(x^n) + \sum_{x^n \in \mathcal{S}_n \cap \mathcal{F}_n(\delta)} P_{X^n}(x^n)$$

$$\leq P_{X^n}(\mathcal{F}_n^c(\delta)) + |\mathcal{S}_n \cap \mathcal{F}_n(\delta)| \cdot \max_{x^n \in \mathcal{F}_n(\delta)} P_{X^n}(x^n)$$

$$< \delta + |\mathcal{S}_n| \cdot \max_{x^n \in \mathcal{F}_n(\delta)} P_{X^n}(x^n)$$

$$< \frac{\varepsilon}{2} + 2^{n(H(X)-2\delta)} \cdot 2^{-n(H(X)-\delta)}$$

$$< \frac{\varepsilon}{2} + 2^{-n\delta}$$

$$< \varepsilon,$$

which is equivalent to $P_e(\mathcal{C}_n) > 1 - \varepsilon$ for $n > N$.   □

Observation 3.7 The results of the above theorem are illustrated in Fig. 3.3, where $\bar{R} = \limsup_{n \to \infty}(1/n)\log_D M_n$ is usually called the *asymptotic* code rate of block codes for compressing the source. It is clear from the figure that the (asymptotic) rate of any block code with arbitrarily small decoding error probability must be greater than or equal to the source entropy.[8] Conversely, the probability of decoding error for any block code of rate smaller than entropy ultimately approaches 1 (and hence is bounded away from zero). Thus for a DMS, the source entropy $H_D(X)$ is the infimum of all "achievable" source (block) coding rates; i.e., it is the infimum of all rates for

---

[8]Note that it is clear from the statement and proof of the forward part of Theorem 3.6 that the source entropy can be achieved as an asymptotic compression rate as long as $(1/n)\log_D M_n$ approaches it from *above* with increasing n. Furthermore, the asymptotic compression rate is defined as the limsup of $(1/n)\log_D M_n$ in order to guarantee reliable compression for n sufficiently large (analogously, in channel coding, the asymptotic transmission rate is defined via the liminf of $(1/n)\log_D M_n$ to ensure reliable communication for all sufficiently large n, see Chap. 4).

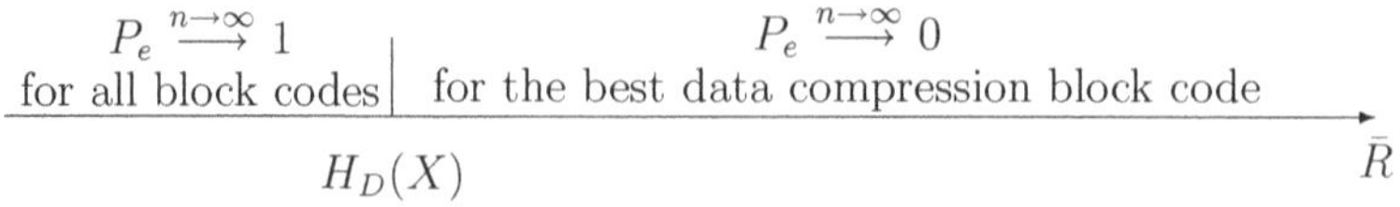

$$\underbrace{P_e \overset{n\to\infty}{\longrightarrow} 1}_{\text{for all block codes}} \quad \underbrace{P_e \overset{n\to\infty}{\longrightarrow} 0}_{\text{for the best data compression block code}}$$

$$\underbrace{}_{H_D(X)} \xrightarrow{} \bar{R}$$

Fig. 3.3 Asymptotic compression rate $\bar{R}$ versus source entropy $H_D(X)$ and behavior of the probability of block decoding error as blocklength n goes to infinity for a discrete memoryless source

which there exists a sequence of D-ary block codes with asymptotically vanishing (as the blocklength goes to infinity) probability of decoding error. Indeed to prove that $H_D(X)$ is such infimum, we decomposed the above theorem in two parts as per the properties of the infimum; see Observation A.11.

For a source with (statistical) memory, the Shannon–McMillan–Breiman theorem cannot be directly applied in its original form, and thereby Shannon's source coding theorem appears restricted to only memoryless sources. However, by exploring the concept behind these theorems, one can find that the key for the validity of Shannon's source coding theorem is actually the existence of a set $\mathcal{A}_n = \{x_1^n, x_2^n, \ldots, x_M^n\}$ with $M \approx D^{nH_D(X)}$ and $P_{X^n}(\mathcal{A}_n^c) \to 0$, namely, the existence of a "typical-like" set $\mathcal{A}_n$ whose size is prohibitively small and whose probability mass is asymptotically large. Thus, if we can find such typical-like set for a source with memory, the source coding theorem for block codes can be extended for this source. Indeed, with appropriate modifications, the Shannon–McMillan–Breiman theorem can be generalized for the class of stationary ergodic sources and hence a block source coding theorem for this class can be established; this is considered in the next section. The block source coding theorem for general (e.g., nonstationary non-ergodic) sources in terms of a generalized "spectral" entropy measure is studied in [73, 172, 175] (see also the end of the next section for a brief description).

3.2.2 Block Codes for Stationary Ergodic Sources

In practice, a stochastic source used to model data often exhibits *memory* or statistical dependence among its random variables; its joint distribution is hence not a product of its marginal distributions. In this section, we consider the asymptotic lossless data compression theorem for the class of stationary ergodic sources.[9]

Before proceeding to generalize the block source coding theorem, we need to first generalize the "entropy" measure for a sequence of dependent random variables X_n (which certainly should be backward compatible to the discrete memoryless cases). A straightforward generalization is to examine the limit of the normalized block entropy of a source sequence, resulting in the concept of *entropy rate*.

---

[9]The definitions of stationarity and ergodicity can be found in Sect. B.3 of Appendix B.

Definition 3.8 (Entropy rate) The *entropy rate* for a source $\{X_n\}_{n=1}^{\infty}$ is denoted by $H(\mathcal{X})$ and defined by

$$H(\mathcal{X}) := \lim_{n \to \infty} \frac{1}{n} H(X^n)$$

provided the limit exists, where $X^n = (X_1, \ldots, X_n)$.

Next, we will show that the entropy rate exists for *stationary* sources (here, we do not need ergodicity for the existence of entropy rate).

Lemma 3.9 *For a stationary source* $\{X_n\}_{n=1}^{\infty}$, *the conditional entropy*

$$H(X_n | X_{n-1}, \ldots, X_1)$$

is nonincreasing in n and also bounded from below by zero. Hence, by Lemma A.20, the limit

$$\lim_{n \to \infty} H(X_n | X_{n-1}, \ldots, X_1)$$

exists.

Proof We have

$$
\begin{aligned}
H(X_n | X_{n-1}, \ldots, X_1) &\leq H(X_n | X_{n-1}, \ldots, X_2) & (3.2.4) \\
&= H(X_n, \ldots, X_2) - H(X_{n-1}, \ldots, X_2) \\
&= H(X_{n-1}, \ldots, X_1) - H(X_{n-2}, \ldots, X_1) & (3.2.5) \\
&= H(X_{n-1} | X_{n-2}, \ldots, X_1),
\end{aligned}
$$

where (3.2.4) follows since conditioning never increases entropy, and (3.2.5) holds because of the stationarity assumption. Finally, recall that each conditional entropy $H(X_n | X_{n-1}, \ldots, X_1)$ is nonnegative. $\qquad\square$

Lemma 3.10 (Cesaro-mean theorem) *If* $a_n \to a$ *as* $n \to \infty$ *and* $b_n = \frac{1}{n} \sum_{i=1}^{n} a_i$, *then* $b_n \to a$ *as* $n \to \infty$.

Proof $a_n \to a$ implies that for any $\varepsilon > 0$, there exists an N such that for all $n > N$, $|a_n - a| < \varepsilon$. Then

$$
\begin{aligned}
|b_n - a| &= \left| \frac{1}{n} \sum_{i=1}^{n} (a_i - a) \right| \\
&\leq \frac{1}{n} \sum_{i=1}^{n} |a_i - a|
\end{aligned}
$$

$$= \frac{1}{n} \sum_{i=1}^{N} |a_i - a| + \frac{1}{n} \sum_{i=N+1}^{n} |a_i - a|$$

$$\leq \frac{1}{n} \sum_{i=1}^{N} |a_i - a| + \frac{n-N}{n} \varepsilon.$$

Hence, $\lim_{n \to \infty} |b_n - a| \leq \varepsilon$. Since ε can be made arbitrarily small, the lemma holds. $\square$

Theorem 3.11 *The entropy rate of a stationary source* $\{X_n\}_{n=1}^{\infty}$ *always exists and is equal to*

$$H(\mathcal{X}) = \lim_{n \to \infty} H(X_n | X_{n-1}, \ldots, X_1).$$

Proof The result directly follows by writing

$$\frac{1}{n} H(X^n) = \frac{1}{n} \sum_{i=1}^{n} H(X_i | X_{i-1}, \ldots, X_1) \quad \text{(chain rule for entropy)}$$

and applying the Cesaro-mean theorem. $\square$

Observation 3.12 It can also be shown that for a stationary source, $(1/n)H(X^n)$ is nonincreasing in n and $(1/n)H(X^n) \geq H(X_n | X_{n-1}, \ldots, X_1)$ for all $n \geq 1$. (The proof is left as an exercise; see Problem 3.)

It is obvious that when $\{X_n\}_{n=1}^{\infty}$ is a discrete memoryless source, $H(X^n) = n \cdot H(X)$ for every n. Hence,

$$H(\mathcal{X}) = \lim_{n \to \infty} \frac{1}{n} H(X^n) = H(X).$$

For a stationary Markov source (of order one),[10]

$$H(\mathcal{X}) = \lim_{n \to \infty} \frac{1}{n} H(X^n) = \lim_{n \to \infty} H(X_n | X_{n-1}, \ldots, X_1) = H(X_2 | X_1),$$

where

$$H(X_2 | X_1) = - \sum_{x_1 \in \mathcal{X}} \sum_{x_2 \in \mathcal{X}} \pi(x_1) P_{X_2 | X_1}(x_2 | x_1) \log P_{X_2 | X_1}(x_2 | x_1),$$

and $\pi(\cdot)$ is a stationary distribution for the Markov source (note that $\pi(\cdot)$ is unique if the Markov source is irreducible[11]). For example, for the stationary binary Markov

---

[10] If a Markov source is mentioned without specifying its order, it is understood that it is a first-order Markov source; see Appendix B for a brief overview on Markov sources and their properties.

[11] See Sect. B.3 of Appendix B for the definition of irreducibility for Markov sources.

source with transition probabilities $P_{X_2|X_1}(0|1) = \alpha$ and $P_{X_2|X_1}(1|0) = \beta$, where $0 < \alpha, \beta < 1$, we have

$$H(\mathcal{X}) = \frac{\beta}{\alpha + \beta} h_{\mathrm{b}}(\alpha) + \frac{\alpha}{\alpha + \beta} h_{\mathrm{b}}(\beta),$$

where $h_{\mathrm{b}}(\alpha) := -\alpha \log_2 \alpha - (1 - \alpha) \log_2(1 - \alpha)$ is the binary entropy function.

Observation 3.13 (Divergence rate for sources with memory) We briefly note that analogously to the notion of entropy rate, the *divergence rate* (or *Kullback–Leibler divergence rate*) can also be defined for sources with memory. More specifically, given two discrete sources $\{X_i\}_{i=1}^{\infty}$ and $\{\hat{X}_i\}_{i=1}^{\infty}$ defined on a common finite alphabet $\mathcal{X}$, with respective sequences of n-fold distributions $\{P_{X^n}\}$ and $\{P_{\hat{X}^n}\}$, then the divergence rate between $\{X_i\}_{i=1}^{\infty}$ and $\{\hat{X}_i\}_{i=1}^{\infty}$ is defined by

$$\lim_{n \to \infty} \frac{1}{n} D(P_{X^n} \| P_{\hat{X}^n})$$

provided the limit exists.[12] The divergence rate is not guaranteed to exist in general; in [350], two examples of non-Markovian ergodic sources are given for which the divergence rate does not exist. However, if the source $\{\hat{X}_i\}$ is time-invariant Markov and $\{X_i\}$ is stationary, then the divergence rate exists and is given in terms of the entropy rate of $\{X_i\}$ and another quantity depending on the (second-order) statistics of $\{X_i\}$ and $\{\hat{X}_i\}$ as follows [157, p. 40]:

$$\lim_{n \to \infty} \frac{1}{n} D(P_{X^n} \| P_{\hat{X}^n}) = -\lim_{n \to \infty} \frac{1}{n} H(X^n) - \sum_{x_1 \in \mathcal{X}} \sum_{x_2 \in \mathcal{X}} P_{X_1 X_2}(x_1, x_2) \log_2 P_{\hat{X}_2|\hat{X}_1}(x_2|x_1).$$

$$(3.2.6)$$

Furthermore, if both $\{X_i\}$ and $\{\hat{X}_i\}$ are time-invariant irreducible Markov sources, then their divergence rate exists and admits the following expression [312, Theorem 1]:

$$\lim_{n \to \infty} \frac{1}{n} D(P_{X^n} \| P_{\hat{X}^n}) = \sum_{x_1 \in \mathcal{X}} \sum_{x_2 \in \mathcal{X}} \pi_X(x_1) P_{X_2|X_1}(x_2|x_1) \log_2 \frac{P_{X_2|X_1}(x_2|x_1)}{P_{\hat{X}_2|\hat{X}_1}(x_2|x_1)},$$

where $\pi_X(\cdot)$ is the stationary distribution of $\{X_i\}$. The above result can also be generalized using the theory of nonnegative matrices and Perron–Frobenius theory for $\{X_i\}$ and $\{\hat{X}_i\}$ being arbitrary (not necessarily irreducible, stationary, etc.) time-invariant Markov chains; see the explicit computable expression in [312, Theorem 2]. A direct consequence of the later result is a formula for the entropy rate of an arbitrary not necessarily stationary time-invariant Markov source [312, Corollary 2].[13]

---

[12] Another notation for the divergence rate is $\lim_{n \to \infty} \frac{1}{n} D(X^n \| \hat{X}^n)$.

[13] More generally, the Rényi entropy rate of order α, $\lim_{n \to \infty} \frac{1}{n} H_\alpha(X^n)$, as well as the Rényi divergence rate of order α, $\lim_{n \to \infty} \frac{1}{n} D_\alpha(P_{X^n} \| P_{\hat{X}^n})$, for arbitrary time-invariant Markov sources

Finally, note that all the above results also hold with the proper modifications if the Markov chains are replaced with kth-order Markov chains (for any integer $k > 1$) [312].

Theorem 3.14 (Generalized AEP or Shannon–McMillan–Breiman Theorem [58, 83, 267, 340]) *If* $\{X_n\}_{n=1}^{\infty}$ *is a stationary ergodic source, then*

$$-\frac{1}{n} \log_2 P_{X^n}(X_1, \ldots, X_n) \xrightarrow{a.s.} H(\mathcal{X}).$$

Since the AEP theorem (law of large numbers) is valid for stationary ergodic sources, all consequences of the AEP will follow, including Shannon's lossless source coding theorem.

Theorem 3.15 (Shannon's source coding theorem for stationary ergodic sources) *Given integer* $D > 1$, *let* $\{X_n\}_{n=1}^{\infty}$ *be a stationary ergodic source with entropy rate (in base D)*

$$H_D(\mathcal{X}) := \lim_{n \to \infty} \frac{1}{n} H_D(X^n).$$

Then the following hold.

- *Forward part (achievability)*: For any $0 < \varepsilon < 1$, there exists a δ with $0 < \delta < \varepsilon$ and a sequence of D-ary block codes $\{\mathcal{C}_n = (n, M_n)\}_{n=1}^{\infty}$ with

$$\limsup_{n \to \infty} \frac{1}{n} \log_D M_n < H_D(\mathcal{X}) + \delta,$$

and probability of decoding error satisfying

$$P_e(\mathcal{C}_n) < \varepsilon$$

for all sufficiently large n.
- *Strong converse part*: For any $0 < \varepsilon < 1$, any sequence of D-ary block codes $\{\mathcal{C}_n = (n, M_n)\}_{n=1}^{\infty}$ with

$$\limsup_{n \to \infty} \frac{1}{n} \log_D M_n < H_D(\mathcal{X})$$

satisfies

$$P_e(\mathcal{C}_n) > 1 - \varepsilon$$

for all n sufficiently large.

$\{X_i\}$ and $\{\hat{X}_i\}$ exist and admit closed-form expressions [311] (see also the earlier work in [279], where the results hold for more restricted classes of Markov sources).

A discrete memoryless (i.i.d.) source is stationary and ergodic (so Theorem 3.6 is clearly a special case of Theorem 3.15). In general, it is hard to check whether a stationary process is ergodic or not. It is known though that if a stationary process is a mixture of two or more stationary ergodic processes, i.e., its n-fold distribution can be written as the mean (with respect to some distribution) of the n-fold distributions of stationary ergodic processes, then it is *not ergodic*.[14]

For example, let P and Q be two distributions on a finite alphabet $\mathcal{X}$ such that the process $\{X_n\}_{n=1}^{\infty}$ is i.i.d. with distribution P and the process $\{Y_n\}_{n=1}^{\infty}$ is i.i.d. with distribution Q. Flip a biased coin (with Heads probability equal to θ, $0 < \theta < 1$) *once* and let

$$Z_n = \begin{cases} X_n, & \text{if Heads,} \\ Y_n, & \text{if Tails,} \end{cases}$$

for $n = 1, 2, \ldots$. Then, the resulting process $\{Z_n\}_{n=1}^{\infty}$ has its n-fold distribution as a mixture of the n-fold distributions of $\{X_n\}_{n=1}^{\infty}$ and $\{Y_n\}_{n=1}^{\infty}$:

$$P_{Z^n}(a^n) = \theta P_{X^n}(a^n) + (1 - \theta) P_{Y^n}(a^n) \tag{3.2.7}$$

for all $a^n \in \mathcal{X}^n$, $n = 1, 2, \ldots$. Then, the process $\{Z_n\}_{n=1}^{\infty}$ is stationary but *not ergodic*.

A specific case for which ergodicity can be easily verified (other than the case of i.i.d. sources) is the case of stationary Markov sources. Specifically, if a (finite alphabet) stationary Markov source is irreducible, then it is ergodic (e.g., see [30, p. 371] and [349, Prop. I.2.9]), and hence a generalized AEP holds for this source. Note that irreducibility can be verified in terms of the source's transition probability matrix.

The following are two examples of stationary processes based on Polya's contagion urn scheme [304], one of which is non-ergodic and the other ergodic.

Example 3.16 (Polya contagion process [304]) We consider a binary process with memory $\{Z_n\}_{n=1}^{\infty}$, which is obtained by the following Polya contagion urn sampling mechanism [304–306] (see also [119, 120]).

An urn initially contains T balls, of which R are red and B are black ($T = R + B$). Successive draws are made from the urn, where after each draw $1 + \Delta$ balls of the same color just drawn are returned to the urn ($\Delta > 0$). The process $\{Z_n\}_{n=1}^{\infty}$ is generated according to the outcome of the draws:

$$Z_n = \begin{cases} 1, & \text{if the } n\text{th ball drawn is red,} \\ 0, & \text{if the } n\text{th ball drawn is black.} \end{cases}$$

In this model, a red ball in the urn can represent an infected person in the population and a black ball can represent a healthy person. Since the number of balls of the color just drawn increases (while the number of balls of the opposite color is unchanged),

---

[14]The converse is also true; i.e., if a stationary process cannot be represented as a mixture of stationary ergodic processes, then it is ergodic.

the likelihood that a ball of the same color as the ball just drawn will be picked in the next draw increases. Hence, the occurrence of an "unfavorable" event (say an infection) increases the probability of future unfavorable events (the same applies for favorable events) and as a result the model provides a basic template for characterizing contagious phenomena.

For any $n \geq 1$, the n-fold distribution of the binary process $\{Z_n\}_{n=1}^{\infty}$ can be derived in closed form as follows:

$$
\begin{aligned}
\Pr[Z^n = a^n] &= \frac{\rho(\rho + \delta) \cdots (\rho + (d-1)\delta)\sigma(\sigma + \delta) \cdots (\sigma + (n-d-1)\delta)}{(1 + \delta)(1 + 2\delta) \cdots (1 + (n-1)\delta)} \\
&= \frac{\Gamma(\frac{1}{\delta}) \, \Gamma(\frac{\rho}{\delta} + d) \, \Gamma(\frac{\sigma}{\delta} + n - d)}{\Gamma(\frac{\rho}{\delta}) \, \Gamma(\frac{\sigma}{\delta}) \, \Gamma(\frac{1}{\delta} + n)}
\end{aligned}
\tag{3.2.8}
$$

for all $a^n = (a_1, a_2, \ldots, a_n) \in \{0, 1\}^n$, where $d = a_1 + a_2 + \cdots + a_n$, $\rho := \frac{R}{T}$, $\sigma := 1 - \rho = \frac{B}{T}$, $\delta := \frac{\Delta}{T}$, and $\Gamma(\cdot)$ is the gamma function given by

$$
\Gamma(x) = \int_0^{\infty} t^{x-1} e^{-t} dt \qquad \text{for } x > 0.
$$

To obtain the last equation in (3.2.8), we use the identity

$$
\prod_{j=0}^{n-1} (\alpha + j\beta) = \beta^n \frac{\Gamma(\frac{\alpha}{\beta} + n)}{\Gamma(\frac{\alpha}{\beta})}
$$

which is obtained using the fact that $\Gamma(x + 1) = x \, \Gamma(x)$. We remark from expression (3.2.8) for the joint distribution that the process $\{Z_n\}$ is *exchangeable*[15] and is thus stationary. Furthermore, it can be shown [120, 306] that the process sample average $\frac{1}{n}(Z_1 + Z_2 + \cdots + Z_n)$ converges almost surely as $n \to \infty$ to a random variable Z, whose distribution is given by the beta distribution with parameters $\rho/\delta = R/\Delta$ and $\sigma/\delta = B/\Delta$. This directly implies that the process $\{Z_n\}_{n=1}^{\infty}$ is *not ergodic* since its sample average does not converge to a constant. It is also shown in [12] that the entropy rate of $\{Z_n\}_{n=1}^{\infty}$ is given by

$$
H(\mathcal{Z}) = E_Z[h_b(Z)] = \int_0^1 h_b(z) f_Z(z) dz
$$

where $h_b(\cdot)$ is the binary entropy function and

---

[15] A process $\{Z_n\}_{n=1}^{\infty}$ is called *exchangeable* (or *symmetrically dependent*) if for every finite positive integer n, the random variables $Z_1, Z_2, \ldots, Z_n$ have the property that their joint distribution is invariant with respect to all permutations of the indices $1, 2, \ldots, n$ (e.g., see [120]). The notion of exchangeability is originally due to de Finetti [90]. It directly follows from the definition that an exchangeable process is stationary.

$$f_Z(z) = \begin{cases} \frac{\Gamma(\frac{1}{\delta})}{\Gamma(\frac{\rho}{\delta})\Gamma(\frac{\sigma}{\delta})} z^{\rho/\delta-1}(1-z)^{\sigma/\delta-1}, & \text{if } 0 < z < 1, \\ 0, & \text{otherwise,} \end{cases}$$

is the beta probability density function with parameters ρ/δ and σ/δ. Note that Theorem 3.15 does not hold for the contagion source $\{Z_n\}$ since it is not ergodic.

Finally, letting $0 \le R_n \le 1$ denote the proportion of red balls in the urns after the nth draw, we can write

$$R_n = \frac{R + (Z_1 + Z_2 + \cdots + Z_n)\Delta}{T + n\Delta} \tag{3.2.9}$$

$$= \frac{R_{n-1}(T + (n-1)\Delta) + Z_n\Delta}{T + n\Delta}. \tag{3.2.10}$$

Now using (3.2.10), we have

$$E[R_n|R_{n-1}, \ldots, R_1] = E[R_n|R_{n-1}]$$
$$= \frac{R_{n-1}(T + (n-1)\Delta) + \Delta}{T + n\Delta} \cdot R_{n-1}$$
$$+ \frac{R_{n-1}(T + (n-1)\Delta)}{T + n\Delta} \cdot (1 - R_{n-1})$$
$$= R_{n-1}$$

almost surely, and thus $\{R_n\}$ is a martingale (e.g., [120, 162]). Since $\{R_n\}$ is bounded, we obtain by the martingale convergence theorem that R_n converges almost surely to some limiting random variable. But from (3.2.9), we note that the asymptotic behavior of R_n is identical to that of $\frac{1}{n}(Z_1 + Z_2 + \cdots + Z_n)$. Thus, R_n also converges almost surely to the above beta-distributed random variable Z.

In [12], a binary additive noise channel, whose noise is the above Polya contagion process $\{Z_n\}_{n=1}^{\infty}$, is investigated as a model for a non-ergodic communication channel with memory. Polya's urn scheme has also been applied and generalized in a wide range of contexts, including genetics [210], evolution and epidemiology [257, 289], image segmentation [35], and network epidemics [182] (see also the survey in [289]).

Example 3.17 (Finite-memory Polya contagion process [12]) The above Polya model has "infinite" memory in the sense that the very first ball drawn from the urn has an identical effect (that does not vanish as the number of draws grows without bound) as the 999 999th ball drawn from the urn on the outcome of the millionth draw. In the context of modeling contagious phenomena, this is not reasonable as one would assume that the effects of an infection dissipate in time. We herein consider a more realistic urn model with finite memory [12].

Consider again an urn originally containing $T = R + B$ balls, of which R are red and B are black. At the nth draw, $n = 1, 2, \ldots$, a ball is selected at random from the urn and replaced with $1 + \Delta$ balls of the same color just drawn ($\Delta > 0$). Then, M draws later, i.e., after the $(n + M)$th draw, Δ balls of the color picked at the nth draw are retrieved from the urn.

Note that in this model, the total number of balls in the urn is constant $(T + M\Delta)$ after an initialization period of M draws. Also, in this scheme, the effect of any draw is limited to M draws in the future. The process $\{Z_n\}_{n=1}^\infty$ again corresponds to the outcome of the draws:

$$Z_n = \begin{cases} 1, & \text{if the } n\text{th ball drawn is red} \\ 0, & \text{if the } n\text{th ball drawn is black.} \end{cases}$$

We have that for $n \geq M + 1$,

$$\Pr\left[Z_n = 1 | Z_{n-1} = z_{n-1}, \ldots, Z_1 = z_1\right] = \frac{R + (z_{n-1} + \cdots + z_{n-M})\Delta}{T + M\Delta}$$

$$= \Pr\left[Z_n = 1 | Z_{n-1} = z_{n-1}, \ldots, Z_{n-M} = z_{n-M}\right]$$

for any $z_i \in \{0, 1\}$, $i = 1, \ldots, n$. Thus, $\{Z_n\}_{n=1}^\infty$ is a Markov process of memory order M. It is also shown in [12] that $\{Z_n\}$ is stationary and its stationary distribution as well as its n-fold distribution and its entropy rate, which is given by

$$H(\mathcal{Z}) = H(Z_{M+1} | Z_M, Z_{M-1}, \ldots, Z_1),$$

are derived in closed form in terms of R/T, Δ/T, and M. Furthermore, it is shown that $\{Z_n\}_{n=1}^\infty$ is irreducible, and hence ergodic. Thus, Theorem 3.15 applies for this finite-memory Polya contagion process. In [420], a generalized version of this process is introduced via a ball sampling mechanism involving a large urn and a finite queue.

Observation 3.18 In complicated situations such as when the source is nonstationary (with time-varying statistics) and/or non-ergodic (such as the non-ergodic processes in (3.2.7) or in Example 3.16), the source entropy rate $H(\mathcal{X})$ (if the limit exists; otherwise, one can look at the lim inf/lim sup of $(1/n)H(X^n)$) has no longer an operational meaning as the smallest possible block compression rate. This causes the need to establish new entropy measures that appropriately characterize the operational limits of an arbitrary stochastic system with memory. This is achieved in [175] where Han and Verdú introduce the spectral notions of *inf/sup-entropy rates* and illustrate the key role these entropy measures play in proving a general lossless block source coding theorem. More specifically, they demonstrate that for an arbitrary (not necessarily stationary and ergodic) finite-alphabet source $\mathcal{X} := \{X^n = (X_1, X_2, \ldots, X_n)\}_{n=1}^\infty$, the expression for the minimum achievable (block) source coding rate is given by the *sup-entropy rate* $\bar{H}(\mathcal{X})$, defined by

$$\bar{H}(\mathcal{X}) := \inf_{\beta \in \mathbb{R}} \left\{ \beta : \limsup_{n \to \infty} \Pr\left[-\frac{1}{n} \log P_{X^n}(X^n) > \beta \right] = 0 \right\}.$$

More details are provided in [73, 172, 175].

3.2.3 Redundancy for Lossless Block Data Compression

Shannon's block source coding theorem establishes that the smallest data compression rate for achieving arbitrarily small error probability for stationary ergodic sources is given by the entropy rate. Thus, one can define the source *redundancy* as the reduction in coding rate one can achieve via asymptotically lossless block source coding versus just using uniquely decodable (completely lossless for any value of the sourceword blocklength n) block source coding. In light of the fact that the former approach yields a source coding rate equal to the entropy rate while the latter approach provides a rate of $\log_2 |\mathcal{X}|$, we therefore define the *total block source coding redundancy* ρ_t (in bits/source symbol) for a stationary ergodic source $\{X_n\}_{n=1}^{\infty}$ as

$$\rho_t := \log_2 |\mathcal{X}| - H(\mathcal{X}).$$

Hence, ρ_t represents the amount of "useless" (or superfluous) statistical source information one can eliminate via binary[16] block source coding.

   If the source is i.i.d. and uniformly distributed, then its entropy rate is equal to $\log_2 |\mathcal{X}|$ and as a result its redundancy is $\rho_t = 0$. This means that the source is *incompressible*, as expected, since in this case *every* sourceword x^n will belong to the δ-typical set $\mathcal{F}_n(\delta)$ for every $n > 0$ and $\delta > 0$ (i.e., $\mathcal{F}_n(\delta) = \mathcal{X}^n$), and hence there are no superfluous sourcewords that can be dispensed of via source coding. If the source has memory or has a nonuniform marginal distribution, then its redundancy is strictly positive and can be classified into two parts:

- Source redundancy due to the *nonuniformity of the source marginal distribution* ρ_d:

$$\rho_d := \log_2 |\mathcal{X}| - H(X_1).$$

- Source redundancy due to the *source memory* ρ_m:

$$\rho_m := H(X_1) - H(\mathcal{X}).$$

As a result, the source total redundancy ρ_t can be decomposed in two parts:

$$\rho_t = \rho_d + \rho_m.$$

   We can summarize the *redundancy* of some typical stationary ergodic sources in the following table.

---

[16]Since we are measuring ρ_t in code bits/source symbol, all logarithms in its expression are in base 2, and hence this redundancy can be eliminated via asymptotically lossless binary block codes (one can also change the units to D-ary code symbol/source symbol using base-D logarithms for the case of D-ary block codes).

Source	ρ_d	ρ_m	ρ_t
i.i.d. uniform	0	0	0
i.i.d. nonuniform	$\log_2 |\mathcal{X}| - H(X_1)$	0	ρ_d
First-order symmetric Markov[a]	0	$H(X_1) - H(X_2|X_1)$	ρ_m
First-order non symmetric Markov	$\log_2 |\mathcal{X}| - H(X_1)$	$H(X_1) - H(X_2|X_1)$	$\rho_d + \rho_m$

[a]A first-order Markov process is symmetric if for any x_1 and $\hat{x}_1$,
$\{a : a = P_{X_2|X_1}(y|x_1)$ for some $y\} = \{a : a = P_{X_2|X_1}(y|\hat{x}_1)$ for some $y\}$.

3.3 Variable-Length Codes for Lossless Data Compression

3.3.1 Non-singular Codes and Uniquely Decodable Codes

We next study variable-length (completely) lossless data compression codes.

Definition 3.19 Consider a discrete source $\{X_n\}_{n=1}^{\infty}$ with finite alphabet $\mathcal{X}$ along with a D-ary code alphabet $\mathcal{B} = \{0, 1, \ldots, D - 1\}$, where $D > 1$ is an integer. Fix integer $n \geq 1$; then a *D-ary nth-order variable-length code* (VLC) is a function

$$f : \mathcal{X}^n \to \mathcal{B}^*$$

mapping (fixed-length) sourcewords of length n to D-ary codewords in $\mathcal{B}^*$ of variable lengths, where $\mathcal{B}^*$ denotes the set of all finite-length strings from $\mathcal{B}$ (i.e., $c \in \mathcal{B}^* \iff \exists$ integer $l \geq 1$ such that $c \in \mathcal{B}^l$).

The *codebook* $\mathcal{C}$ of a VLC is the set of all codewords:

$$\mathcal{C} = f(\mathcal{X}^n) = \{f(x^n) \in \mathcal{B}^* : x^n \in \mathcal{X}^n\}.$$

A variable-length lossless data compression code is a code in which the source symbols can be completely reconstructed without distortion. In order to achieve this goal, the source symbols have to be encoded unambiguously in the sense that any two different source symbols (with positive probabilities) are represented by different codewords. Codes satisfying this property are called *non-singular codes*. In practice, however, the encoder often needs to encode a sequence of source symbols, which results in a concatenated sequence of codewords. If any concatenation of codewords can also be unambiguously reconstructed without punctuation, then the code is said to be *uniquely decodable*. In other words, a VLC is uniquely decodable if all finite sequences of sourcewords ($x^n \in \mathcal{X}^n$) are mapped onto distinct strings of codewords: for any m and m', $(x_1^n, x_2^n, \ldots, x_m^n) \neq (y_1^n, y_2^n, \ldots, y_{m'}^n)$ implies that

$$(f(x_1^n), f(x_2^n), \ldots, f(x_m^n)) \neq (f(y_1^n), f(y_2^n), \ldots, f(y_{m'}^n)),$$

or equivalently,

$$(f(x_1^n), f(x_2^n), \ldots, f(x_m^n)) = (f(y_1^n), f(y_2^n), \ldots, f(y_{m'}^n))$$

implies that

$$m = m' \text{ and } x_j^n = y_j^m \text{ for } j = 1, \ldots, m.$$

Note that a non-singular VLC is not necessarily uniquely decodable. For example, consider a binary (first-order) code for the source with alphabet

$$\mathcal{X} = \{A, B, C, D, E, F\}$$

given by

$$\text{code of } A = 0,$$
$$\text{code of } B = 1,$$
$$\text{code of } C = 00,$$
$$\text{code of } D = 01,$$
$$\text{code of } E = 10,$$
$$\text{code of } F = 11.$$

The above code is clearly non-singular; it is, however, not uniquely decodable because the codeword sequence, 010, can be reconstructed as ABA, DA, or AE (i.e., $(f(A), f(B), f(A)) = (f(D), f(A)) = (f(A), f(E))$ even though (A, B, A), (D, A), and (A, E) are all non-equal).

One important objective is to find out how "efficiently" we can represent a given discrete source via a uniquely decodable nth-order VLC and provide a construction technique that (at least asymptotically, as $n \to \infty$) attains the optimal "efficiency." In other words, we want to determine what is the smallest possible average code rate (or equivalently, the smallest average codeword length) that an nth-order uniquely decodable VLC can have when (losslessly) representing a given source, and we want to give an explicit code construction that can attain this smallest possible rate (at least asymptotically in the sourceword length n).

Definition 3.20 Let $\mathcal{C}$ be a D-ary nth-order VLC

$$f : \mathcal{X}^n \to \{0, 1, \ldots, D - 1\}^*$$

for a discrete source $\{X_n\}_{n=1}^\infty$ with alphabet $\mathcal{X}$ and distribution $P_{X^n}(x^n)$, $x^n \in \mathcal{X}^n$. Setting $\ell(c_{x^n})$ as the length of the codeword $c_{x^n} = f(x^n)$ associated with sourceword x^n, then the *average codeword length* for $\mathcal{C}$ is given by

$$\bar{\ell} := \sum_{x^n \in \mathcal{X}^n} P_{X^n}(x^n)\ell(c_{x^n})$$

and its *average code rate* (in D-ary code symbols/source symbol) is given by

$$\overline{R}_n := \frac{\overline{\ell}}{n} = \frac{1}{n} \sum_{x^n \in \mathcal{X}^n} P_{X^n}(x^n) \ell(c_{x^n}).$$

The following theorem provides a strong condition that a uniquely decodable code must satisfy.[17]

Theorem 3.21 (Kraft inequality for uniquely decodable codes) *Let $\mathcal{C}$ be a uniquely decodable D-ary nth-order VLC for a discrete source $\{X_n\}_{n=1}^{\infty}$ with alphabet $\mathcal{X}$. Let the $M = |\mathcal{X}|^n$ codewords of $\mathcal{C}$ have lengths $\ell_1, \ell_2, \ldots, \ell_M$, respectively. Then, the following inequality must hold:*

$$\sum_{m=1}^{M} D^{-\ell_m} \leq 1.$$

Proof Suppose that we use the codebook $\mathcal{C}$ to encode N sourcewords ($x_k^n \in \mathcal{X}^n$, $k = 1, \ldots, N$) arriving in a sequence; this yields a concatenated codeword sequence

$$c_1 c_2 c_3 \ldots c_N.$$

Let the lengths of the codewords be respectively denoted by

$$\ell(c_1), \ell(c_2), \ldots, \ell(c_N).$$

Consider

$$\left(\sum_{c_1 \in \mathcal{C}} \sum_{c_2 \in \mathcal{C}} \cdots \sum_{c_N \in \mathcal{C}} D^{-[\ell(c_1) + \ell(c_2) + \cdots + \ell(c_N)]} \right).$$

It is obvious that the above expression is equal to

$$\left(\sum_{c \in \mathcal{C}} D^{-\ell(c)} \right)^N = \left(\sum_{m=1}^{M} D^{-\ell_m} \right)^N.$$

(Note that $|\mathcal{C}| = M$.) On the other hand, all the code sequences with length

$$i = \ell(c_1) + \ell(c_2) + \cdots + \ell(c_N)$$

contribute equally to the sum of the identity, which is D^{-i}. Let A_i denote the number of N-codeword sequences that have length i. Then, the above identity can be rewritten as

$$\left(\sum_{m=1}^{M} D^{-\ell_m} \right)^N = \sum_{i=1}^{LN} A_i D^{-i},$$

---

[17]This theorem is also attributed to McMillan.

where

$$L := \max_{c \in C} \ell(c).$$

Since C is by assumption a uniquely decodable code, the codeword sequence must be unambiguously decodable. Observe that a code sequence with length i has at most D^i unambiguous combinations. Therefore, $A_i \leq D^i$, and

$$\left(\sum_{m=1}^{M} D^{-\ell_m} \right)^N = \sum_{i=1}^{LN} A_i D^{-i} \leq \sum_{i=1}^{LN} D^i D^{-i} = LN,$$

which implies that

$$\sum_{m=1}^{M} D^{-\ell_m} \leq (LN)^{1/N}.$$

The proof is completed by noting that the above inequality holds for every N, and the upper bound $(LN)^{1/N}$ goes to 1 as N goes to infinity. □

The Kraft inequality is a very useful tool, especially for showing that the fundamental lower bound of the average rate of uniquely decodable VLCs for discrete memoryless sources is given by the source entropy.

Theorem 3.22 *The average rate of every uniquely decodable D-ary nth-order VLC for a discrete memoryless source $\{X_n\}_{n=1}^{\infty}$ is lower bounded by the source entropy $H_D(X)$ (measured in D-ary code symbols/source symbol).*

Proof Consider a uniquely decodable D-ary nth-order VLC code for the source $\{X_n\}_{n=1}^{\infty}$

$$f : \mathcal{X}^n \to \{0, 1, \ldots, D-1\}^*$$

and let $\ell(c_{x^n})$ denote the length of the codeword $c_{x^n} = f(x^n)$ for sourceword x^n. Then

$$\overline{R}_n - H_D(X) = \frac{1}{n} \sum_{x^n \in \mathcal{X}^n} P_{X^n}(x^n) \ell(c_{x^n}) - \frac{1}{n} H_D(X^n)$$

$$= \frac{1}{n} \left[\sum_{x^n \in \mathcal{X}^n} P_{X^n}(x^n) \ell(c_{x^n}) - \sum_{x^n \in \mathcal{X}^n} \left(-P_{X^n}(x^n) \log_D P_{X^n}(x^n) \right) \right]$$

$$= \frac{1}{n} \sum_{x^n \in \mathcal{X}^n} P_{X^n}(x^n) \log_D \frac{P_{X^n}(x^n)}{D^{-\ell(c_{x^n})}}$$

$$\geq \frac{1}{n} \left[\sum_{x^n \in \mathcal{X}^n} P_{X^n}(x^n) \right] \log_D \frac{\left[\sum_{x^n \in \mathcal{X}^n} P_{X^n}(x^n) \right]}{\left[\sum_{x^n \in \mathcal{X}^n} D^{-\ell(c_{x^n})} \right]}$$

(log-sum inequality)

$$= -\frac{1}{n} \log \left[\sum_{x^n \in \mathcal{X}'^n} D^{-\ell(c_{x^n})} \right]$$
$$\geq 0,$$

where the last inequality follows from the Kraft inequality for uniquely decodable codes and the fact that the logarithm is a strictly increasing function. □

By examining the above proof, we observe that

$$\overline{R}_n = H_D(X) \quad \text{iff} \quad P_{X^n}(x^n) = D^{-l(c_{x^n})};$$

i.e., the source symbol probabilities are (negative) integer powers of D. Such a source is called *D-adic* [83]. In this case, the code is called *absolutely optimal* as it achieves the source entropy lower bound (it is thus optimal in terms of yielding a minimal average code rate for any given n).

Furthermore, we know from the above theorem that the average code rate is no smaller than the source entropy. Indeed, a lossless data compression code, whose average code rate achieves entropy, should be optimal (note that if a code's average rate is below entropy, then the Kraft inequality is violated and the code is no longer uniquely decodable). In summary, we have

- Uniquely decodability $\implies$ the Kraft inequality holds.
- Uniquely decodability $\implies$ average code rate of VLCs for memoryless sources is lower bounded by the source entropy.

Exercise 3.23

1. Find a non-singular and also non-uniquely decodable code that violates the Kraft inequality. (Hint: The answer is already provided in this section.)
2. Find a non-singular and also non-uniquely decodable code that beats the entropy lower bound.

3.3.2 Prefix or Instantaneous Codes

A *prefix* code[18] is a VLC which is self-punctuated in the sense that there is no need to append extra symbols for differentiating adjacent codewords. A more precise definition follows:

Definition 3.24 *(Prefix code)* A VLC is called a *prefix code* or an *instantaneous code* if no codeword is a prefix of any other codeword.

A prefix code is also named an *instantaneous code* because the codeword sequence can be decoded *instantaneously* (it is immediately recognizable) without the reference to future codewords in the same sequence. Note that a uniquely decodable

---

[18] Another name for prefix codes is *prefix-free* codes.

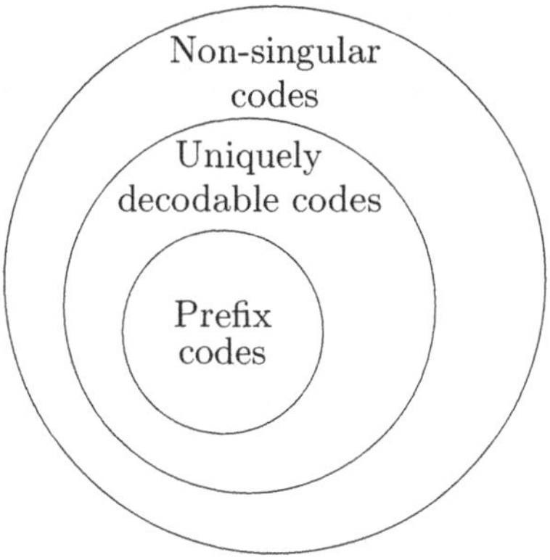

Fig. 3.4 Classification of
variable-length codes

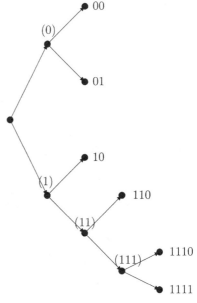

Fig. 3.5 Tree structure of a
binary prefix code. The
codewords are those residing
on the leaves, which in this
case are 00, 01, 10, 110,
1110, and 1111

code is not necessarily prefix-free and may not be decoded instantaneously. The
relationship between different codes encountered thus far is depicted in Fig. 3.4.

A D-ary prefix code can be represented graphically as an initial segment of a
D-ary tree. An example of a tree representation for a binary ($D = 2$) prefix code is
shown in Fig. 3.5.

Theorem 3.25 (Kraft inequality for prefix codes) *There exists a D-ary nth-order
prefix code for a discrete source $\{X_n\}_{n=1}^{\infty}$ with alphabet $\mathcal{X}$ iff the codewords of length
ℓ_m, $m = 1, \ldots, M$, satisfy the Kraft inequality, where $M = |\mathcal{X}|^n$.*

Proof Without loss of generality, we provide the proof for the case of $D = 2$ (binary
codes).

Forward part: *Prefix codes satisfy the Kraft inequality.*

The codewords of a prefix code can always be put on the leaves of a tree. Pick up a length

$$\ell_{\max} := \max_{1 \le m \le M} \ell_m.$$

A tree has originally $2^{\ell_{\max}}$ nodes on level $\ell_{\max}$. Each codeword of length ℓ_m obstructs $2^{\ell_{\max} - \ell_m}$ nodes on level $\ell_{\max}$. In other words, when any node is chosen as a codeword, all its children will be excluded from being codewords (as for a prefix code, no codeword can be a prefix of any other code). There are exactly $2^{\ell_{\max} - \ell_m}$ excluded nodes on level $\ell_{\max}$ of the tree. Note that no two codewords obstruct the same nodes on level $\ell_{\max}$. Hence, the number of totally obstructed codewords on level $\ell_{\max}$ should be less than $2^{\ell_{\max}}$, i.e.,

$$\sum_{m=1}^{M} 2^{\ell_{\max} - \ell_m} \le 2^{\ell_{\max}},$$

which immediately implies the Kraft inequality:

$$\sum_{m=1}^{M} 2^{-\ell_m} \le 1.$$

(This part can also be proven by stating the fact that a prefix code is a uniquely decodable code. The objective of adding this proof is to illustrate the characteristics of a tree-like prefix code.)

Converse part: *Kraft inequality implies the existence of a prefix code.*

Suppose that $\ell_1, \ell_2, \ldots, \ell_M$ satisfy the Kraft inequality. We will show that there exists a binary tree with M selected nodes where the ith node resides on level ℓ_i.

Let n_i be the number of nodes (among the M nodes) residing on level i (namely, n_i is the number of codewords with length i or $n_i = |\{m : \ell_m = i\}|$), and let

$$\ell_{\max} := \max_{1 \le m \le M} \ell_m.$$

Then, from the Kraft inequality, we have

$$n_1 2^{-1} + n_2 2^{-2} + \cdots + n_{\ell_{\max}} 2^{-\ell_{\max}} \le 1.$$

The above inequality can be rewritten in a form that is more suitable for this proof as follows:

$$n_1 2^{-1} \le 1$$
$$n_1 2^{-1} + n_2 2^{-2} \le 1$$
$$\vdots$$
$$n_1 2^{-1} + n_2 2^{-2} + \cdots + n_{\ell_{\max}} 2^{-\ell_{\max}} \le 1.$$

Hence,

$$n_1 \leq 2$$
$$n_2 \leq 2^2 - n_1 2^1$$
$$\vdots$$
$$n_{\ell_{\max}} \leq 2^{\ell_{\max}} - n_1 2^{\ell_{\max}-1} - \cdots - n_{\ell_{\max}-1} 2^1,$$

which can be interpreted in terms of a tree model as follows: the first inequality says that the number of codewords of length 1 is less than the available number of nodes on the first level, which is 2. The second inequality says that the number of codewords of length 2 is less than the total number of nodes on the second level, which is 2^2, minus the number of nodes obstructed by the first-level nodes already occupied by codewords. The succeeding inequalities demonstrate the availability of a sufficient number of nodes at each level after the nodes blocked by shorter length codewords have been removed. Because this is true at every codeword length up to the maximum codeword length, the assertion of the theorem is proved. □

Theorems 3.21 and 3.25 unveil the following relation between a variable-length uniquely decodable code and a prefix code.

Corollary 3.26 *A uniquely decodable D-ary nth-order code can always be replaced by a D-ary nth-order prefix code with the same average codeword length (and hence the same average code rate).*

The following theorem interprets the relationship between the average code rate of a prefix code and the source entropy.

Theorem 3.27 *Consider a discrete memoryless source $\{X_n\}_{n=1}^{\infty}$.*

1. *For any D-ary nth-order prefix code for the source, the average code rate is no less than the source entropy $H_D(X)$.*
2. *There must exist a D-ary nth-order prefix code for the source whose average code rate is no greater than $H_D(X) + \frac{1}{n}$, namely,*

$$\overline{R}_n := \frac{1}{n} \sum_{x^n \in \mathcal{X}^n} P_{X^n}(x^n)\ell(c_{x^n}) \leq H_D(X) + \frac{1}{n}, \tag{3.3.1}$$

where c_{x^n} is the codeword for sourceword x^n, and $\ell(c_{x^n})$ is the length of codeword c_{x^n}.

Proof A prefix code is uniquely decodable, and hence it directly follows from Theorem 3.22 that its average code rate is no less than the source entropy.

To prove the second part, we can design a prefix code satisfying both (3.3.1) and the Kraft inequality, which immediately implies the existence of the desired code by Theorem 3.25. Choose the codeword length for sourceword x^n as

$$\ell(c_{x^n}) = \lfloor -\log_D P_{X^n}(x^n) \rfloor + 1. \tag{3.3.2}$$

Then
$$D^{-\ell(c_{x^n})} \leq P_{X^n}(x^n).$$

Summing both sides over all source symbols, we obtain

$$\sum_{x^n \in \mathcal{X}^n} D^{-\ell(c_{x^n})} \leq 1,$$

which is exactly the Kraft inequality. On the other hand, (3.3.2) implies

$$\ell(c_{x^n}) \leq -\log_D P_{X^n}(x^n) + 1,$$

which in turn implies

$$\sum_{x^n \in \mathcal{X}^n} P_{X^n}(x^n)\ell(c_{x^n}) \leq \sum_{x^n \in \mathcal{X}^n} \left[-P_{X^n}(x^n) \log_D P_{X^n}(x^n) \right] + \sum_{x^n \in \mathcal{X}^n} P_{X^n}(x^n)$$
$$= H_D(X^n) + 1 = n H_D(X) + 1,$$

where the last equality holds since the source is memoryless. □

We note that nth-order prefix codes (which encode sourcewords of length n) for memoryless sources can yield an average code rate arbitrarily close to the source entropy when allowing n to grow without bound. For example, a memoryless source with alphabet

$$\{A, B, C\}$$

and probability distribution

$$P_X(A) = 0.8, \quad P_X(B) = P_X(C) = 0.1$$

has an entropy equal to

$$-0.8 \cdot \log_2 0.8 - 0.1 \cdot \log_2 0.1 - 0.1 \cdot \log_2 0.1 = 0.92 \text{ bits.}$$

One optimal binary first-order or single-letter encoding (with $n = 1$) prefix codes for this source is given by $c(A) = 0$, $c(B) = 10$ and $c(C) = 11$, where $c(\cdot)$ is the encoding function. Then, the resultant average code rate for this code is

$$0.8 \times 1 + 0.2 \times 2 = 1.2 \text{ bits} \geq 0.92 \text{ bits.}$$

Now if we consider a second-order (with $n = 2$) prefix code by encoding two consecutive source symbols at a time, the new source alphabet becomes

$$\{AA, AB, AC, BA, BB, BC, CA, CB, CC\},$$

and the resultant probability distribution is calculated by

$$(\forall\, x_1, x_2 \in \{A, B, C\}) \quad P_{X^2}(x_1, x_2) = P_X(x_1)P_X(x_2)$$

as the source is memoryless. Then, an optimal binary prefix codes for the source is given by

$$c(AA) = 0$$
$$c(AB) = 100$$
$$c(AC) = 101$$
$$c(BA) = 110$$
$$c(BB) = 111100$$
$$c(BC) = 111101$$
$$c(CA) = 1110$$
$$c(CB) = 111110$$
$$c(CC) = 111111.$$

The average code rate of this code now becomes

$$\frac{0.64(1 \times 1) + 0.08(3 \times 3 + 4 \times 1) + 0.01(6 \times 4)}{2} = 0.96 \text{ bits},$$

which is closer to the source entropy of 0.92 bits. As n increases, the average code rate will be brought closer to the source entropy.

From Theorems 3.22 and 3.27, we obtain Shannon's lossless variable-length source coding theorem for discrete memoryless sources.

Theorem 3.28 (Shannon's lossless variable-length source coding theorem: DMS) *Fix integer $D > 1$ and consider a DMS $\{X_n\}_{n=1}^{\infty}$ with distribution P_X and entropy $H_D(X)$ (measured in D-ary units). Then the following hold.*

- *Forward part (achievability): For any $\varepsilon > 0$, there exists a D-ary nth-order prefix (hence uniquely decodable) code*

$$f: \mathcal{X}^n \to \{0, 1, \ldots, D - 1\}^*$$

for the source with an average code rate $\overline{R}_n$ satisfying

$$\overline{R}_n < H_D(X) + \varepsilon$$

for n sufficiently large.
- *Converse part: Every uniquely decodable code*

$$f: \mathcal{X}^n \to \{0, 1, \ldots, D - 1\}^*$$

for the source has an average code rate $\overline{R}_n \geq H_D(X)$.

Thus, for a discrete memoryless source, its entropy $H_D(X)$ (measured in D-ary units) represents the *smallest* variable-length lossless compression rate for n sufficiently large.

Proof The forward part follows directly from Theorem 3.27 by choosing n large enough such that $1/n < \varepsilon$, and the converse part is already given by Theorem 3.22. □

Observation 3.29 (Shannon's lossless variable-length coding theorem: stationary sources) Theorem 3.28 actually also holds for the class of *stationary sources* by replacing the source entropy $H_D(X)$ with the source entropy rate

$$H_D(\mathcal{X}) := \lim_{n \to \infty} \frac{1}{n} H_D(X^n),$$

measured in D-ary units. The proof is very similar to the proofs of Theorems 3.22 and 3.27 with slight modifications (such as using the fact that $\frac{1}{n} H_D(X^n)$ is nonincreasing with n for stationary sources).

Observation 3.30 (Rényi's entropy and lossless data compression) In the lossless variable-length source coding theorem, we have chosen the criterion of minimizing the average codeword length. Implicit in the use of average codeword length as a performance criterion is the assumption that the cost of compression varies *linearly* with codeword length. This is not always the case as in some applications, where the processing cost of decoding may be elevated and buffer overflows caused by long codewords can cause problems, an *exponential* cost/penalty function for codeword lengths can be more appropriate than a linear cost function [54, 67, 206]. Naturally, one would desire to choose a generalized function with exponential costs such that the familiar linear cost function (given by the average codeword length) is a special limiting case.

Indeed in [67], given a D-ary nth-order VLC $\mathcal{C}$

$$f: \mathcal{X}^n \to \{0, 1, \ldots, D-1\}^*$$

for a discrete source $\{X_i\}_{i=1}^{\infty}$ with alphabet $\mathcal{X}$ and distribution P_{X^n}, Campbell considers the following exponential cost function, called the *average codeword length of order* t:

$$L_n(t) := \frac{1}{t} \log_D \left(\sum_{x^n \in \mathcal{X}^n} P_{X^n}(x^n) D^{t \cdot \ell(c_{x^n})} \right),$$

where t is a chosen positive constant, $c_{x^n} = f(x^n)$ is the codeword associated with sourceword x^n, and $\ell(\cdot)$ is the length of c_{x^n}. Similarly, $\frac{1}{n} L_n(t)$ denotes the *average code rate of order* t. The criterion for optimality now becomes that an nth-order code is said to be *optimal if its cost $L_n(t)$ is the smallest among all possible uniquely decodable codes.*

In the limiting case when $t \to 0$,

$$L_n(t) \to \sum_{x \in \mathcal{X}} P_X(x)\ell(c_x) = \overline{\ell}$$

and we recover the average codeword length, as desired. Also, when $t \to \infty$, $L_n(t) \to \max_{x \in \mathcal{X}} \ell(c_x)$, which is the maximum codeword length for all codewords in $\mathcal{C}$. Note that for a fixed $t > 0$, minimizing $L_n(t)$ is equivalent to minimizing $\sum_{x^n \in \mathcal{X}^n} P_{X^n}(x^n) D^{t\ell(c_{x^n})}$. In this sum, the weight for codeword c_{x^n} is $D^{t\ell(c_{x^n})}$, and hence smaller codeword lengths are favored.

For this source coding setup with an exponential cost function, Campbell established in [67] an operational characterization for Rényi's entropy by proving the following lossless variable-length coding theorem for memoryless sources.

Theorem 3.31 (Lossless source coding theorem under exponential costs) *Consider a DMS $\{X_n\}$ with Rényi entropy in D-ary units and of order α given by*

$$H_\alpha(X) = \frac{1}{1 - \alpha} \log_D \left(\sum_{x \in \mathcal{X}} P_X^\alpha(x) \right).$$

Fixing $t > 0$ and setting $\alpha = \frac{1}{1+t}$, the following hold.

- *For any $\varepsilon > 0$, there exists a D-ary nth-order uniquely decodable code $f : \mathcal{X}^n \to \{0, 1, \ldots, D - 1\}^*$ for the source with an average code rate of order t satisfying*

$$\frac{1}{n} L_n(t) \le H_\alpha(X) + \varepsilon$$

for n sufficiently large.
- *Conversely, it is not possible to find a uniquely decodable code whose average code rate of order t is less than $H_\alpha(X)$.*

Noting (by Lemma 2.52) that the Rényi entropy of order α reduces to the Shannon entropy (in D-ary units) as $\alpha \to 1$, the above theorem reduces to Theorem 3.28 as $\alpha \to 1$ (or equivalently, as $t \to 0$). Finally, in [309, Sect. 4.4], [310, 311], the above source coding theorem is extended for time-invariant Markov sources in terms of the Rényi entropy rate, $\lim_{n \to \infty} \frac{1}{n} H_\alpha(X^n)$ with $\alpha = (1 + t)^{-1}$, which exists and can be calculated in closed form for such sources.

3.3.3 Examples of Binary Prefix Codes

(A) Huffman Codes: Optimal Variable-Length Codes

Given a discrete source with alphabet $\mathcal{X}$, we next construct an optimal binary first-order (single-letter) uniquely decodable variable-length code

$$f : \mathcal{X} \to \{0, 1\}^*,$$

where optimality is in the sense that the code's average codeword length (or equivalently, its average code rate) is minimized over the class of all binary uniquely decodable codes for the source. Note that finding optimal nth-order codes with $n > 1$ follows directly by considering $\mathcal{X}^n$ as a new source with expanded alphabet (i.e., by mapping n source symbols at a time).

By Corollary 3.26, we remark that in our search for optimal uniquely decodable codes, we can restrict our attention to the (smaller) class of optimal prefix codes. We thus proceed by observing the following necessary conditions of optimality for binary prefix codes.

Lemma 3.32 *Let $\mathcal{C}$ be an optimal binary prefix code with codeword lengths ℓ_i, $i = 1, \ldots, M$, for a source with alphabet $\mathcal{X} = \{a_1, \ldots, a_M\}$ and symbol probabilities $p_1, \ldots, p_M$. We assume, without loss of generality, that*

$$p_1 \geq p_2 \geq p_3 \geq \cdots \geq p_M,$$

and that any group of source symbols with identical probability is listed in order of increasing codeword length (i.e., if $p_i = p_{i+1} = \cdots = p_{i+s}$, then $\ell_i \leq \ell_{i+1} \leq \cdots \leq \ell_{i+s}$). Then the following properties hold.

1. *Higher probability source symbols have shorter codewords: $p_i > p_j$ implies $\ell_i \leq \ell_j$, for $i, j = 1, \ldots, M$.*
2. *The two least probable source symbols have codewords of equal length: $\ell_{M-1} = \ell_M$.*
3. *Among the codewords of length ℓ_M, two of the codewords are identical except in the last digit.*

Proof

(1) If $p_i > p_j$ and $\ell_i > \ell_j$, then it is possible to construct a better code $\mathcal{C}'$ by interchanging ("swapping") codewords i and j of $\mathcal{C}$, since

$$\begin{aligned}\bar{\ell}(\mathcal{C}') - \bar{\ell}(\mathcal{C}) &= p_i \ell_j + p_j \ell_i - (p_i \ell_i + p_j \ell_j) \\ &= (p_i - p_j)(\ell_j - \ell_i) \\ &< 0.\end{aligned}$$

Hence, code $\mathcal{C}'$ is better than code $\mathcal{C}$, contradicting the fact that $\mathcal{C}$ is optimal.

(2) We first know that $\ell_{M-1} \leq \ell_M$, since

- If $p_{M-1} > p_M$, then $\ell_{M-1} \leq \ell_M$ by result 1 above.
- If $p_{M-1} = p_M$, then $\ell_{M-1} \leq \ell_M$ by our assumption about the ordering of codewords for source symbols with identical probability.

Now, if $\ell_{M-1} < \ell_M$, we may delete the last digit of codeword M, and the deletion cannot result in another codeword since $\mathcal{C}$ is a prefix code. Thus, the deletion forms a new prefix code with a better average codeword length than C, contradicting the fact that C is optimal. Hence, we must have that $\ell_{M-1} = \ell_M$.

(3) Among the codewords of length ℓ_M, if no two codewords agree in all digits except the last, then we may delete the last digit in all such codewords to obtain a better codeword.   □

The above observation suggests that if we can construct an optimal code for the entire source except for its two least likely symbols, then we can construct an optimal overall code. Indeed, the following lemma due to Huffman [195] follows from Lemma 3.32.

Lemma 3.33 (Huffman) *Consider a source with alphabet $\mathcal{X} = \{a_1, \ldots, a_M\}$ and symbol probabilities $p_1, \ldots, p_M$ such that $p_1 \geq p_2 \geq \cdots \geq p_M$. Consider the reduced source alphabet $\mathcal{Y} = \{a_1, \ldots, a_{M-2}, a_{M-1,M}\}$ obtained from $\mathcal{X}$, where the first $M - 2$ symbols of $\mathcal{Y}$ are identical to those in $\mathcal{X}$ and symbol $a_{M-1,M}$ has probability $p_{M-1} + p_M$ and is obtained by combining the two least likely source symbols a_{M-1} and a_M of $\mathcal{X}$. Suppose that C', given by $f' : \mathcal{Y} \rightarrow \{0, 1\}^*$, is an optimal prefix code for the reduced source $\mathcal{Y}$. We now construct a prefix code C, $f : \mathcal{X} \rightarrow \{0, 1\}^*$, for the original source $\mathcal{X}$ as follows:*

- *The codewords for symbols $a_1, a_2, \ldots, a_{M-2}$ are exactly the same as the corresponding codewords in C':*

$$f(a_1) = f'(a_1), f(a_2) = f'(a_2), \ldots, f(a_{M-2}) = f'(a_{M-2}).$$

- *The codewords associated with symbols a_{M-1} and a_M are formed by appending a "0" and a "1", respectively, to the codeword $f'(a_{M-1,M})$ associated with the letter $a_{M-1,M}$ in C':*

$$f(a_{M-1}) = [f'(a_{M-1,M})\, 0] \quad and \quad f(a_M) = [f'(a_{M-1,M})\, 1].$$

Then, code C is optimal for the original source $\mathcal{X}$.

Hence, the problem of finding the optimal code for a source of alphabet size M is reduced to the problem of finding an optimal code for the reduced source of alphabet size $M - 1$. In turn, we can reduce the problem to that of size $M - 2$ and so on. Indeed, the above lemma yields a recursive algorithm for constructing optimal binary prefix codes.

Huffman encoding algorithm: Repeatedly, apply the above lemma until one is left with a reduced source with two symbols. An optimal binary prefix code for this source consists of the codewords 0 and 1. Then proceed backward, constructing (as outlined in the above lemma) optimal codes for each reduced source until one arrives at the original source.

Example 3.34 Consider a source with alphabet

$$\mathcal{X} = \{1, 2, 3, 4, 5, 6\}$$

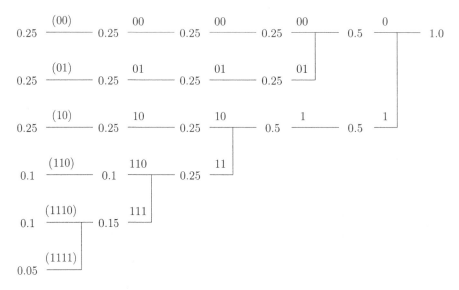

Fig. 3.6 Example of the Huffman encoding

and symbol probabilities 0.25, 0.25, 0.25, 0.1, 0.1, and 0.05, respectively. By following the Huffman encoding procedure as shown in Fig. 3.6, we obtain the Huffman
code as

$$00, 01, 10, 110, 1110, 1111.$$

Observation 3.35

- Huffman codes are not unique for a given source distribution; e.g., by inverting all
  the code bits of a Huffman code, one gets another Huffman code, or by resolving
  ties in different ways in the Huffman algorithm, one also obtains different Huffman
  codes (but all of these codes have the same minimal $\overline{R}_n$).
- One can obtain optimal codes that are not Huffman codes; e.g., by interchanging
  two codewords of the same length of a Huffman code, one can get another non-
  Huffman (but optimal) code. Furthermore, one can construct an optimal *suffix* code
  (i.e., a code in which no codeword can be a suffix of another codeword) from a
  Huffman code (which is a prefix code) by reversing the Huffman codewords.
- Binary Huffman codes always satisfy the Kraft inequality with equality (their code
  tree is "saturated"); e.g., see [87, p. 72].
- Any nth-order binary Huffman code $f : \mathcal{X}^n \to \{0, 1\}^*$ for a stationary source
  $\{X_n\}_{n=1}^\infty$ with finite alphabet $\mathcal{X}$ satisfies

$$H(\mathcal{X}) \le \frac{1}{n} H(X^n) \le \overline{R}_n < \frac{1}{n} H(X^n) + \frac{1}{n}.$$

Thus, as n increases to infinity, $\overline{R}_n \to H(\mathcal{X})$; but while the encoding–decoding delay increases only linearly with n, the storage complexity grows exponentially with n.

- Note that *nonbinary* (i.e., for $D > 2$) Huffman codes can also be constructed in a mostly similar way as for the case of binary Huffman codes by designing a D-ary tree and iteratively applying Lemma 3.33, where now the D least likely source symbols are combined at each stage. The only difference from the case of binary Huffman codes is that we have to ensure that we are ultimately left with D symbols at the last stage of the algorithm to guarantee the code's optimality. This is remedied by expanding the original source alphabet $\mathcal{X}$ by adding "dummy" symbols (each with zero probability) so that the alphabet size of the expanded source $|\mathcal{X}'|$ is the smallest positive integer greater than or equal to $|\mathcal{X}|$ with

$$|\mathcal{X}'| = 1 \quad (\text{modulo } D - 1).$$

For example, if $|\mathcal{X}| = 6$ and $D = 3$ (ternary codes), we obtain that $|\mathcal{X}'| = 7$, meaning that we need to enlarge the original source $\mathcal{X}$ by adding one dummy (zero probability) source symbol.

We thus obtain that the necessary conditions for optimality of Lemma 3.32 also hold for D-ary prefix codes when replacing $\mathcal{X}$ with the expanded source $\mathcal{X}'$ and replacing "two" with "D" in the statement of the lemma. The resulting D-ary Huffman code will be an optimal code for the original source $\mathcal{X}$ (e.g., see [135, Chap. 3] and [266, Chap. 11]).

- *Generalized Huffman codes under exponential costs*: When the lossless compression problem allows for exponential costs, as discussed in Observation 3.30 and formalized in Theorem 3.31, a straightforward generalization of Huffman's algorithm, which minimizes the average code rate of order t, $\frac{1}{n} L_n(t)$, can be obtained [192, Theorem 1']. More specifically, while in Huffman's algorithm, each new node (for a combined or equivalent symbol) is assigned weight $p_i + p_j$, where p_i and p_j are the lowest weights (probabilities) among the available nodes, in the generalized algorithm, each new node is assigned weight $2^t(p_i + p_j)$. With this simple modification, one can directly construct such generalized Huffman codes (e.g., see [310] for examples of codes designed for Markov sources).

(B) Shannon–Fano–Elias Code

Assume $\mathcal{X} = \{1, \ldots, M\}$ and $P_X(x) > 0$ for all $x \in \mathcal{X}$. Define

$$F(x) := \sum_{a \leq x} P_X(a),$$

and

$$\bar{F}(x) := \sum_{a < x} P_X(a) + \frac{1}{2} P_X(x).$$

Encoder: For any $x \in \mathcal{X}$, express $\bar{F}(x)$ in decimal binary form, say

$$\bar{F}(x) = .c_1 c_2 \ldots c_k \ldots,$$

and take the first k (fractional) bits as the codeword of source symbol x, i.e.,

$$(c_1, c_2, \ldots, c_k),$$

where $k := \lceil \log_2(1/P_X(x)) \rceil + 1$.
Decoder: Given codeword $(c_1, \ldots, c_k)$, compute the cumulative sum of $F(\cdot)$ starting from the smallest element in $\{1, 2, \ldots, M\}$ until the first x satisfying

$$F(x) \geq .c_1 \ldots c_k.$$

Then, x should be the original source symbol.
Proof of decodability: For any number $a \in [0, 1]$, let $[a]_k$ denote the operation that chops the binary representation of a after k bits (i.e., removing the $(k+1)$th bit, the $(k+2)$th bit, etc.). Then

$$\bar{F}(x) - \left[\bar{F}(x)\right]_k < \frac{1}{2^k}.$$

Since $k = \lceil \log_2(1/P_X(x)) \rceil + 1$,

$$\frac{1}{2^k} \leq \frac{1}{2} P_X(x)$$

$$= \left[\sum_{a<x} P_X(a) + \frac{P_X(x)}{2}\right] - \sum_{a \leq x-1} P_X(a)$$

$$= \bar{F}(x) - F(x-1).$$

Hence,

$$F(x-1) = \left[F(x-1) + \frac{1}{2^k}\right] - \frac{1}{2^k} \leq \bar{F}(x) - \frac{1}{2^k} < \left[\bar{F}(x)\right]_k.$$

In addition,

$$F(x) > \bar{F}(x) \geq \left[\bar{F}(x)\right]_k.$$

Consequently, x is the first element satisfying

$$F(x) \geq .c_1 c_2 \ldots c_k.$$

$\square$

Average codeword length:

$$\bar{\ell} = \sum_{x \in \mathcal{X}} P_X(x) \left(\left\lceil \log_2 \frac{1}{P_X(x)} \right\rceil + 1 \right)$$

$$< \sum_{x \in \mathcal{X}} P_X(x) \left(\log_2 \frac{1}{P_X(x)} + 2 \right)$$

$$= H(X) + 2 \text{ bits.}$$

Observation 3.36 The Shannon–Fano–Elias code is a prefix code.

3.3.4 Examples of Universal Lossless Variable-Length Codes

In Sect. 3.3.3, we assume that the source distribution is known. Thus, we can use either Huffman codes or Shannon–Fano–Elias codes to compress the source. What if the source distribution is not a known priori? Is it still possible to establish a completely lossless data compression code which is *universally* good (or asymptotically optimal) for all sources of interest? The answer is affirmative. Examples of such universal codes are adaptive Huffman codes [136], arithmetic codes [242, 243, 322] (which are based on the Shannon–Fano–Elias code), and Lempel–Ziv codes [404, 430, 431], which are efficiently employed in various forms in many multimedia compression packages and standards. We herein give a brief and basic description of adaptive Huffman and Lempel–Ziv codes.

(A) Adaptive Huffman Codes

A straightforward universal coding scheme is to use the empirical distribution (or relative frequencies) as the true distribution, and then apply the optimal Huffman code according to the empirical distribution. If the source is i.i.d., the relative frequencies will converge to its true marginal probability. Therefore, such universal codes should be good for all i.i.d. sources. However, in order to get an accurate estimation of the true distribution, one must observe a sufficiently long source sequence under which the coder will suffer a long delay. This can be improved using *adaptive universal Huffman codes* [136].

The working procedure of an adaptive Huffman code is as follows. Start with an initial guess of the source distribution (based on the assumption that the source is DMS). As a new source symbol arrives, encode the data in terms of the Huffman coding scheme according to the current estimated distribution, and then update the estimated distribution and the Huffman codebook according to the newly arrived source symbol.

To be specific, let the source alphabet be $\mathcal{X} := \{a_1, \ldots, a_M\}$. Define

$$N(a_i|x^n) := \text{number of } a_i \text{ occurrence in } x_1, x_2, \ldots, x_n.$$

Then, the (current) relative frequency of a_i is $N(a_i|x^n)/n$. Let $c_n(a_i)$ denote the Huffman codeword of source symbol a_i with respect to the distribution

$$\left[\frac{N(a_1|x^n)}{n}, \frac{N(a_2|x^n)}{n}, \dots, \frac{N(a_M|x^n)}{n} \right].$$

Now suppose that $x_{n+1} = a_j$. The codeword $c_n(a_j)$ is set as output, and the relative frequency for each source outcome becomes

$$\frac{N(a_j|x^{n+1})}{n+1} = \frac{n \cdot (N(a_j|x^n)/n) + 1}{n+1}$$

and

$$\frac{N(a_i|x^{n+1})}{n+1} = \frac{n \cdot (N(a_i|x^n)/n)}{n+1} \quad \text{for } i \neq j.$$

This observation results in the following distribution update policy:

$$P_{\hat{X}}^{(n+1)}(a_j) = \frac{n P_{\hat{X}}^{(n)}(a_j) + 1}{n+1}$$

and

$$P_{\hat{X}}^{(n+1)}(a_i) = \frac{n}{n+1} P_{\hat{X}}^{(n)}(a_i) \quad \text{for } i \neq j,$$

where $P_{\hat{X}}^{(n+1)}$ represents the estimate of the true distribution P_X at time $(n+1)$.

Note that in the adaptive Huffman coding scheme, the encoder and decoder need not be redesigned at every time, but only when a sufficient change in the estimated distribution occurs such that the so-called *sibling property* is violated.

Definition 3.37 *(Sibling property)* A binary prefix code is said to have the *sibling property* if its code tree satisfies

1. every node in the code tree (except for the root node) has a sibling (i.e., the code tree is saturated), and
2. the node can be listed in nondecreasing order of probabilities with each node being adjacent to its sibling.

The next observation indicates the fact that the Huffman code is the only prefix code satisfying the sibling property.

Observation 3.38 A binary prefix code is a Huffman code iff it satisfies the sibling property.

An example for a code tree satisfying the sibling property is shown in Fig. 3.7. The first requirement is satisfied since the tree is saturated. The second requirement can be checked by the node list in Fig. 3.7.

If the next observation (say at time $n = 17$) is a_3, then its codeword 100 is set as output (using the Huffman code corresponding to $P_{\hat{X}}^{(16)}$). The estimated distribution is updated as follows:

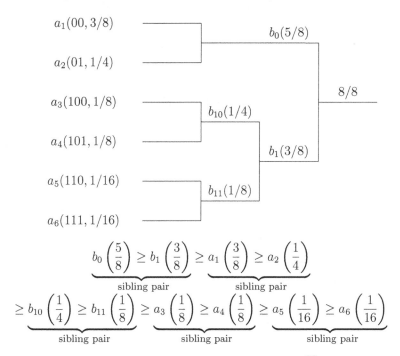

Fig. 3.7 Example of the sibling property based on the code tree from $P_{\hat{X}}^{(16)}$. The arguments inside the parenthesis following a_j respectively indicate the codeword and the probability associated with a_j. Here, "b" is used to denote the internal nodes of the tree with the assigned (partial) code as its subscript. The number in the parenthesis following b is the probability sum of all its children

$$P_{\hat{X}}^{(17)}(a_1) = \frac{16 \times (3/8)}{17} = \frac{6}{17}, \quad P_{\hat{X}}^{(17)}(a_2) = \frac{16 \times (1/4)}{17} = \frac{4}{17}$$

$$P_{\hat{X}}^{(17)}(a_3) = \frac{16 \times (1/8) + 1}{17} = \frac{3}{17}, \quad P_{\hat{X}}^{(17)}(a_4) = \frac{16 \times (1/8)}{17} = \frac{2}{17}$$

$$P_{\hat{X}}^{(17)}(a_5) = \frac{16 \times [1/(16)]}{17} = \frac{1}{17}, \quad P_{\hat{X}}^{(17)}(a_6) = \frac{16 \times [1/(16)]}{17} = \frac{1}{17}.$$

The sibling property is then violated (cf. Fig. 3.8). Hence, codebook needs to be updated according to the new estimated distribution, and the observation at $n = 18$ shall be encoded using the new codebook in Fig. 3.9. Details about adaptive Huffman codes can be found in [136].

(B) Lempel–Ziv Codes

We now introduce a well-known and high-performing universal coding scheme, which is named after its inventors, Lempel and Ziv [430, 431] (it is also named as the Lempel–Ziv–Welch compression algorithm after Welch developed an efficient

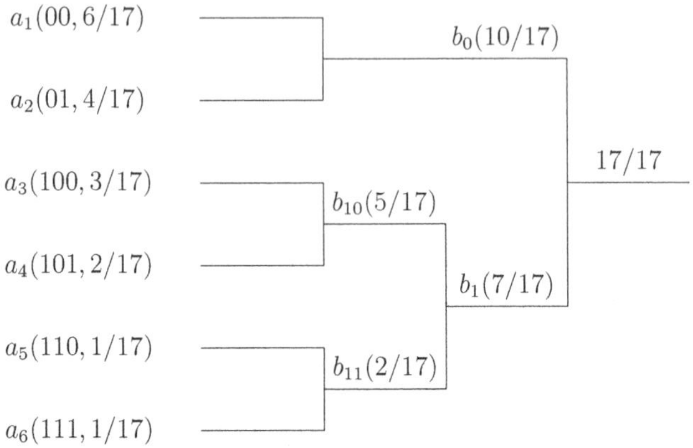

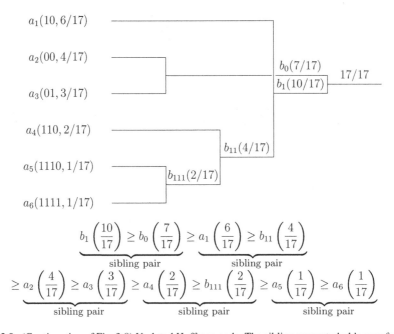

$$\underbrace{b_0\left(\frac{10}{17}\right) \geq b_1\left(\frac{7}{17}\right)}_{\text{sibling pair}} \geq a_1\left(\frac{6}{17}\right) \geq b_{10}\left(\frac{5}{17}\right)$$

$$\geq a_2\left(\frac{4}{17}\right) \geq \underbrace{a_3\left(\frac{3}{17}\right) \geq a_4\left(\frac{2}{17}\right)}_{\text{sibling pair}} \geq b_{11}\left(\frac{2}{17}\right) \geq \underbrace{a_5\left(\frac{1}{17}\right) \geq a_6\left(\frac{1}{17}\right)}_{\text{sibling pair}}$$

Fig. 3.8 (Continuation of Fig. 3.7) Example of violation of the sibling property after observing a new symbol a_3 at $n = 17$. Note that node a_1 is not adjacent to its sibling a_2

$$\underbrace{b_1\left(\frac{10}{17}\right) \geq b_0\left(\frac{7}{17}\right)}_{\text{sibling pair}} \geq a_1\left(\frac{6}{17}\right) \geq \underbrace{b_{11}\left(\frac{4}{17}\right)}_{\text{sibling pair}}$$

$$\geq \underbrace{a_2\left(\frac{4}{17}\right) \geq a_3\left(\frac{3}{17}\right)}_{\text{sibling pair}} \geq \underbrace{a_4\left(\frac{2}{17}\right) \geq b_{111}\left(\frac{2}{17}\right)}_{\text{sibling pair}} \geq \underbrace{a_5\left(\frac{1}{17}\right) \geq a_6\left(\frac{1}{17}\right)}_{\text{sibling pair}}$$

Fig. 3.9 (Continuation of Fig. 3.8) Updated Huffman code. The sibling property holds now for the new code

version of the original Lempel–Ziv technique [404]). These codes, unlike Huffman
and Shannon–Fano–Elias codes, map variable-length sourcewords (as opposed to
fixed-length codewords) onto codewords.

Suppose the source alphabet is binary. Then, the Lempel–Ziv encoder can be
described as follows.

Encoder:

1. Parse the input sequence into strings that have never appeared before. For example, if the input sequence is 1011010100010..., the algorithm first grabs the
   first letter 1 and finds that it has never appeared before. So 1 is the *first string*.
   Then, the algorithm scoops the second letter 0 and also determines that it has not
   appeared before, and hence, put it to be the *next string*. The algorithm moves on
   to the next letter 1 and finds that this string has appeared. Hence, it hits another
   letter 1 and yields a new string 11, and so on. Under this procedure, the source
   sequence is parsed into the strings

$$1, 0, 11, 01, 010, 00, 10.$$

2. Let L be the number of distinct strings of the parsed source. Then, we need
   $\lfloor \log_2 L \rfloor + 1$ bits to index these strings (starting from one). In the above example,
   the indices are

$$\begin{aligned} \text{parsed source:} \quad & 1 \quad 0 \quad 11 \quad 01 \quad 010 \quad 00 \quad 10 \\ \text{index:} \quad & 001 \; 010 \; 011 \; 100 \; 101 \; 110 \; 111 \end{aligned}$$

The codeword of each string is then the index of its prefix concatenated with the
last bit in its source string. For example, the codeword of source string 010 will
be the index of 01, i.e., 100, concatenated with the last bit of the source string,
i.e., 0. Through this procedure, encoding the above-parsed strings with $L = 3$
yields the codeword sequence

$$(000, 1)(000, 0)(001, 1)(010, 1)(100, 0)(010, 0)(001, 0)$$

or equivalently,

$$0001000000110101100001000010.$$

Note that the conventional Lempel–Ziv encoder requires two passes: the first pass
to decide L, and the second pass to generate the codewords. The algorithm, however,
can be modified so that it requires only one pass over the entire source string. Also,
note that the above algorithm uses an *equal* number of bits ($\lfloor \log_2 L \rfloor + 1$) to all the
location indices, which can also be relaxed by proper modification.

Decoder: The decoding is straightforward from the encoding procedure.

Theorem 3.39 *The above algorithm asymptotically achieves the entropy rate of any
stationary ergodic source (with unknown statistics).*

Proof Refer to [83, Sect. 13.5].   □

Problems

1. A binary discrete memoryless source $\{X_n\}_{n=1}^\infty$ has distribution $P_X(1) = 0.005$. A binary codeword is provided for every sequence of 100 source digits containing three or fewer ones. In other words, the set of sourcewords of length 100 that are encoded to distinct block codewords is

$$\mathcal{A} := \{x^{100} \in \{0, 1\}^{100} : \text{number of } 1's \text{ in } x^{100} \le 3\}.$$

   (a) Show that $\mathcal{A}$ is indeed a typical set $\mathcal{F}_{100}(0.2)$ defined using the base-2 logarithm.
   (b) Find the minimum codeword blocklength in bits for the block coding scheme.
   (c) Find the probability for sourcewords not in $\mathcal{A}$.
   (d) Use Chebyshev's inequality to bound the probability of observing a sourceword outside $\mathcal{A}$. Compare this bound with the actual probability computed in part (c).
   Hint: Let X_i represent the binary random digit at instance i, and let $S_n = X_1 + \cdots + X_n$. Note that $\Pr[S_{100} \ge 4]$ is equal to

$$\Pr\left[\left|\frac{1}{100} S_{100} - 0.005\right| \ge 0.035\right].$$

2. *Weak Converse to the Fixed-Length Source Coding Theorem*: Recall (see Observation 3.3) that an (n, M) fixed-length source code for a discrete memoryless source (DMS) $\{X_n\}_{n=1}^\infty$ with finite alphabet $\mathcal{X}$ consists of an encoder $f: \mathcal{X}^n \to \{1, 2, \ldots, M\}$, and a decoder $g: \{1, 2, \ldots, M\} \to \mathcal{X}^n$. The rate of the code is

$$R_n := \frac{1}{n} \log_2 M \text{ bits/source symbol,}$$

and its probability of decoding error is

$$P_e = \Pr[X^n \ne \hat{X}^n],$$

where $\hat{X}^n = g(f(X^n))$.

   (a) Show that any fixed-length source code (n, M) for a DMS satisfies

$$P_e \ge \frac{H(X) - R_n}{\log_2 |\mathcal{X}|} - \frac{1}{n \log_2 |\mathcal{X}|},$$

where $H(X)$ is the source entropy.

Hint: Show that $\log_2 M \geq I(X^n; \hat{X}^n)$, and use Fano's inequality.

(b) Deduce the (weak) converse to the fixed-length source coding theorem for DMSs by proving that for any (n, M) source code with $\limsup_{n\to\infty} R_n < H(X)$, its P_e is bounded away from zero for n sufficiently large.

3. For a stationary source $\{X_n\}_{n=1}^{\infty}$, show that for any integer $n > 1$,

   (a) $\frac{1}{n}H(X^n) \leq \frac{1}{n-1}H(X^{n-1})$
   (b) $\frac{1}{n}H(X^n) \geq H(X_n|X^{n-1})$.
   Hint: Use the chain rule for entropy and the fact that

   $$H(X_i|X_{i-1}, \ldots, X_1) = H(X_n|X_{n-1}, \ldots, X_{n-i+1})$$

   for every i.

4. *Randomized random walk*: An ant walks randomly on a line of integers. At time instance i, it may move forward with probability $1 - Z_{i-1}$, or it may move backward with probability Z_{i-1}, where $\{Z_i\}_{i=0}^{\infty}$ are identically distributed random variables with finite alphabet $\mathcal{Z} \subset [0, 1]$. Let X_i be the number on which the ant stands at time instance i, and let $X_0 = 0$ (with probability one).

   (a) Show that

   $$H(X_1, X_2, \ldots, X_n|Z_0, Z_1, \ldots, Z_{n-1}) = nE[h_b(Z)],$$

   where $h_b(\cdot)$ is the binary entropy function.
   (b) If $\Pr[Z_0 = 0] = \Pr[Z_0 = 1] = \frac{1}{2}$, determine $H(X_1, X_2, \ldots, X_n)$.
   (c) Find the entropy rate of the process $\{X_n\}_{n=1}^{\infty}$ in (b).

5. A source with binary alphabet $\mathcal{X} = \{0, 1\}$ emits a sequence of random variables $\{X_n\}_{n=1}^{\infty}$. Let $\{Z_n\}_{n=1}^{\infty}$ be a binary independent and identically distributed (i.i.d.) sequence of random variables such that $\Pr\{Z_n = 1\} = \Pr\{Z_n = 0\}$. We assume that $\{X_n\}_{n=1}^{\infty}$ is generated according to the equation

   $$X_n = X_{n-1} \oplus X_{n-2} \oplus Z_n, \quad n = 1, 2, \ldots$$

   where $\oplus$ denotes addition modulo-2, and $X_0 = X_{-1} = 0$. Find the entropy rate of $\{X_n\}_{n=1}^{\infty}$.

6. For each of the following codes, either prove unique decodability or give an ambiguous concatenated sequence of codewords:

   (a) $\{1, 0, 00\}$.
   (b) $\{1, 01, 00\}$.
   (c) $\{1, 10, 00\}$.
   (d) $\{1, 10, 01\}$.
   (e) $\{0, 01\}$.
   (f) $\{00, 01, 10, 11\}$.

7. We know the fact that the average code rate of all nth-order uniquely decodable codes for a DMS must be no less than the source entropy. But this is not necessarily true for non-singular codes. Give an example of a non-singular code in which the average code rate is less than the source entropy.

8. Under what condition does the average code rate of a uniquely decodable binary first-order variable-length code for a DMS equal the source entropy?
   Hint: See the discussion after Theorem 3.22.

9. *Binary Markov Source*: Consider the binary homogeneous Markov source: $\{X_n\}_{n=1}^{\infty}$, $X_n \in \mathcal{X} = \{0, 1\}$, with

$$\Pr\{X_{n+1} = j | X_n = i\} = \begin{cases} \frac{\rho}{1+\delta}, & \text{if } i = 0 \text{ and } j = 1, \\ \frac{\rho+\delta}{1+\delta}, & \text{if } i = 1 \text{ and } j = 1, \end{cases}$$

where $n \geq 1$, $0 \leq \rho \leq 1$ and $\delta \geq 0$.

   (a) Find the initial state distribution $(\Pr\{X_1 = 0\}, \Pr\{X_1 = 1\})$ required to make the source $\{X_n\}_{n=1}^{\infty}$ stationary.
       Assume in the next questions that the source is stationary.
   (b) Find the entropy rate of $\{X_n\}_{n=1}^{\infty}$ in terms of ρ and δ.
   (c) For $\delta = 1$ and $\rho = 1/2$, compute the source redundancies ρ_d, ρ_m, and ρ_t.
   (d) Suppose that $\rho = 1$. Is $\{X_n\}_{n=1}^{\infty}$ irreducible? What is the value of the entropy rate in this case?
   (e) For $\delta = 0$, show that $\{X_n\}_{n=1}^{\infty}$ is a discrete memoryless source and compute its entropy rate in terms of ρ.
   (f) If $\rho = 1/2$ and $\delta = 3/2$, design first-, second-, and third-order binary Huffman codes for this source. Determine in each case the average code rate and compare it to the entropy rate.

10. *Polya contagion process of memory two*: Consider the finite-memory Polya contagion source presented in Example 3.17 with $M = 2$.

    (a) Find the transition distribution of this binary Markov process and determine its stationary distribution in terms of the source parameters.
    (b) Find the source entropy rate.

11. Suppose random variables Z_1 and Z_2 are independent from each other and have the same distribution as Z with

$$\begin{cases} \Pr[Z = e_1] = 0.4, \\ \Pr[Z = e_2] = 0.3, \\ \Pr[Z = e_3] = 0.2, \\ \Pr[Z = e_4] = 0.1. \end{cases}$$

    (a) Design a first-order binary Huffman code $f: \{e_1, e_2, e_3, e_4\} \to \{0, 1\}^*$ for Z.

(b) Applying the Huffman code in (a) to Z_1 and Z_2 and concatenating $f(Z_1)$ with $f(Z_2)$ yields an overall codeword for the pair (Z_1, Z_2) given by

$$f(Z_1, Z_2) := (f(Z_1), f(Z_2)) = (U_1, U_2, \ldots, U_k),$$

where k ranges from 2 to 6, depending on the outcomes of Z_1 and Z_2. Are U_1 and U_2 independent? Justify your answer.
Hint: Examine $\Pr[U_2 = 0|U_1 = u_1]$ for different values of u_1.

(c) Is the average code rate equal to the entropy given by

$$0.4\log_2 \frac{1}{0.4} + 0.3\log_2 \frac{1}{0.3} + 0.2\log_2 \frac{1}{0.2}$$

$$+0.1\log_2 \frac{1}{0.1} = 1.84644 \text{ bits/letter?}$$

Justify your answer.

(d) Now if we apply the Huffman code in (a) sequentially to the i.i.d. sequence $Z_1, Z_2, Z_3, \ldots$ with the same marginal distribution as Z, and yield the output $U_1, U_2, U_3, \ldots$, can $U_1, U_2, U_3, \ldots$ be further compressed?
If your answer to this question is NO, prove the i.i.d. uniformity of $U_1, U_2, U_3, \ldots$. If your answer to this question is YES, then explain why the optimal Huffman code does not give an i.i.d. uniform output.
Hint: Examine whether the average code rate can achieve the source entropy.

12. In the second part of Theorem 3.27, it is shown that there exists a D-ary prefix code with

$$\bar{R}_n = \frac{1}{n} \sum_{x \in \mathcal{X}} P_X(x)\ell(c_x) \le H_D(X) + \frac{1}{n},$$

where c_x is the codeword for the source symbol x and $\ell(c_x)$ is the length of codeword c_x. Show that the upper bound can be improved to

$$\bar{R}_n < H_D(X) + \frac{1}{n}.$$

Hint: Replace $\ell(c_x) = \lfloor -\log_D P_X(x) \rfloor + 1$ by a new assignment.

13. Let $X_1, X_2, X_3, \ldots$ be an i.i.d. random variables with common infinite alphabet $\mathcal{X} = \{x_1, x_2, x_3, \ldots\}$, and assume that $P_X(x_i) > 0$ for every i.

(a) Prove that the average code rate of the first-order (single-letter) **binary** Huffman code is equal to $H(X)$ iff $P_X(x_i)$ is equal to 2^{-n_i} for every i, where $\{n_i\}_{i\ge 1}$ is a sequence of positive integers.
Hint: The if-part can be proved by the new bound in Problem 12, and the only-if-part can be proved by modifying the proof of Theorem 3.22.

(b) What is the sufficient and necessary condition under which the average code rate of the first-order (single-letter) **ternary** Huffman code equals $H_3(X)$?

(c) Prove that the average code rate of the second-order (two-letter) **binary** Huffman code cannot be equal to $H(X) + 1/2$ bits?

Hint: Use the new bound in Problem 12.

14. Decide whether each of the following statements is *true* or *false*. Prove the validity of those that are true and give counterexamples or arguments based on known facts to disprove those that are false.

(a) Every Huffman code for a discrete memoryless source (DMS) has a corresponding suffix code with the same average code rate.

(b) Consider a DMS $\{X_n\}_{n=1}^{\infty}$ with alphabet $\mathcal{X} = \{a_1, a_2, a_3, a_4, a_5, a_6\}$ and probability distribution

$$[p_1, p_2, p_3, p_4, p_5, p_6] = \left[\frac{1}{4}, \frac{1}{4}, \frac{1}{4}, \frac{1}{8}, \frac{1}{16}, \frac{1}{16}\right],$$

where $p_i := \Pr\{X = a_i\}$, $i = 1, \ldots, 6$. The Shannon–Fano–Elias code $f: \mathcal{X} \to \{0, 1\}^*$ for the source is optimal.

15. Consider a discrete memoryless source $\{X_i\}_{i=1}^{\infty}$ with alphabet $\mathcal{X} = \{a, b, c\}$ and distribution $P[X = a] = 1/2$ and $P[X = b] = P[X = c] = 1/4$.

(a) Design an optimal first-order binary prefix code for this source (i.e., for $n = 1$).

(b) Design an optimal second-order binary prefix code for this source (i.e., for $n = 2$).

(c) Compare the codes in terms of both performance and complexity. Which code would you recommend? Justify your answer.

16. Let $\{(X_i, Y_i)\}_{i=1}^{\infty}$ be a two-dimensional DMS with alphabet $\mathcal{X} \times \mathcal{Y} = \{0, 1\} \times \{0, 1\}$ and common distribution $P_{X,Y}$ given by

$$P_{X,Y}(0, 0) = P_{X,Y}(1, 1) = \frac{1 - \epsilon}{2}$$

and

$$P_{X,Y}(0, 1) = P_{X,Y}(1, 0) = \frac{\epsilon}{2},$$

where $0 < \epsilon < 1$.

(a) Find the limit of the random variable $[P_{X^n}(X^n)]^{\frac{1}{2n}}$ as $n \to \infty$.

(b) Find the limit of the random variable

$$\frac{1}{n} \log_2 \frac{P_{X^n, Y^n}(X^n, Y^n)}{P_{X^n}(X^n) P_{Y^n}(Y^n)}$$

as $n \to \infty$.

17. Consider a discrete memoryless source $\{X_i\}_{i=1}^{\infty}$ with alphabet $\mathcal{X}$ and distribution p_X. Let $\mathcal{C} = f(\mathcal{X})$ be a uniquely decodable binary code

$$f: \mathcal{X} \to \{0, 1\}^*$$

that maps single source letters onto binary strings such that its average code rate $\overline{R}_C$ satisfies

$$\overline{R}_C = H(X) \qquad \text{bits/source symbol.}$$

In other words, $\mathcal{C}$ is *absolutely optimal*.
Now consider a second binary code $\mathcal{C}' = f'(\mathcal{X}^n)$ for the source that maps source n-tuples onto binary strings:

$$f': \mathcal{X}^n \to \{0, 1\}^*.$$

Provide a construction for the map f' such that the code $\mathcal{C}'$ is also absolutely optimal.

18. Consider two random variables X and Y with values in finite sets $\mathcal{X}$ and $\mathcal{Y}$, respectively. Let $\bar{l}_X$, $\bar{l}_Y$, and $\bar{l}_{XY}$ denote the average codeword lengths of the optimal (first-order) prefix codes

$$f: \mathcal{X} \to \{0, 1\}^*,$$

$$g: \mathcal{Y} \to \{0, 1\}^*$$

and

$$h: \mathcal{X} \times \mathcal{Y} \to \{0, 1\}^*,$$

respectively; i.e., $\bar{l}_X = E[l(f(X))], \bar{l}_Y = E[l(g(Y))],$ and $\bar{l}_{XY} = E[l(h(X, Y))]$.

Prove that

(a) $\bar{l}_X + \bar{l}_Y - \bar{l}_{XY} < I(X; Y) + 2.$
(b) $\bar{l}_{XY} \le \bar{l}_X + \bar{l}_Y.$

19. *Entropy rate*: Consider a stationary source $\{X_n\}_{n=1}^{\infty}$ with finite alphabet $\mathcal{X}$ and entropy rate $H(\mathcal{X})$.

(a) Show that the normalized conditional entropy $\frac{1}{n} H(X_{n+1}^{2n} | X^n)$ is nonincreasing in n. where $X^n = (X_1, \ldots, X_n)$ and $X_{n+1}^{2n} = (X_{n+1}, \ldots, X_{2n})$.
(b) Show that

$$H(X_{2n} | X^{2n-1}) \le \frac{1}{n} H(X_{n+1}^{2n} | X^n) \le \frac{1}{n} H(X^n).$$

(c) Find the limits of $\frac{1}{n} H(X_{n+1}^{2n} | X^n)$ and $\frac{1}{n} I(X_{n+1}^{2n}; X^n)$ as $n \to \infty$.
(d) Given that the source $\{X_n\}$ is stationary Markov, compare $\frac{1}{n} H(X_{n+1}^{2n} | X^n)$ to $H(\mathcal{X})$.

20. *Divergence rate*: Prove the expression in (3.2.6) for the divergence rate between a stationary source $\{X_i\}$ and a time-invariant Markov source $\{\hat{X}_i\}$, with both sources having a common finite alphabet $\mathcal{X}$. Generalize the result if the source $\{\hat{X}_i\}$ is a time-invariant kth-order Markov chain.
21. Prove Observation 3.29.
22. Prove Lemma 3.33.

Chapter 4
Data Transmission and Channel Capacity

4.1 Principles of Data Transmission

A noisy communication channel is an input–output medium in which the output is not completely or deterministically specified by the input. The channel is indeed stochastically modeled, where given channel input x, the channel output y is governed by a transition (conditional) probability distribution denoted by $P_{Y|X}(y|x)$. Since two different inputs may give rise to the same output, the receiver, upon receipt of an output, needs to *guess* the most probable sent input. In general, words of length n are sent and received over the channel; in this case, the channel is characterized by a sequence of n-dimensional transition distributions $P_{Y^n|X^n}(y^n|x^n)$, for $n = 1, 2, \ldots$. A block diagram depicting a data transmission or channel coding system with no (output) feedback is given in Fig. 4.1.

The designer of a data transmission (or channel) code needs to carefully select *codewords* from the set of channel input words (of a given length) so that a minimal ambiguity is obtained at the channel receiver. For example, suppose that a channel has binary input and output alphabets and that its transition probability distribution induces the following conditional probability on its output symbols given that input words of length 2 are sent:

$$P_{Y|X^2}(y = 0|x^2 = 00) = 1$$
$$P_{Y|X^2}(y = 0|x^2 = 01) = 1$$
$$P_{Y|X^2}(y = 1|x^2 = 10) = 1$$
$$P_{Y|X^2}(y = 1|x^2 = 11) = 1,$$

which can be graphically depicted as

© Springer Nature Singapore Pte Ltd. 2018
F. Alajaji and P.-N. Chen, *An Introduction to Single-User Information Theory*,
Springer Undergraduate Texts in Mathematics and Technology,
https://doi.org/10.1007/978-981-10-8001-2_4

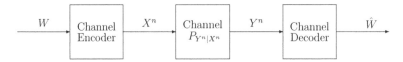

Fig. 4.1 A data transmission system, where W represents the message for transmission, X^n denotes the codeword corresponding to message W, Y^n represents the received word due to channel input X^n, and $\hat{W}$ denotes the reconstructed message from Y^n

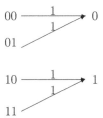

and a binary message (either event A or event B) is required to be transmitted from the sender to the receiver. Then the data transmission code with (codeword 00 for event A, codeword 10 for event B) obviously induces less ambiguity at the receiver than the code with (codeword 00 for event A, codeword 01 for event B).

In short, the objective in designing a data transmission (or channel) code is to transform a noisy channel into a reliable medium for sending messages and recovering them at the receiver with minimal loss. To achieve this goal, the designer of a data transmission code needs to take advantage of the common parts between the sender and the receiver sites that are least affected by the channel noise. We will see that these common parts are probabilistically captured by the mutual information between the channel input and the channel output.

As illustrated in the previous example, if a "least-noise-affected" subset of the channel input words is appropriately selected as the set of codewords, the messages intended to be transmitted can be reliably sent to the receiver with arbitrarily small error. One then raises the question:

What is the maximum amount of information (per channel use) that can be reliably transmitted over a given noisy channel?

In the above example, we can transmit a binary message error-free, and hence, the amount of information that can be reliably transmitted is at least 1 bit per channel use (or channel symbol). It can be expected that the amount of information that can be reliably transmitted for a highly noisy channel should be less than that for a less noisy channel. But such a comparison requires a good measure of the "*noisiness*" of channels.

From an information-theoretic viewpoint, "*channel capacity*" provides a good measure of the *noisiness* of a channel; it represents the maximal amount of informational messages (per channel use) that can be transmitted via a data transmission code over the channel and recovered with arbitrarily small probability of error at the receiver. In addition to its dependence on the channel transition distribution, channel capacity also depends on the coding constraint imposed on the channel input, such as "only block (fixed-length) codes are allowed." In this chapter, we will study channel capacity for block codes (namely, only block transmission code can be used).[1] Throughout the chapter, the noisy channel is assumed to be *memoryless* (as defined in the next section).

4.2 Discrete Memoryless Channels

Definition 4.1 (*Discrete channel*) A discrete communication channel is characterized by

- A finite input alphabet $\mathcal{X}$.
- A finite output alphabet $\mathcal{Y}$.
- A sequence of n-dimensional transition distributions

$$\{P_{Y^n|X^n}(y^n|x^n)\}_{n=1}^{\infty}$$

such that $\sum_{y^n \in \mathcal{Y}^n} P_{Y^n|X^n}(y^n|x^n) = 1$ for every $x^n \in \mathcal{X}^n$, where $x^n = (x_1, \ldots, x_n) \in \mathcal{X}^n$ and $y^n = (y_1, \ldots, y_n) \in \mathcal{Y}^n$. We assume that the above sequence of n-dimensional distribution is consistent, i.e.,

$$
\begin{aligned}
P_{Y^i|X^i}(y^i|x^i) &= \frac{\sum_{x_{i+1} \in \mathcal{X}} \sum_{y_{i+1} \in \mathcal{Y}} P_{X^{i+1}}(x^{i+1}) P_{Y^{i+1}|X^{i+1}}(y^{i+1}|x^{i+1})}{\sum_{x_{i+1} \in \mathcal{X}} P_{X^{i+1}}(x^{i+1})} \\
&= \sum_{x_{i+1} \in \mathcal{X}} \sum_{y_{i+1} \in \mathcal{Y}} P_{X_{i+1}|X^i}(x_{i+1}|x^i) P_{Y^{i+1}|X^{i+1}}(y^{i+1}|x^{i+1})
\end{aligned}
$$

for every x^i, y^i, $P_{X_{i+1}|X^i}$ and $i = 1, 2, \ldots$.

In general, real-world communications channels exhibit statistical memory in the sense that current channel outputs statistically depend on past outputs as well as past, current, and (possibly) future inputs. However, for the sake of simplicity, we restrict our attention in this chapter to the class of memoryless channels (see Problem 4.27 for a brief discussion of channels with memory).

Definition 4.2 (*Discrete memoryless channel*) A discrete memoryless channel (DMC) is a channel whose sequence of transition distributions $P_{Y^n|X^n}$ satisfies

$$P_{Y^n|X^n}(y^n|x^n) = \prod_{i=1}^{n} P_{Y|X}(y_i|x_i) \tag{4.2.1}$$

---

[1] See [397] for recent results regarding channel capacity when no coding constraints are applied to the channel input (so that variable-length codes can be employed).

for every $n = 1, 2, \ldots, x^n \in \mathcal{X}^n$ and $y^n \in \mathcal{Y}^n$. In other words, a DMC is fully described by the channel's transition distribution matrix $\mathbb{Q} := [p_{x,y}]$ of size $|\mathcal{X}| \times |\mathcal{Y}|$, where

$$p_{x,y} := P_{Y|X}(y|x)$$

for $x \in \mathcal{X}, y \in \mathcal{Y}$. Furthermore, the matrix $\mathbb{Q}$ is *stochastic*; i.e., the sum of the entries in each of its rows is equal to 1 $\left(\text{since } \sum_{y \in \mathcal{Y}} p_{x,y} = 1 \text{ for all } x \in \mathcal{X}\right)$.

Observation 4.3 We note that the DMC's condition (4.2.1) is actually *equivalent* to the following two sets of conditions [29]:

$$\begin{cases} P_{Y_n|X^n, Y^{n-1}}(y_n|x^n, y^{n-1}) = P_{Y|X}(y_n|x_n) \ \forall n = 1, 2, \ldots, x^n, y^n; \\ \qquad\qquad\qquad\qquad\qquad\qquad\qquad\qquad\qquad\qquad\qquad\qquad (4.2.2a) \\ P_{Y^{n-1}|X^n}(y^{n-1}|x^n) = P_{Y^{n-1}|X^{n-1}}(y^{n-1}|x^{n-1}) \ \forall n = 2, 3, \ldots, x^n, y^{n-1}. \\ \qquad\qquad\qquad\qquad\qquad\qquad\qquad\qquad\qquad\qquad\qquad\qquad (4.2.2b) \end{cases}$$

$$\begin{cases} P_{Y_n|X^n, Y^{n-1}}(y_n|x^n, y^{n-1}) = P_{Y|X}(y_n|x_n) \ \forall n = 1, 2, \ldots, x^n, y^n; \\ \qquad\qquad\qquad\qquad\qquad\qquad\qquad\qquad\qquad\qquad\qquad\qquad (4.2.3a) \\ P_{X_n|X^{n-1}, Y^{n-1}}(x_n|x^{n-1}, y^{n-1}) = P_{X_n|X^{n-1}}(x_n|x^{n-1}) \ \forall n = 1, 2, \ldots, x^n, y^{n-1}. \\ \qquad\qquad\qquad\qquad\qquad\qquad\qquad\qquad\qquad\qquad\qquad\qquad (4.2.3b) \end{cases}$$

Condition (4.2.2a) [also (4.2.3a)] implies that the current output Y_n only depends on the current input X_n but not on past inputs X^{n-1} and outputs Y^{n-1}. Condition (4.2.2b) indicates that the past outputs Y^{n-1} do not depend on the current input X_n. These two conditions together give

$$\begin{aligned} P_{Y^n|X^n}(y^n|x^n) &= P_{Y^{n-1}|X^n}(y^{n-1}|x^n) P_{Y_n|X^n, Y^{n-1}}(y_n|x^n, y^{n-1}) \\ &= P_{Y^{n-1}|X^{n-1}}(y^{n-1}|x^{n-1}) P_{Y|X}(y_n|x_n); \end{aligned}$$

hence, (4.2.1) holds recursively for $n = 1, 2, \ldots$ The converse [i.e., (4.2.1) implies both (4.2.2a) and (4.2.2b)] is a direct consequence of

$$P_{Y_n|X^n, Y^{n-1}}(y_n|x^n, y^{n-1}) = \frac{P_{Y^n|X^n}(y^n|x^n)}{\sum_{y_n \in \mathcal{Y}} P_{Y^n|X^n}(y^n|x^n)}$$

and

$$P_{Y^{n-1}|X^n}(y^{n-1}|x^n) = \sum_{y_n \in \mathcal{Y}} P_{Y^n|X^n}(y^n|x^n).$$

Similarly, (4.2.3b) states that the current input X_n is independent of past outputs Y^{n-1}, which together with (4.2.3a) implies again

$$P_{Y^n|X^n}(y^n|x^n)$$

$$= \frac{P_{X^n,Y^n}(x^n, y^n)}{P_{X^n}(x^n)}$$

$$= \frac{P_{X^{n-1},Y^{n-1}}(x^{n-1}, y^{n-1})P_{X_n|X^{n-1},Y^{n-1}}(x_n|x^{n-1}, y^{n-1})P_{Y_n|X^n,Y^{n-1}}(y_n|x^n, y^{n-1})}{P_{X^{n-1}}(x^{n-1})P_{X_n|X^{n-1}}(x_n|x^{n-1})}$$

$$= P_{Y^{n-1}|X^{n-1}}(y^{n-1}|x^{n-1})P_{Y|X}(y_n|x_n),$$

hence, recursively yielding (4.2.1). The converse for (4.2.3b)—i.e., (4.2.1) implying (4.2.3b)—can be analogously proved by noting that

$$P_{X_n|X^{n-1},Y^{n-1}}(x_n|x^{n-1}, y^{n-1}) = \frac{P_{X^n}(x^n)\sum_{y_n \in \mathcal{Y}} P_{Y^n|X^n}(y^n|x^n)}{P_{X^{n-1}}(x^{n-1})P_{Y^{n-1}|X^{n-1}}(y^{n-1}|x^{n-1})}.$$

Note that the above definition of DMC in (4.2.1) prohibits the use of channel feedback, as feedback allows the current channel input to be a function of past channel outputs (therefore, conditions (4.2.2b) and (4.2.3b) cannot hold with feedback). Instead, a causality condition generalizing (4.2.2a) (e.g., see Problem 4.28 or [415, Definition 7.4]) will be needed for a channel with feedback.

Examples of DMCs:

1. *Identity (noiseless) channels*: An identity channel has equal size input and output alphabets ($|\mathcal{X}| = |\mathcal{Y}|$) and channel transition probability satisfying

$$P_{Y|X}(y|x) = \begin{cases} 1 & \text{if } y = x \\ 0 & \text{if } y \neq x. \end{cases}$$

This is a noiseless or perfect channel as the channel input is received error-free at the channel output.

2. *Binary symmetric channels*: A binary symmetric channel (BSC) is a channel with binary input and output alphabets such that each input has a (conditional) probability given by ε for being received inverted at the output, where $\varepsilon \in [0, 1]$ is called the channel's *crossover probability* or *bit error rate*. The channel's transition distribution matrix is given by

$$\mathbb{Q} = [p_{x,y}] = \begin{bmatrix} p_{0,0} & p_{0,1} \\ p_{1,0} & p_{1,1} \end{bmatrix}$$

$$= \begin{bmatrix} P_{Y|X}(0|0) & P_{Y|X}(1|0) \\ P_{Y|X}(0|1) & P_{Y|X}(1|1) \end{bmatrix} = \begin{bmatrix} 1 - \varepsilon & \varepsilon \\ \varepsilon & 1 - \varepsilon \end{bmatrix} \qquad (4.2.4)$$

and can be graphically represented via a transition diagram as shown in Fig. 4.2.

If we set $\varepsilon = 0$, then the BSC reduces to the binary identity (noiseless) channel. The channel is called "symmetric" since $P_{Y|X}(1|0) = P_{Y|X}(0|1)$; i.e.,

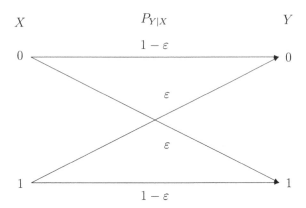

Fig. 4.2 Binary symmetric channel (BSC)

it has the same probability for flipping an input bit into a 0 or a 1. A detailed discussion of DMCs with various symmetry properties will be discussed later in this chapter.

Despite its simplicity, the BSC is rich enough to capture most of the complexity of coding problems over more general channels. For example, it can exactly model the behavior of practical channels with additive memoryless Gaussian noise used in conjunction of binary symmetric modulation and hard-decision demodulation (e.g., see [407, p. 240]). It is also worth pointing out that the BSC can be explicitly represented via a binary modulo-2 additive noise channel whose output at time i is the modulo-2 sum of its input and noise variables:

$$Y_i = X_i \oplus Z_i \quad \text{for} \quad i = 1, 2, \ldots, \tag{4.2.5}$$

where $\oplus$ denotes addition modulo-2, Y_i, X_i, and Z_i are the channel output, input, and noise, respectively, at time i, the alphabets $\mathcal{X} = \mathcal{Y} = \mathcal{Z} = \{0, 1\}$ are all binary. It is assumed in (4.2.5) that X_i and Z_j are independent of each other for any $i, j = 1, 2, \ldots$, and that the noise process is a Bernoulli(ε) process—i.e., a binary i.i.d. process with $\Pr[Z = 1] = \varepsilon$.

3. *Binary erasure channels*: In the BSC, some input bits are received perfectly and others are received corrupted (flipped) at the channel output. In some channels, however, some input bits are lost during transmission instead of being received corrupted (for example, packets in data networks may get dropped or blocked due to congestion or bandwidth constraints). In this case, the receiver knows the exact location of these bits in the received bitstream or codeword, but not their actual value. Such bits are then declared as "erased" during transmission and are called "erasures." This gives rise to the so-called binary erasure channel (BEC) as illustrated in Fig. 4.3, with input alphabet $\mathcal{X} = \{0, 1\}$ and output alphabet $\mathcal{Y} = \{0, E, 1\}$, where E represents an erasure (we may assume that E is a real number strictly greater than one), and channel transition matrix given by

Fig. 4.3 Binary erasure channel (BEC)

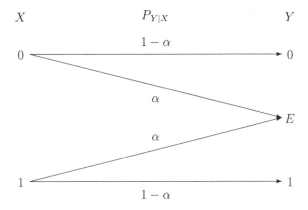

$$Q = [p_{x,y}] = \begin{bmatrix} p_{0,0} & p_{0,E} & p_{0,1} \\ p_{1,0} & p_{1,E} & p_{1,1} \end{bmatrix}$$

$$= \begin{bmatrix} P_{Y|X}(0|0) & P_{Y|X}(E|0) & P_{Y|X}(1|0) \\ P_{Y|X}(0|1) & P_{Y|X}(E|1) & P_{Y|X}(1|1) \end{bmatrix}$$

$$= \begin{bmatrix} 1-\alpha & \alpha & 0 \\ 0 & \alpha & 1-\alpha \end{bmatrix}, \tag{4.2.6}$$

where $0 \le \alpha \le 1$ is called the channel's *erasure probability*. We also observe that, like the BSC, the BEC can be explicitly expressed as follows:

$$Y_i = X_i \cdot \mathbf{1}\{Z_i \ne E\} + E \cdot \mathbf{1}\{Z_i = E\} \quad \text{for} \quad i = 1, 2, \ldots, \tag{4.2.7}$$

where

$$\mathbf{1}\{Z_i \ne E\} := \begin{cases} 1 & \text{if } Z_i \ne E \\ 0 & \text{if } Z_i = E \end{cases}$$

is the indicator function of the set $\{Z_i \ne E\}$, Y_i, X_i, and Z_i are the channel output, input, and *erasure*, respectively, at time i and the alphabets are $\mathcal{X} = \{0, 1\}$, $\mathcal{Z} = \{0, E\}$ and $\mathcal{Y} = \{0, 1, E\}$. Indeed, when the erasure variable $Z_i = E$, $Y_i = E$ and an erasure occurs in the channel; also, when $Z_i = 0$, $Y_i = X_i$ and the input is received perfectly. In the BEC functional representation in (4.2.7), it is assumed that X_i and Z_j are independent of each other for any i, j and that the erasure process $\{Z_i\}$ is i.i.d. with $\Pr[Z = E] = \alpha$.

4. *Binary channels with errors and erasures*: One can combine the BSC with the BEC to obtain a binary channel with both errors and erasures, as shown in Fig. 4.4. We will call such channel the binary symmetric erasure channel (BSEC). In this case, the channel's transition matrix is given by

Fig. 4.4 Binary symmetric erasure channel (BSEC)

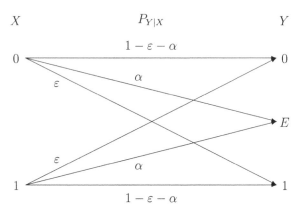

$$Q = [p_{x,y}] = \begin{bmatrix} p_{0,0} & p_{0,E} & p_{0,1} \\ p_{1,0} & p_{1,E} & p_{1,1} \end{bmatrix} = \begin{bmatrix} 1 - \varepsilon - \alpha & \alpha & \varepsilon \\ \varepsilon & \alpha & 1 - \varepsilon - \alpha \end{bmatrix}, \qquad (4.2.8)$$

where $\varepsilon, \alpha \in [0, 1]$ are the channel's crossover and erasure probabilities, respectively, with $\varepsilon + \alpha \le 1$. Clearly, setting $\alpha = 0$ reduces the BSEC to the BSC, and setting $\varepsilon = 0$ reduces the BSEC to the BEC. Analogously to the BSC and the BEC, the BSEC admits an explicit expression in terms of a *noise-erasure process*:

$$Y_i = \begin{cases} X_i \oplus Z_i & \text{if } Z_i \ne E \\ E & \text{if } Z_i = E \end{cases}$$

$$= (X_i \oplus Z_i) \cdot \mathbf{1}\{Z_i \ne E\} + E \cdot \mathbf{1}\{Z_i = E\} \qquad (4.2.9)$$

for $i = 1, 2, \ldots$, where $\oplus$ is modulo-2 addition,[2] $\mathbf{1}\{\cdot\}$ is the indicator function, Y_i, X_i, and Z_i are the channel output, input, and *noise-erasure* variable, respectively, at time i and the alphabets are $\mathcal{X} = \{0, 1\}$ and $\mathcal{Y} = \mathcal{Z} = \{0, 1, E\}$. Indeed, when the noise-erasure variable $Z_i = E$, $Y_i = E$ and an erasure occurs in the channel; when $Z_i = 0$, $Y_i = X_i$ and the input is received perfectly; finally when $Z_i = 1$, $Y_i = X_i \oplus 1$ and the input bit is received in error. In the BSEC functional characterization (4.2.9), it is assumed that X_i and Z_j are independent of each other for any i, j and that the noise-erasure process $\{Z_i\}$ is i.i.d. with $\Pr[Z = E] = \alpha$ and $\Pr[Z = 1] = \varepsilon$.

More generally, the channel needs not have a symmetric property in the sense of having identical transition distributions when inputs bits 0 or 1 are sent. For example, the channel's transition matrix can be given by

---

[2] Strictly speaking, note that $X \oplus Z$ in not defined in (4.2.9) when $Z = E$. However, as $\mathbf{1}\{Z \ne E\} = 0$ when $Z = E$, this is remedied by using the convention that an undefined quantity multiplied by zero is equal to zero.

$$\mathbb{Q} = [p_{x,y}] = \begin{bmatrix} p_{0,0} & p_{0,E} & p_{0,1} \\ p_{1,0} & p_{1,E} & p_{1,1} \end{bmatrix} = \begin{bmatrix} 1 - \varepsilon - \alpha & \alpha & \varepsilon \\ \varepsilon' & \alpha' & 1 - \varepsilon' - \alpha' \end{bmatrix}, \quad (4.2.10)$$

where the probabilities $\varepsilon \neq \varepsilon'$ and $\alpha \neq \alpha'$ in general. We call such channel, an *asymmetric* channel with errors and erasures (this model might be useful to represent practical channels using asymmetric or nonuniform modulation constellations).

5. *q-ary symmetric channels:* Given an integer $q \geq 2$, the q-ary symmetric channel is a nonbinary extension of the BSC; it has alphabets $\mathcal{X} = \mathcal{Y} = \{0, 1, \ldots, q-1\}$ of size q and channel transition matrix given by

$$\mathbb{Q} = [p_{x,y}]$$

$$= \begin{bmatrix} p_{0,0} & p_{0,1} & \cdots & p_{0,q-1} \\ p_{1,0} & p_{1,1} & \cdots & p_{1,q-1} \\ \vdots & \vdots & \vdots & \vdots \\ p_{q-1,0} & p_{q-1,1} & \cdots & p_{q-1,q-1} \end{bmatrix}$$

$$= \begin{bmatrix} 1 - \varepsilon & \frac{\varepsilon}{q-1} & \cdots & \frac{\varepsilon}{q-1} \\ \frac{\varepsilon}{q-1} & 1 - \varepsilon & \cdots & \frac{\varepsilon}{q-1} \\ \vdots & \vdots & \vdots & \vdots \\ \frac{\varepsilon}{q-1} & \frac{\varepsilon}{q-1} & \cdots & 1 - \varepsilon \end{bmatrix}, \quad (4.2.11)$$

where $0 \leq \varepsilon \leq 1$ is the channel's *symbol error rate (or probability)*. When $q = 2$, the channel reduces to the BSC with bit error rate ε, as expected.

As the BSC, the q-ary symmetric channel can be expressed as a modulo-q additive noise channel with common input, output and noise alphabets $\mathcal{X} = \mathcal{Y} = \mathcal{Z} = \{0, 1, \ldots, q-1\}$ and whose output Y_i at time i is given by $Y_i = X_i \oplus_q Z_i$, for $i = 1, 2, \ldots$, where $\oplus_q$ denotes addition modulo-q, and X_i and Z_i are the channel's input and noise variables, respectively, at time i. Here, the noise process $\{Z_n\}_{n=1}^{\infty}$ is assumed to be an i.i.d. process with distribution

$$\Pr[Z = 0] = 1 - \varepsilon \quad \text{and} \quad \Pr[Z = a] = \frac{\varepsilon}{q-1} \quad \forall a \in \{1, \ldots, q-1\}.$$

It is also assumed that the input and noise processes are independent of each other.

6. *q-ary erasure channels*: Given an integer $q \geq 2$, one can also consider a nonbinary extension of the BEC, yielding the so-called q-ary erasure channel. Specifically, this channel has input and output alphabets given by $\mathcal{X} = \{0, 1, \ldots, q-1\}$ and $\mathcal{Y} = \{0, 1, \ldots, q-1, E\}$, respectively, where E denotes an erasure, and channel transition distribution given by

$$P_{Y|X}(y|x) = \begin{cases} 1 - \alpha & \text{if } y = x, x \in \mathcal{X} \\ \alpha & \text{if } y = E, x \in \mathcal{X} \\ 0 & \text{if } y \neq x, x \in \mathcal{X}, \end{cases} \qquad (4.2.12)$$

where $0 \leq \alpha \leq 1$ is the erasure probability. As expected, setting $q = 2$ reduces the channel to the BEC.

Note that the same functional representation (4.2.7) also holds for this channel, where $\{Z_i\}_{i=1}^{\infty}$ is an i.i.d. (input independent) erasure process with alphabet $\{0, E\}$. Finally, a nonbinary extension to the BSEC can be similarly obtained.

4.3 Block Codes for Data Transmission Over DMCs

Definition 4.4 (*Fixed-length data transmission code*) Given positive integers n and M (where $M = M_n$), and a discrete channel with input alphabet $\mathcal{X}$ and output alphabet $\mathcal{Y}$, a fixed-length data transmission code (or block code) for this channel with blocklength n and rate $\frac{1}{n} \log_2 M$ message bits per channel symbol (or channel use) is denoted by $\mathcal{C}_n = (n, M)$ and consists of:

1. M information messages intended for transmission.
2. An encoding function

$$f : \{1, 2, \ldots, M\} \to \mathcal{X}^n$$

   yielding codewords $f(1), f(2), \ldots, f(M) \in \mathcal{X}^n$, each of length n. The set of these M codewords is called the codebook and we also usually write $\mathcal{C}_n = \{f(1), f(2), \ldots, f(M)\}$ to list the codewords.
3. A decoding function $g : \mathcal{Y}^n \to \{1, 2, \ldots, M\}$.

The set $\{1, 2, \ldots, M\}$ is called the *message set* and we assume that a message W follows a uniform distribution over the set of messages: $\Pr[W = w] = \frac{1}{M}$ for all $w \in \{1, 2, \ldots, M\}$. A block diagram for the channel code is given at the beginning of this chapter; see Fig. 4.1. As depicted in the diagram, to convey message W over the channel, the encoder sends its corresponding codeword $X^n = f(W)$ at the channel input. Finally, Y^n is received at the channel output (according to the memoryless channel distribution $P_{Y^n|X^n}$) and the decoder yields $\hat{W} = g(Y^n)$ as the message estimate.

Definition 4.5 (*Average probability of error*) The average probability of error for a channel block code $\mathcal{C}_n = (n, M)$ code with encoder $f(\cdot)$ and decoder $g(\cdot)$ used over a channel with transition distribution $P_{Y^n|X^n}$ is defined as

$$P_e(\mathcal{C}_n) := \frac{1}{M} \sum_{w=1}^{M} \lambda_w(\mathcal{C}_n),$$

where

$$\lambda_w(\mathcal{C}_n) := \Pr[\hat{W} \neq W | W = w] = \Pr[g(Y^n) \neq w | X^n = f(w)]$$
$$= \sum_{y^n \in \mathcal{Y}^n \,:\, g(y^n) \neq w} P_{Y^n | X^n}(y^n | f(w))$$

is the code's conditional probability of decoding error given that message w is sent over the channel.

Note that, since we have assumed that the message W is drawn uniformly from the set of messages, we have that

$$P_e(\mathcal{C}_n) = \Pr[\hat{W} \neq W].$$

Observation 4.6 (Maximal probability of error) Another more conservative error criterion is the so-called *maximal probability of error*

$$\lambda(\mathcal{C}_n) := \max_{w \in \{1, 2, \dots, M\}} \lambda_w(\mathcal{C}_n).$$

Clearly, $P_e(\mathcal{C}_n) \leq \lambda(\mathcal{C}_n)$; so one would expect that $P_e(\mathcal{C}_n)$ behaves differently than $\lambda(\mathcal{C}_n)$. However, it can be shown that from a code $\mathcal{C}_n = (n, M)$ with arbitrarily small $P_e(\mathcal{C}_n)$, one can construct (by throwing away from $\mathcal{C}_n$ half of its codewords with largest conditional probability of error) a code $\mathcal{C}'_n = (n, \frac{M}{2})$ with arbitrarily small $\lambda(\mathcal{C}'_n)$ at essentially the same code rate as n grows to infinity (e.g., see [83, p. 204], [415, p. 163]).[3] Hence, for simplicity, we will only use $P_e(\mathcal{C}_n)$ as our criterion when evaluating the "goodness" or reliability[4] of channel block codes; but one must keep in mind that our results hold under $\lambda(\mathcal{C}_n)$ as well, in particular the channel coding theorem below.

Our target is to find a good channel block code (or to show the existence of a good channel block code). From the perspective of the (weak) law of large numbers, a good choice is to draw the code's codewords based on the *jointly typical* set between the input and the output of the channel, since all the probability mass is ultimately placed on the jointly typical set. The decoding failure then occurs only when the channel input–output pair does not lie in the jointly typical set, which implies that the probability of decoding error is ultimately small. We next define the jointly typical set.

---

[3]Note that this fact holds for single-user channels with known transition distributions (as given in Definition 4.1) that remain constant throughout the transmission of a codeword. It does not however hold for single-user channels whose statistical descriptions may vary in an unknown manner from symbol to symbol during a codeword transmission; such channels, which include the class of "arbitrarily varying channels" (see [87, Chap. 2, Sect. 6]), will not be considered in this textbook.

[4]We interchangeably use the terms "goodness" or "reliability" for a block code to mean that its (average) probability of error asymptotically vanishes with increasing blocklength.

Definition 4.7 *(Jointly typical set)* The set $\mathcal{F}_n(\delta)$ of jointly δ-typical n-tuple pairs (x^n, y^n) with respect to the memoryless distribution $P_{X^n, Y^n}(x^n, y^n) = \prod_{i=1}^{n} P_{X,Y}(x_i, y_i)$ is defined by

$$
\mathcal{F}_n(\delta) := \Bigg\{ (x^n, y^n) \in \mathcal{X}^n \times \mathcal{Y}^n :
$$

$$
\left| -\frac{1}{n} \log_2 P_{X^n}(x^n) - H(X) \right| < \delta, \quad \left| -\frac{1}{n} \log_2 P_{Y^n}(y^n) - H(Y) \right| < \delta,
$$

$$
\text{and} \quad \left| -\frac{1}{n} \log_2 P_{X^n, Y^n}(x^n, y^n) - H(X, Y) \right| < \delta \Bigg\}.
$$

In short, a pair (x^n, y^n) generated by independently drawing n times under $P_{X,Y}$ is jointly δ-typical if its joint and marginal empirical entropies are, respectively, δ-close to the true joint and marginal entropies.

With the above definition, we directly obtain the joint AEP theorem.

Theorem 4.8 (Joint AEP) *If* (X_1, Y_1), (X_2, Y_2), ..., (X_n, Y_n), ... *are i.i.d., i.e.,* $\{(X_i, Y_i)\}_{i=1}^{\infty}$ *is a dependent pair of DMSs, then*

$$
-\frac{1}{n} \log_2 P_{X^n}(X_1, X_2, \ldots, X_n) \rightarrow H(X) \quad \text{in probability,}
$$

$$
-\frac{1}{n} \log_2 P_{Y^n}(Y_1, Y_2, \ldots, Y_n) \rightarrow H(Y) \quad \text{in probability,}
$$

and

$$
-\frac{1}{n} \log_2 P_{X^n, Y^n}((X_1, Y_1), \ldots, (X_n, Y_n)) \rightarrow H(X, Y) \quad \text{in probability}
$$

as $n \rightarrow \infty$.

Proof By the weak law of large numbers, we have the desired result.   □

Theorem 4.9 (Shannon–McMillan–Breiman theorem for pairs) *Given a dependent pair of DMSs with joint entropy* $H(X, Y)$ *and any* δ *greater than zero, we can choose* n *big enough so that the jointly* δ-*typical set satisfies:*

1. $P_{X^n, Y^n}(\mathcal{F}_n^c(\delta)) < \delta$ *for sufficiently large n.*
2. *The number of elements in* $\mathcal{F}_n(\delta)$ *is at least* $(1 - \delta)2^{n(H(X,Y)-\delta)}$ *for sufficiently large n, and at most* $2^{n(H(X,Y)+\delta)}$ *for every n.*
3. *If* $(x^n, y^n) \in \mathcal{F}_n(\delta)$, *its probability of occurrence satisfies*

$$
2^{-n(H(X,Y)+\delta)} < P_{X^n, Y^n}(x^n, y^n) < 2^{-n(H(X,Y)-\delta)}.
$$

Proof The proof is quite similar to that of the Shannon–McMillan–Breiman theorem for a single memoryless source presented in the previous chapter; we hence leave it as an exercise. □

We next introduce the notion of *operational capacity* for a channel as the largest coding transmission rate that can be conveyed reliably (i.e., with asymptotically decaying error probability) over the channel when the coding blocklength is allowed to grow without bound.

Definition 4.10 *(Operational capacity)* A rate R is said to be *achievable* for a discrete channel if there exists a sequence of (n, M_n) channel codes $\mathscr{C}_n$ with

$$\liminf_{n \to \infty} \frac{1}{n} \log_2 M_n \geq R \quad \text{and} \quad \lim_{n \to \infty} P_e(\mathscr{C}_n) = 0.$$

The channel's *operational capacity*, C_{op}, is the supremum of all achievable rates:

$$C_{op} = \sup\{R : R \text{ is achievable}\}.$$

We herein arrive at the main result of this chapter, Shannon's channel coding theorem for DMCs. It states that for a DMC, its operational capacity C_{op} is actually equal to a quantity C, conveniently termed as channel *capacity* (or *information capacity*) and defined as the maximum of the channel's mutual information over the set of its input distributions (see below). In other words, the quantity C is indeed the supremum of all achievable channel code rates, and this is shown in two parts in the theorem in light of the properties of the supremum; see Observation A.5. As a result, for a given DMC, its quantity C, which can be calculated by solely using the channel's transition matrix $\mathbb{Q}$, constitutes the largest rate at which one can reliably transmit information via a block code over this channel. Thus, it is possible to communicate reliably over an inherently noisy DMC at a *fixed* rate (without decreasing it) as long as this rate is below C and the code's blocklength is allowed to be large.

Theorem 4.11 (Shannon's channel coding theorem) *Consider a DMC with finite input alphabet* $\mathcal{X}$, *finite output alphabet* $\mathcal{Y}$ *and transition distribution probability* $P_{Y|X}(y|x)$, $x \in \mathcal{X}$ *and* $y \in \mathcal{Y}$. *Define the channel capacity[5]*

---

[5]First note that the mutual information $I(X; Y)$ is actually a function of the input statistics P_X and the channel statistics $P_{Y|X}$. Hence, we may write it as

$$I(P_X, P_{Y|X}) = \sum_{x \in \mathcal{X}} \sum_{y \in \mathcal{Y}} P_X(x) P_{Y|X}(y|x) \log_2 \frac{P_{Y|X}(y|x)}{\sum_{x' \in \mathcal{X}} P_X(x') P_{Y|X}(y|x')}.$$

Such an expression is more suitable for calculating the channel capacity.

Note also that the channel capacity C is well-defined since, for a fixed $P_{Y|X}$, $I(P_X, P_{Y|X})$ is concave and continuous in P_X (with respect to both the variational distance and the Euclidean distance (i.e., $\mathcal{L}_2$-distance) [415, Chap. 2]), and since the set of all input distributions P_X is a compact (closed and bounded) subset of $\mathbb{R}^{|\mathcal{X}|}$ due to the finiteness of $\mathcal{X}$. Hence, there exists a P_X that achieves the supremum of the mutual information and the maximum is attainable.

$$C := \max_{P_X} I(X; Y) = \max_{P_X} I(P_X, P_{Y|X}),$$

where the maximum is taken over all input distributions P_X. Then, the following hold.

- *Forward part (achievability): For any $0 < \varepsilon < 1$, there exist $\gamma > 0$ and a sequence of data transmission block codes $\{\mathscr{C}_n = (n, M_n)\}_{n=1}^{\infty}$ with*

$$\liminf_{n \to \infty} \frac{1}{n} \log_2 M_n \geq C - \gamma$$

and

$$P_e(\mathscr{C}_n) < \varepsilon \quad \text{for sufficiently large } n,$$

where $P_e(\mathscr{C}_n)$ denotes the (average) probability of error for block code $\mathscr{C}_n$.
- *Converse part: For any $0 < \varepsilon < 1$, any sequence of data transmission block codes $\{\mathscr{C}_n = (n, M_n)\}_{n=1}^{\infty}$ with*

$$\liminf_{n \to \infty} \frac{1}{n} \log_2 M_n > C$$

satisfies

$$P_e(\mathscr{C}_n) > (1 - \epsilon)\mu \quad \text{for sufficiently large } n, \tag{4.3.1}$$

where

$$\mu = 1 - \frac{C}{\liminf_{n \to \infty} \frac{1}{n} \log_2 M_n} > 0,$$

i.e., the codes' probability of error is bounded away from zero for all n sufficiently large.[6]

Proof of the forward part: It suffices to prove the *existence* of a good block code sequence (satisfying the rate condition, i.e., $\liminf_{n \to \infty} (1/n) \log_2 M_n \geq C - \gamma$ for some $\gamma > 0$) whose average error probability is ultimately less than ε. Since the forward part holds trivially when $C = 0$ by setting $M_n = 1$, we assume in the sequel that $C > 0$.

We will use Shannon's original *random coding* proof technique in which the good block code sequence is not deterministically constructed; instead, its existence is *implicitly* proven by showing that for a class (ensemble) of block code sequences $\{\mathscr{C}_n\}_{n=1}^{\infty}$ and a code-selecting distribution $\Pr[\mathscr{C}_n]$ over these block code sequences, the expectation value of the average error probability, evaluated under the code-selecting distribution on these block code sequences, can be made smaller than ε for n sufficiently large:

---

[6]Note that (4.3.1) actually implies that $\liminf_{n \to \infty} P_e(\mathscr{C}_n) \geq \lim_{\epsilon \downarrow 0} (1 - \epsilon)\mu = \mu$, where the error probability lower bound has nothing to do with ϵ. Here, we state the converse of Theorem 4.11 in a form in parallel to the converse statements in Theorems 3.6 and 3.15.

$$E_{\mathcal{C}_n}[P_e(\mathcal{C}_n)] = \sum_{\mathcal{C}_n} \Pr[\mathcal{C}_n] P_e(\mathcal{C}_n) \to 0 \quad \text{as } n \to \infty.$$

Hence, there must exist at least one such a desired good code sequence $\{\mathcal{C}_n^*\}_{n=1}^{\infty}$ among them (with $P_e(\mathcal{C}_n^*) \to 0$ as $n \to \infty$).

Fix $\varepsilon \in (0, 1)$ and some γ in $(0, \min\{4\varepsilon, C\})$. Observe that there exists N_0 such that for $n > N_0$, we can choose an integer M_n with

$$C - \frac{\gamma}{2} \geq \frac{1}{n} \log_2 M_n > C - \gamma. \tag{4.3.2}$$

(Since we are only concerned with the case of "sufficient large n," it suffices to consider only those n's satisfying $n > N_0$, and ignore those n's for $n \leq N_0$.)

Define $\delta := \gamma/8$. Let $P_{\hat{X}}$ be a probability distribution that achieves the channel capacity:

$$C := \max_{P_X} I(P_X, P_{Y|X}) = I(P_{\hat{X}}, P_{Y|X}).$$

Denote by $P_{\hat{Y}^n}$ the channel output distribution due to channel input product distribution $P_{\hat{X}^n}$ (with $P_{\hat{X}^n}(x^n) = \prod_{i=1}^{n} P_{\hat{X}}(x_i)$), i.e.,

$$P_{\hat{Y}^n}(y^n) = \sum_{x^n \in \mathcal{X}^n} P_{\hat{X}^n, \hat{Y}^n}(x^n, y^n)$$

where

$$P_{\hat{X}^n, \hat{Y}^n}(x^n, y^n) := P_{\hat{X}^n}(x^n) P_{Y^n|X^n}(y^n|x^n)$$

for all $x^n \in \mathcal{X}^n$ and $y^n \in \mathcal{Y}^n$. Note that since $P_{\hat{X}^n}(x^n) = \prod_{i=1}^{n} P_{\hat{X}}(x_i)$ and the channel is memoryless, the resulting joint input–output process $\{(\hat{X}_i, \hat{Y}_i)\}_{i=1}^{\infty}$ is also memoryless with

$$P_{\hat{X}^n, \hat{Y}^n}(x^n, y^n) = \prod_{i=1}^{n} P_{\hat{X}, \hat{Y}}(x_i, y_i)$$

and

$$P_{\hat{X}, \hat{Y}}(x, y) = P_{\hat{X}}(x) P_{Y|X}(y|x) \quad \text{for } x \in \mathcal{X}, y \in \mathcal{Y}.$$

We next present the proof in three steps.

Step 1: Code construction.

For any blocklength n, independently select M_n channel inputs with replacement[7] from $\mathcal{X}^n$ according to the distribution $P_{\hat{X}^n}(x^n)$. For the selected M_n channel inputs yielding codebook $\mathcal{C}_n := \{c_1, c_2, \ldots, c_{M_n}\}$, define the encoder $f_n(\cdot)$ and decoder $g_n(\cdot)$, respectively, as follows:

---

[7]Here, the channel inputs are selected with replacement. That means it is possible and acceptable that all the selected M_n channel inputs are identical.

$$f_n(m) = c_m \quad \text{for } 1 \le m \le M_n,$$

and

$$g_n(y^n) = \begin{cases} m, & \text{if } c_m \text{ is the only codeword in } \mathscr{C}_n \\ & \text{satisfying } (c_m, y^n) \in \mathcal{F}_n(\delta); \\ \text{any one in } \{1, 2, \ldots, M_n\}, & \text{otherwise,} \end{cases}$$

where $\mathcal{F}_n(\delta)$ is defined in Definition 4.7 with respect to distribution $P_{\hat{X}^n, \hat{Y}^n}$. (We evidently assume that the codebook $\mathscr{C}_n$ and the channel distribution $P_{Y|X}$ are known at both the encoder and the decoder.) Hence, the code $\mathscr{C}_n$ operates as follows. A message W is chosen according to the uniform distribution from the set of messages. The encoder f_n then transmits the Wth codeword c_W in $\mathscr{C}_n$ over the channel. Then, Y^n is received at the channel output and the decoder guesses the sent message via $\hat{W} = g_n(Y^n)$.

Note that there is a total $|\mathcal{X}|^{nM_n}$ possible randomly generated codebooks $\mathscr{C}_n$ and the probability of selecting each codebook is given by

$$\Pr[\mathscr{C}_n] = \prod_{m=1}^{M_n} P_{\hat{X}^n}(c_m).$$

Step 2: Conditional error probability.

For each (randomly generated) data transmission code $\mathscr{C}_n$, the conditional probability of error given that message m was sent, $\lambda_m(\mathscr{C}_n)$, can be upper bounded by

$$\lambda_m(\mathscr{C}_n) \le \sum_{y^n \in \mathcal{Y}^n \,:\, (c_m, y^n) \notin \mathcal{F}_n(\delta)} P_{Y^n|X^n}(y^n | c_m)$$

$$+ \sum_{\substack{m'=1 \\ m' \ne m}}^{M_n} \sum_{y^n \in \mathcal{Y}^n \,:\, (c_{m'}, y^n) \in \mathcal{F}_n(\delta)} P_{Y^n|X^n}(y^n | c_m), \qquad (4.3.3)$$

where the first term in (4.3.3) considers the case that the received channel output y^n is not jointly δ-typical with c_m, (and hence, the decoding rule $g_n(\cdot)$ would possibly result in a wrong guess), and the second term in (4.3.3) reflects the situation when y^n is jointly δ-typical with not only the transmitted codeword c_m but also with another codeword $c_{m'}$ (which may cause a decoding error).

By taking expectation in (4.3.3) with respect to the mth codeword-selecting distribution $P_{\hat{X}^n}(c_m)$, we obtain

$$\sum_{c_m \in \mathcal{X}^n} P_{\hat{X}^n}(c_m) \lambda_m(\mathcal{C}_n) \le \sum_{c_m \in \mathcal{X}^n} \sum_{y^n \notin \mathcal{F}_n(\delta|c_m)} P_{\hat{X}^n}(c_m) P_{Y^n|X^n}(y^n|c_m)$$

$$+ \sum_{\substack{c_m \in \mathcal{X}^n \\ }} \sum_{\substack{m'=1 \\ m' \neq m}}^{M_n} \sum_{y^n \in \mathcal{F}_n(\delta|c_{m'})} P_{\hat{X}^n}(c_m) P_{Y^n|X^n}(y^n|c_m)$$

$$= P_{\hat{X}^n,\hat{Y}^n}\left(\mathcal{F}_n^c(\delta)\right)$$

$$+ \sum_{\substack{m'=1 \\ m' \neq m}}^{M_n} \sum_{c_m \in \mathcal{X}^n} \sum_{y^n \in \mathcal{F}_n(\delta|c_{m'})} P_{\hat{X}^n,\hat{Y}^n}(c_m, y^n),$$

$$(4.3.4)$$

where

$$\mathcal{F}_n(\delta|x^n) := \left\{ y^n \in \mathcal{Y}^n : (x^n, y^n) \in \mathcal{F}_n(\delta) \right\}.$$

Step 3: Average error probability.

We now can analyze the expectation of the average error probability

$$E_{\mathcal{C}_n}[P_e(\mathcal{C}_n)]$$

over the ensemble of all codebooks $\mathcal{C}_n$ generated at random according to $\Pr[\mathcal{C}_n]$ and show that it asymptotically vanishes as n grows without bound. We obtain the following series of inequalities:

$$E_{\mathcal{C}_n}[P_e(\mathcal{C}_n)] = \sum_{\mathcal{C}_n} \Pr[\mathcal{C}_n] P_e(\mathcal{C}_n)$$

$$= \sum_{c_1 \in \mathcal{X}^n} \cdots \sum_{c_{M_n} \in \mathcal{X}^n} P_{\hat{X}^n}(c_1) \cdots P_{\hat{X}^n}(c_{M_n}) \left(\frac{1}{M_n} \sum_{m=1}^{M_n} \lambda_m(\mathcal{C}_n) \right)$$

$$= \frac{1}{M_n} \sum_{m=1}^{M_n} \sum_{c_1 \in \mathcal{X}^n} \cdots \sum_{c_{m-1} \in \mathcal{X}^n} \sum_{c_{m+1} \in \mathcal{X}^n} \cdots \sum_{c_{M_n} \in \mathcal{X}^n}$$

$$P_{\hat{X}^n}(c_1) \cdots P_{\hat{X}^n}(c_{m-1}) P_{\hat{X}^n}(c_{m+1}) \cdots P_{\hat{X}^n}(c_{M_n})$$

$$\times \left(\sum_{c_m \in \mathcal{X}^n} P_{\hat{X}^n}(c_m) \lambda_m(\mathcal{C}_n) \right)$$

$$\le \frac{1}{M_n} \sum_{m=1}^{M_n} \sum_{c_1 \in \mathcal{X}^n} \cdots \sum_{c_{m-1} \in \mathcal{X}^n} \sum_{c_{m+1} \in \mathcal{X}^n} \cdots \sum_{c_{M_n} \in \mathcal{X}^n}$$

$$P_{\hat{X}^n}(c_1) \cdots P_{\hat{X}^n}(c_{m-1}) P_{\hat{X}^n}(c_{m+1}) \cdots P_{\hat{X}^n}(c_{M_n})$$

$$\times P_{\hat{X}^n,\hat{Y}^n}\left(\mathcal{F}_n^c(\delta)\right)$$

$$+ \frac{1}{M_n} \sum_{m=1}^{M_n} \sum_{c_1 \in \mathcal{X}^n} \cdots \sum_{c_{m-1} \in \mathcal{X}^n} \sum_{c_{m+1} \in \mathcal{X}^n} \cdots \sum_{c_{M_n} \in \mathcal{X}^n}$$

$$P_{\hat{X}^n}(c_1) \cdots P_{\hat{X}^n}(c_{m-1}) P_{\hat{X}^n}(c_{m+1}) \cdots P_{\hat{X}^n}(c_{M_n})$$

$$\times \sum_{\substack{m'=1 \\ m' \neq m}}^{M_n} \sum_{c_m \in \mathcal{X}^n} \sum_{y^n \in \mathcal{F}_n(\delta|c_{m'})} P_{\hat{X}^n, \hat{Y}^n}(c_m, y^n) \qquad (4.3.5)$$

$$= P_{\hat{X}^n, \hat{Y}^n}\left(\mathcal{F}_n^c(\delta)\right)$$

$$+ \frac{1}{M_n} \sum_{m=1}^{M_n} \left\{ \sum_{\substack{m'=1 \\ m' \neq m}}^{M_n} \left[\sum_{c_1 \in \mathcal{X}^n} \cdots \sum_{c_{m-1} \in \mathcal{X}^n} \sum_{c_{m+1} \in \mathcal{X}^n} \cdots \sum_{c_{M_n} \in \mathcal{X}^n} \right. \right.$$

$$P_{\hat{X}^n}(c_1) \cdots P_{\hat{X}^n}(c_{m-1}) P_{\hat{X}^n}(c_{m+1}) \cdots P_{\hat{X}^n}(c_{M_n})$$

$$\left. \left. \times \sum_{c_m \in \mathcal{X}^n} \sum_{y^n \in \mathcal{F}_n(\delta|c_{m'})} P_{\hat{X}^n, \hat{Y}^n}(c_m, y^n) \right] \right\},$$

where (4.3.5) follows from (4.3.4), and the last step holds since $P_{\hat{X}^n, \hat{Y}^n}\left(\mathcal{F}_n^c(\delta)\right)$ is a constant independent of $c_1, \ldots, c_{M_n}$ and m. Observe that for $n > N_0$,

$$\sum_{\substack{m'=1 \\ m' \neq m}}^{M_n} \left[\sum_{c_1 \in \mathcal{X}^n} \cdots \sum_{c_{m-1} \in \mathcal{X}^n} \sum_{c_{m+1} \in \mathcal{X}^n} \cdots \sum_{c_{M_n} \in \mathcal{X}^n} \right.$$

$$P_{\hat{X}^n}(c_1) \cdots P_{\hat{X}^n}(c_{m-1}) P_{\hat{X}^n}(c_{m+1}) \cdots P_{\hat{X}^n}(c_{M_n})$$

$$\left. \times \sum_{c_m \in \mathcal{X}^n} \sum_{y^n \in \mathcal{F}_n(\delta|c_{m'})} P_{\hat{X}^n, \hat{Y}^n}(c_m, y^n) \right]$$

$$= \sum_{\substack{m'=1 \\ m' \neq m}}^{M_n} \left[\sum_{c_m \in \mathcal{X}^n} \sum_{c_{m'} \in \mathcal{X}^n} \sum_{y^n \in \mathcal{F}_n(\delta|c_{m'})} P_{\hat{X}^n}(c_{m'}) P_{\hat{X}^n, \hat{Y}^n}(c_m, y^n) \right]$$

$$= \sum_{\substack{m'=1 \\ m' \neq m}}^{M_n} \left[\sum_{c_{m'} \in \mathcal{X}^n} \sum_{y^n \in \mathcal{F}_n(\delta|c_{m'})} P_{\hat{X}^n}(c_{m'}) \left(\sum_{c_m \in \mathcal{X}^n} P_{\hat{X}^n, \hat{Y}^n}(c_m, y^n) \right) \right]$$

$$= \sum_{\substack{m'=1 \\ m' \neq m}}^{M_n} \left[\sum_{c_{m'} \in \mathcal{X}^n} \sum_{y^n \in \mathcal{F}_n(\delta|c_{m'})} P_{\hat{X}^n}(c_{m'}) P_{\hat{Y}^n}(y^n) \right]$$

$$= \sum_{\substack{m'=1 \\ m' \neq m}}^{M_n} \left[\sum_{(c_{m'}, y^n) \in \mathcal{F}_n(\delta)} P_{\hat{X}^n}(c_{m'}) P_{\hat{Y}^n}(y^n) \right]$$

$$\leq \sum_{\substack{m'=1 \\ m' \neq m}}^{M_n} |\mathcal{F}_n(\delta)| 2^{-n(H(\hat{X})-\delta)} 2^{-n(H(\hat{Y})-\delta)}$$

$$\leq \sum_{\substack{m'=1 \\ m' \neq m}}^{M_n} 2^{n(H(\hat{X},\hat{Y})+\delta)} 2^{-n(H(\hat{X})-\delta)} 2^{-n(H(\hat{Y})-\delta)}$$

$$= (M_n - 1) 2^{n(H(\hat{X},\hat{Y})+\delta)} 2^{-n(H(\hat{X})-\delta)} 2^{-n(H(\hat{Y})-\delta)}$$

$$< M_n \cdot 2^{n(H(\hat{X},\hat{Y})+\delta)} 2^{-n(H(\hat{X})-\delta)} 2^{-n(H(\hat{Y})-\delta)}$$

$$\leq 2^{n(C-4\delta)} \cdot 2^{-n(I(\hat{X};\hat{Y})-3\delta)} = 2^{-n\delta},$$

where the first inequality follows from the definition of the jointly typical set $\mathcal{F}_n(\delta)$, the second inequality holds by the Shannon–McMillan–Breiman theorem for pairs (Theorem 4.9), the last inequality follows since $C = I(\hat{X}; \hat{Y})$ by definition of $\hat{X}$ and $\hat{Y}$, and since $(1/n)\log_2 M_n \leq C - (\gamma/2) = C - 4\delta$. Consequently,

$$E_{\mathcal{C}_n}[P_e(\mathcal{C}_n)] \leq P_{\hat{X}^n, \hat{Y}^n}\left(\mathcal{F}_n^c(\delta)\right) + 2^{-n\delta},$$

which for sufficiently large n (and $n > N_0$), can be made smaller than $2\delta = \gamma/4 < \varepsilon$ by the Shannon–McMillan–Breiman theorem for pairs.   □

Before proving the converse part of the channel coding theorem, let us recall Fano's inequality in a channel coding context. Consider an (n, M_n) channel block code $\mathcal{C}_n$ with encoding and decoding functions given by

$$f_n : \{1, 2, \ldots, M_n\} \rightarrow \mathcal{X}^n$$

and

$$g_n : \mathcal{Y}^n \rightarrow \{1, 2, \ldots, M_n\},$$

respectively. Let message W, which is uniformly distributed over the set of messages $\{1, 2, \ldots, M_n\}$, be sent via codeword $X^n(W) = f_n(W)$ over the DMC, and let Y^n be received at the channel output. At the receiver, the decoder estimates the sent message via $\hat{W} = g_n(Y^n)$ and the probability of estimation error is given by the code's average error probability:

$$\Pr[W \neq \hat{W}] = P_e(\mathcal{C}_n)$$

since W is uniformly distributed. Then, Fano's inequality (2.5.2) yields

$$H(W|Y^n) \leq 1 + P_e(\mathcal{C}_n) \log_2(M_n - 1)$$
$$< 1 + P_e(\mathcal{C}_n) \log_2 M_n. \tag{4.3.6}$$

We next proceed with the proof of the converse part.

Proof of the converse part: For any (n, M_n) block channel code $\mathcal{C}_n$ as described above, we have that $W \to X^n \to Y^n$ form a Markov chain; we thus obtain by the data processing inequality that

$$I(W; Y^n) \leq I(X^n; Y^n). \tag{4.3.7}$$

We can also upper bound $I(X^n; Y^n)$ in terms of the channel capacity C as follows:

$$I(X^n; Y^n) \leq \max_{P_{X^n}} I(X^n; Y^n)$$

$$\leq \max_{P_{X^n}} \sum_{i=1}^{n} I(X_i; Y_i) \quad \text{(by Theorem 2.21)}$$

$$\leq \sum_{i=1}^{n} \max_{P_{X^n}} I(X_i; Y_i)$$

$$= \sum_{i=1}^{n} \max_{P_{X_i}} I(X_i; Y_i)$$

$$= nC. \tag{4.3.8}$$

Consequently, code $\mathcal{C}_n$ satisfies the following:

$$\log_2 M_n = H(W) \quad \text{(since W is uniformly distributed)}$$
$$= H(W|Y^n) + I(W; Y^n)$$
$$\leq H(W|Y^n) + I(X^n; Y^n) \quad \text{(by (4.3.7))}$$
$$\leq H(W|Y^n) + nC \quad \text{(by (4.3.8))}$$
$$< 1 + P_e(\mathcal{C}_n) \cdot \log_2 M_n + nC. \quad \text{(by (4.3.6))}$$

This implies that

$$P_e(\mathcal{C}_n) > 1 - \frac{C}{(1/n) \log_2 M_n} - \frac{1}{\log_2 M_n} = 1 - \frac{C + 1/n}{(1/n) \log_2 M_n}.$$

So if $\liminf_{n \to \infty} (1/n) \log_2 M_n = \frac{C}{1-\mu}$, then for any $0 < \varepsilon < 1$, there exists an integer N such that for $n \geq N$,

$$\frac{1}{n} \log_2 M_n \geq \frac{C + 1/n}{1 - (1 - \varepsilon)\mu}, \tag{4.3.9}$$

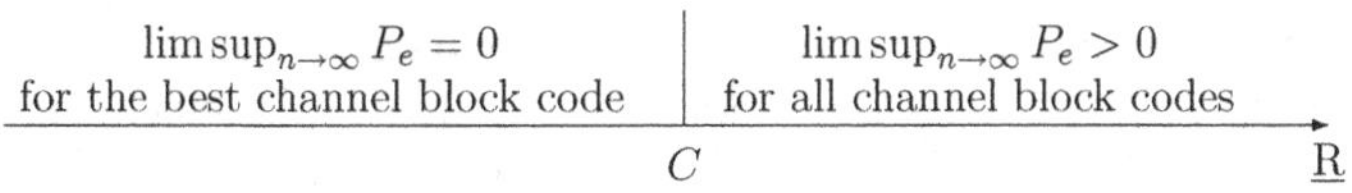

$\limsup_{n\to\infty} P_e = 0$ for the best channel block code	$\limsup_{n\to\infty} P_e > 0$ for all channel block codes

$\qquad\qquad\qquad\qquad\qquad\qquad C \qquad\qquad\qquad\qquad\qquad\qquad\qquad\qquad \underline{R}$

Fig. 4.5 Asymptotic channel coding rate $\underline{R}$ versus channel capacity C and behavior of the probability of error as blocklength n goes to infinity for a DMC

because, otherwise, (4.3.9) would be violated for infinitely many n, implying a contradiction that

$$\liminf_{n\to\infty} \frac{1}{n} \log_2 M_n \leq \liminf_{n\to\infty} \frac{C+1/n}{1-(1-\varepsilon)\mu} = \frac{C}{1-(1-\varepsilon)\mu}.$$

Hence, for $n \geq N$,

$$P_e(\mathcal{C}_n) > 1 - [1-(1-\varepsilon)\mu]\frac{C+1/n}{C+1/n} = (1-\epsilon)\mu > 0;$$

i.e., $P_e(\mathcal{C}_n)$ is bounded away from zero for n sufficiently large.   □

Observation 4.12 The results of the above channel coding theorem, which proves that $C_{op} = C$, are illustrated in Fig. 4.5,[8] where $\underline{R} = \liminf_{n\to\infty} \frac{1}{n} \log_2 M_n$ (measured in message bits/channel use) is usually called the *asymptotic* coding rate of channel block codes. As indicated in the figure, the asymptotic rate of any good block code for the DMC must be smaller than or equal to the channel capacity C.[9] Conversely, any block code with (asymptotic) rate greater than C, will have its probability of error bounded away from zero.

Observation 4.13 (Zero error codes) In the converse part of Theorem 4.11, we showed that

$$\liminf_{n\to\infty} P_e(\mathcal{C}_n) = 0 \quad\Longrightarrow\quad \liminf_{n\to\infty} \frac{1}{n}\log_2 M_n \leq C. \qquad (4.3.10)$$

---

[8]Note that Theorem 4.11 actually implies that $\lim_{n\to\infty} P_e = 0$ for $\underline{R} < C_{op} = C$ and that $\liminf_{n\to\infty} P_e > 0$ for $\underline{R} > C_{op} = C$; these properties, however, might not hold for more general channels than the DMC. For general channels, three partitions instead of two may result, i.e., $\underline{R} < C_{op}$, $C_{op} < \underline{R} < \bar{C}_{op}$ and $\underline{R} > \bar{C}_{op}$, which, respectively, correspond to $\limsup_{n\to\infty} P_e = 0$ for the best block code, $\limsup_{n\to\infty} P_e > 0$ but $\liminf_{n\to\infty} P_e = 0$ for the best block code, and $\liminf_{n\to\infty} P_e > 0$ for all channel codes, where $\bar{C}_{op}$ is called the channel's *optimistic operational capacity* [394, 396]. Since $\bar{C}_{op} = C_{op} = C$ for DMCs, the three regions are reduced to two. A formula for $\bar{C}_{op}$ in terms of a generalized (spectral) mutual information rate is established in [75].
[9]It can be seen from the theorem that C can be achieved as an asymptotic transmission rate as long as $(1/n)\log_2 M_n$ approaches C from *below* with increasing n (see (4.3.2)).

We next briefly examine the situation when we require that all (n, M_n) codes $\mathscr{C}_n$ are to be used with exactly no errors for any value of the blocklength n; i.e., $P_e(\mathscr{C}_n) = 0$ for every n. In this case, we readily obtain that $H(W|Y^n) = 0$, which in turn implies (by invoking the data processing inequality) that for any n,

$$
\begin{aligned}
\log_2 M_n &= H(W|Y^n) + I(W; Y^n) \\
&= I(W; Y^n) \\
&\leq I(X^n; Y^n) \\
&\leq nC.
\end{aligned}
$$

Thus, we have proved that

$$
P_e(\mathscr{C}_n) = 0 \;\forall n \quad \Longrightarrow \quad \limsup_{n \to \infty} \frac{1}{n} \log_2 M_n \leq C,
$$

which is a stronger result than (4.3.10).

Shannon's channel coding theorem, established in 1948 [340], provides the ultimate limit for reliable communication over a noisy channel. However, it does not provide an explicit efficient construction for good codes since searching for a good code from the ensemble of randomly generated codes is prohibitively complex, as its size grows double exponentially with blocklength (see Step 1 of the proof of the forward part). It thus spurred the entire area of *coding theory*, which flourished over the last several decades with the aim of constructing powerful error-correcting codes operating close to the capacity limit. Particular advances were made for the class of *linear codes* (also known as group codes) whose rich[10] yet elegantly simple algebraic structures made them amenable for efficient practically implementable encoding and decoding. Examples of such codes include Hamming, Golay, Bose–Chaudhuri–Hocquenghem (BCH), Reed–Muller, Reed–Solomon and convolutional codes. In 1993, the so-called *Turbo codes* were introduced by Berrou et al. [44, 45] and shown experimentally to perform close to the channel capacity limit for the class of memoryless channels. Similar near-capacity achieving linear codes were later established with the rediscovery of Gallager's *low-density parity-check codes (LDPC)* [133, 134, 251, 252]. A more recent breakthrough was the invention of *polar codes* by Arikan in 2007, when he provided a deterministic construction of codes that can provably achieve channel capacity [22, 23]; see the next section for a brief illustrative example on polar codes for the BEC. Many of the above codes are used with increased sophistication in today's ubiquitous communication, information and multimedia technologies. For detailed studies on channel coding theory, see the following texts [50, 52, 208, 248, 254, 321, 407].

---

[10]Indeed, there exist linear codes that can achieve the capacity of memoryless channels with additive noise (e.g., see [87, p. 114]). Such channels include the BSC and the q-ary symmetric channel.

4.4 Example of Polar Codes for the BEC

As noted above, polar coding is a new channel coding method proposed by Arikan [22, 23], which can provably achieve the capacity of any binary-input memoryless channel $\mathbb{Q}$ whose capacity is realized by a uniform input distribution (e.g., quasi-symmetric channels). The proof technique and code construction, which has low encoding and decoding complexity, are purely based on information-theoretic concepts. For simplicity, we focus solely on a channel $\mathbb{Q}$ given by the BEC with erasure probability ε, which we denote as $BEC(\varepsilon)$ for short.

The main idea behind polar codes is channel "polarization," which transforms many independent uses of $BEC(\varepsilon)$, n uses to be precise (where n is the coding blocklength),[11] into extremal "polarized" channels; i.e., channels which are either perfect (noiseless) or completely noisy. It is shown that as $n \to \infty$, the number of unpolarized channels converges to 0 and the fraction of perfect channels converges to $I(X; Y) = 1 - \varepsilon$ under a uniform input, which is the capacity of the BEC. A polar code can then be naturally obtained by sending information bits directly through those perfect channels and sending known bits (usually called frozen bits) through the completely noisy channels.

We start with the simplest case of $n = 2$. The channel transformation depicted in Fig. 4.6a is usually called the basic transformation. In this figure, we have two independent uses of $BEC(\varepsilon)$, namely, (X_1, Y_1) and (X_2, Y_2), where every bit has ε chance of being erased. In other words, under uniformly distributed X_1 and X_2, we have

$$I(\mathbb{Q}) := I(X_1; Y_1) = I(X_2; Y_2) = 1 - \varepsilon.$$

Now consider the following linear modulo-2 operation shown in Fig. 4.6:

$$X_1 = U_1 \oplus U_2,$$
$$X_2 = U_2,$$

where U_1 and U_2 represent uniformly distributed independent message bits. The decoder performs successive cancellation decoding as follows. It first decodes U_1 from the received (Y_1, Y_2), and then decodes U_2 based on (Y_1, Y_2) and the previously decoded U_1 (assuming the decoding is done correctly). This will create two new channels; namely, the "worse" channel $\mathbb{Q}^-$ and the "better" channel $\mathbb{Q}^+$ given by

$$\mathbb{Q}^- : U_1 \to (Y_1, Y_2),$$
$$\mathbb{Q}^+ : U_2 \to (Y_1, Y_2, U_1),$$

---

[11]Recall that in channel coding, a codeword of length n is typically sent by using the channel n consecutive times (i.e., in series). But in polar coding, an equivalent method is applied, which consists of using n identical and independent copies of the channel in parallel, with each channel being utilized only once.

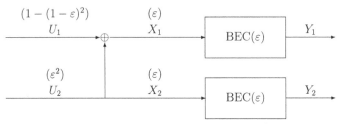

(a) Transformation for two independent uses of BEC(ε).

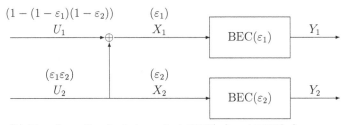

(b) Transformation for independent BEC(ε_1) and BEC(ε_2).

Fig. 4.6 Basic transformation with $n = 2$

respectively (the names of these channels will be justified shortly). Note that correctly receiving $Y_1 = X_1$ alone is not enough for us to determine U_1, since U_2 is a uniform random variable that is independent of U_1. One really needs to have both $Y_1 = X_1$ and $Y_2 = X_2$ for correctly decoding U_1. This observation implies that $\mathbb{Q}^-$ is a BEC with erasure probability[12] $\varepsilon^- := 1 - (1 - \varepsilon)^2$. Also, note that given U_1, either $Y_1 = X_1$ or $Y_2 = X_2$ is sufficient to determine U_2. This implies that $\mathbb{Q}^+$ is a BEC with erasure probability $\varepsilon^+ := \varepsilon^2$.

Overall, we have

$$
\begin{aligned}
I(\mathbb{Q}^+) + I(\mathbb{Q}^-) &= I(U_2; Y_1, Y_2, U_1) + I(U_1; Y_1, Y_2) \\
&= (1 - \varepsilon^2) + [1 - (1 - (1 - \varepsilon)^2)] \\
&= 2(1 - \varepsilon) \\
&= 2I(\mathbb{Q}),
\end{aligned}
\tag{4.4.1}
$$

and

---

[12]More precisely, channel $\mathbb{Q}^-$ has the same behavior as a BEC, and it can be exactly converted to a BEC after relabeling its output pair (y_1, y_2) as an equivalent three-valued symbol $y_{1,2}$ as follows:

$$
y_{1,2} = \begin{cases}
0 & \text{if } (y_1, y_2) \in \{(0, 0), (1, 1)\}, \\
E & \text{if } (y_1, y_2) \in \{(0, E), (1, E), (E, E), (E, 0), (E, 1)\}, \\
1 & \text{if } (y_1, y_2) \in \{(0, 1), (1, 0)\}.
\end{cases}
$$

A similar conversion can be applied to channel $\mathbb{Q}^+$.

$$(1 - \varepsilon)^2 = I(\mathbb{Q}^-)$$
$$\leq I(\mathbb{Q}) = 1 - \varepsilon$$
$$\leq I(\mathbb{Q}^+) = 1 - \varepsilon^2, \tag{4.4.2}$$

with equality iff $\varepsilon(1 - \varepsilon) = 0$ (i.e., $\varepsilon = 0$ or $\varepsilon = 1$). Equation (4.4.1) shows that the basic transformation does not incur any loss in mutual information. Furthermore, (4.4.2) indeed confirms that $\mathbb{Q}^+$ and $\mathbb{Q}^-$ are, respectively, better and worse than $\mathbb{Q}$.[13]

So far, we have talked about how to use the basic transformation to generate a better channel $\mathbb{Q}^+$ and a worse channel $\mathbb{Q}^-$ from two independent uses of $\mathbb{Q} = $ BEC(ε). Now, let us consider the case of $n = 4$ and suppose we perform the basic transformation twice to send (i.i.d. uniform) message bits (U_1, U_2, U_3, U_4), yielding

$$\mathbb{Q}^- : V_1 \to (Y_1, Y_2), \qquad \text{where } X_1 = V_1 \oplus V_2,$$
$$\mathbb{Q}^+ : V_2 \to (Y_1, Y_2, V_1), \quad \text{where } X_2 = V_2,$$
$$\mathbb{Q}^- : V_3 \to (Y_3, Y_4), \qquad \text{where } X_3 = V_3 \oplus V_4,$$
$$\mathbb{Q}^+ : V_4 \to (Y_3, Y_4, V_3), \quad \text{where } X_4 = V_4,$$

where $V_1 = U_1 \oplus U_2$, $V_3 = U_2$, $V_2 = U_3 \oplus U_4$ and $V_4 = U_4$. Since both $\mathbb{Q}^-$ channels have the same erasure probability $\varepsilon^- = 1 - (1 - \varepsilon)^2$, and since both $\mathbb{Q}^+$ channels have the same erasure probability $\varepsilon^+ = \varepsilon^2$, we can now take two $\mathbb{Q}^-$ channels and perform the basic transformation again to generate two new channels: $\mathbb{Q}^{--} : U_1 \to (Y_1, Y_2, Y_3, Y_4)$ with erasure probability $\varepsilon^{--} := 1 - (1 - \varepsilon^-)^2$ and $\mathbb{Q}^{-+} : U_2 \to (Y_1, Y_2, Y_3, Y_4, U_1)$ with erasure probability $\varepsilon^{-+} := (\varepsilon^-)^2$. Similarly, we can use two $\mathbb{Q}^+$ channels to form $\mathbb{Q}^{+-} : U_3 \to (Y_1, Y_2, Y_3, Y_4, U_1, U_2)$ with erasure probability $\varepsilon^{+-} := 1 - (1 - \varepsilon^+)^2$ and $\mathbb{Q}^{++} : U_4 \to (Y_1, Y_2, Y_3, Y_4, U_1, U_3, U_2)$ with erasure probability $\varepsilon^{++} := (\varepsilon^+)^2$.

The key attribute of this technique is that we do not have to stop here; in fact, we can keep exploiting this property until all the channels eventually become either very good (i.e., perfect) or very bad (i.e., completely noisy). In polar coding terminology, the process of using multiple basic transformations to get $X_1, \ldots, X_n$ from $U_1, \ldots, U_n$ (where the U_i's are i.i.d. uniform message random variables) is called channel "combining" and that of using $Y_1, \ldots, Y_n$ and $U_1, \ldots, U_{i-1}$ to obtain U_i for $i \in \{1, \ldots, n\}$ is called channel "splitting." Altogether, the phenomenon is called channel "polarization."

For constructing a polar code with blocklength $n = 2^m$ and 2^k codewords (i.e., with each binary message word having length k), one can perform m stages of channel polarization and transmit uncoded k message bits via the k positions with

---

[13]The same reasoning can be applied to form the basic transformation for two independent but not identically distributed BECs as shown in Fig. 4.6b, where $\mathbb{Q}^+$ and $\mathbb{Q}^-$ become BEC($\varepsilon_1 \varepsilon_2$) and BEC($1 - (1 - \varepsilon_1)(1 - \varepsilon_2)$), respectively. This extension may be useful when combining n independent uses of a channel in a multistage manner (in particular, when the two channels to be combined may become non-identically distributed after the second stage). In Example 4.14, only identically distributed BECs will be combined at each stage, which is a typical design for polar coding.

largest mutual informations. The other $n - k$ positions are stuffed with frozen bits; this encoding process is precisely channel combining. The decoder successively decodes U_i, $i \in \{1, \ldots, n\}$, based on $(Y_1, \ldots, Y_n)$ and the previously decoded $\hat{U}_j$, $j \in \{1, \ldots, i - 1\}$. This decoder is called a successive cancellation decoder and mimics the behavior of channel splitting in the process of channel polarization.

Example 4.14 Consider a BEC with erasure probability $\varepsilon = 0.5$ and let $n = 8$. The channel polarization process for this example is shown in Fig. 4.7. Note that since the mutual information of a BEC(ε) under a uniform input is simply $1 - \varepsilon$; one can equivalently keep tracking the erasure probabilities as we have shown in parentheses in Fig. 4.7. Now, suppose we would like to construct a $(8, 4)$ polar code, we pick the four positions with largest mutual informations (i.e., smallest erasure probabilities). That is, we pick (U_4, U_6, U_7, U_8) to send uncoded bits and the other positions are frozen.

As an example of the computation of the erasure probabilities, 0.5625 for T_2 is obtained from 0.75×0.75, which are the numbers above V_1 and V_3, and combining T_1 and T_2 produces $1 - (1 - 0.9375)(1 - 0.9375) \approx 0.9961$, which is the number above U_1.

Ever since their invention by Arikan [22, 23], polar codes have generated extensive interest; see [25, 26, 226, 228, 329, 371, 372] and the references therein and thereafter. A key reason for their prevalence is that they form the first coding scheme that has an explicit low-complexity construction structure while being capable of achieving channel capacity as code length approaches infinity. More importantly, polar codes do not exhibit the error floor behavior, which Turbo and (to a lesser extent) LDPC codes are prone to. In practice, since one cannot have infinitely many stages of polarization, there will always exist unpolarized channels. The development of effective construction and decoding methods for polar codes with practical blocklengths is an active area of research. Due to their attractive properties, polar codes were adopted in 2016 by the 3rd Generation Partnership Project (3GPP) as error-correcting codes for the control channel of the 5th generation (5G) mobile communication standard [99].

We conclude by noting that the notion of polarization is not unique to channel coding; it can also be applied to source coding and other information-theoretic problems including secrecy and multiuser systems (e.g., cf. [24, 148, 226, 227, 256]).

4.5   Calculating Channel Capacity

Given a DMC with finite input alphabet $\mathcal{X}$, finite output alphabet $\mathcal{Y}$ and channel transition matrix $\mathbb{Q} = [p_{x,y}]$ of size $|\mathcal{X}| \times |\mathcal{Y}|$, where $p_{x,y} := P_{Y|X}(y|x)$, for $x \in \mathcal{X}$ and $y \in \mathcal{Y}$, we would like to calculate

$$C := \max_{P_X} I(X; Y)$$

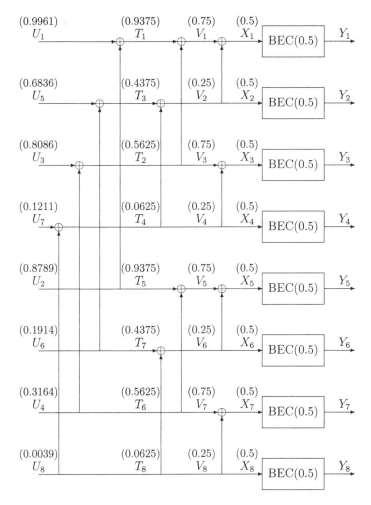

Fig. 4.7 Channel polarization for $\mathbb{Q} = \mathrm{BEC}(0.5)$ with $n = 8$

where the maximization (which is well-defined) is carried over the set of input distributions P_X, and $I(X; Y)$ is the mutual information between the channel's input and output.

Note that C can be determined numerically via nonlinear optimization techniques —such as the iterative algorithms developed by Arimoto [27] and Blahut [49, 51], see also [88] and [415, Chap. 9]. In general, there are no closed-form (single-letter) analytical expressions for C. However, for many "simplified" channels, it is possible to analytically determine C under some "symmetry" properties of their channel transition matrix.

4.5.1 Symmetric, Weakly Symmetric, and Quasi-symmetric Channels

Definition 4.15 A DMC with finite input alphabet $\mathcal{X}$, finite output alphabet $\mathcal{Y}$ and channel transition matrix $\mathbb{Q} = [p_{x,y}]$ of size $|\mathcal{X}| \times |\mathcal{Y}|$ is said to be *symmetric* if the rows of $\mathbb{Q}$ are permutations of each other and the columns of $\mathbb{Q}$ are permutations of each other. The channel is said to be *weakly symmetric* if the rows of $\mathbb{Q}$ are permutations of each other and all the column sums in $\mathbb{Q}$ are equal.

It directly follows from the definition that symmetry implies weak-symmetry. Examples of symmetric DMCs include the BSC, the q-ary symmetric channel and the following ternary channel with $\mathcal{X} = \mathcal{Y} = \{0, 1, 2\}$ and transition matrix:

$$\mathbb{Q} = \begin{bmatrix} P_{Y|X}(0|0) & P_{Y|X}(1|0) & P_{Y|X}(2|0) \\ P_{Y|X}(0|1) & P_{Y|X}(1|1) & P_{Y|X}(2|1) \\ P_{Y|X}(0|2) & P_{Y|X}(1|2) & P_{Y|X}(2|2) \end{bmatrix} = \begin{bmatrix} 0.4 & 0.1 & 0.5 \\ 0.5 & 0.4 & 0.1 \\ 0.1 & 0.5 & 0.4 \end{bmatrix}.$$

The following DMC with $|\mathcal{X}| = |\mathcal{Y}| = 4$ and

$$\mathbb{Q} = \begin{bmatrix} 0.5 & 0.25 & 0.25 & 0 \\ 0.5 & 0.25 & 0.25 & 0 \\ 0 & 0.25 & 0.25 & 0.5 \\ 0 & 0.25 & 0.25 & 0.5 \end{bmatrix} \tag{4.5.1}$$

is weakly symmetric (but not symmetric). Noting that all above channels involve square transition matrices, we emphasize that $\mathbb{Q}$ can be rectangular while satisfying the symmetry or weak-symmetry properties. For example, the DMC with $|\mathcal{X}| = 2$, $|\mathcal{Y}| = 4$ and

$$\mathbb{Q} = \begin{bmatrix} \frac{1-\varepsilon}{2} & \frac{\varepsilon}{2} & \frac{1-\varepsilon}{2} & \frac{\varepsilon}{2} \\ \frac{\varepsilon}{2} & \frac{1-\varepsilon}{2} & \frac{\varepsilon}{2} & \frac{1-\varepsilon}{2} \end{bmatrix} \tag{4.5.2}$$

is symmetric (where $\varepsilon \in [0, 1]$), while the DMC with $|\mathcal{X}| = 2$, $|\mathcal{Y}| = 3$ and

$$\mathbb{Q} = \begin{bmatrix} \frac{1}{3} & \frac{1}{6} & \frac{1}{2} \\ \frac{1}{3} & \frac{1}{2} & \frac{1}{6} \end{bmatrix}$$

is weakly symmetric.

Lemma 4.16 *The capacity of a weakly symmetric channel $\mathbb{Q}$ is achieved by a uniform input distribution and is given by*

$$C = \log_2 |\mathcal{Y}| - H(q_1, q_2, \ldots, q_{|\mathcal{Y}|}) \tag{4.5.3}$$

where $(q_1, q_2, \ldots, q_{|\mathcal{Y}|})$ denotes any row of $\mathbb{Q}$ and

$$H(q_1, q_2, \ldots, q_{|\mathcal{Y}|}) := -\sum_{i=1}^{|\mathcal{Y}|} q_i \log_2 q_i$$

is the row entropy.

Proof The mutual information between the channel's input and output is given by

$$I(X; Y) = H(Y) - H(Y|X)$$
$$= H(Y) - \sum_{x \in \mathcal{X}} P_X(x) H(Y|X = x)$$

where $H(Y|X = x) = -\sum_{y \in \mathcal{Y}} P_{Y|X}(y|x) \log_2 P_{Y|X}(y|x) = -\sum_{y \in \mathcal{Y}} p_{x,y} \log_2 p_{x,y}$.

Noting that every row of $\mathbb{Q}$ is a permutation of every other row, we obtain that $H(Y|X = x)$ is independent of x and can be written as

$$H(Y|X = x) = H(q_1, q_2, \ldots, q_{|\mathcal{Y}|}),$$

where $(q_1, q_2, \ldots, q_{|\mathcal{Y}|})$ is any row of $\mathbb{Q}$. Thus

$$H(Y|X) = \sum_{x \in \mathcal{X}} P_X(x) H(q_1, q_2, \ldots, q_{|\mathcal{Y}|})$$
$$= H(q_1, q_2, \ldots, q_{|\mathcal{Y}|}) \left(\sum_{x \in \mathcal{X}} P_X(x) \right)$$
$$= H(q_1, q_2, \ldots, q_{|\mathcal{Y}|}).$$

This implies

$$I(X; Y) = H(Y) - H(q_1, q_2, \ldots, q_{|\mathcal{Y}|})$$
$$\leq \log_2 |\mathcal{Y}| - H(q_1, q_2, \ldots, q_{|\mathcal{Y}|}),$$

with equality achieved iff Y is uniformly distributed over $\mathcal{Y}$. We next show that choosing a uniform input distribution, $P_X(x) = \frac{1}{|\mathcal{X}|} \ \forall x \in \mathcal{X}$, yields a uniform output distribution, hence maximizing mutual information. Indeed, under a uniform input distribution, we obtain that for any $y \in \mathcal{Y}$,

$$P_Y(y) = \sum_{x \in \mathcal{X}} P_X(x) P_{Y|X}(y|x) = \frac{1}{|\mathcal{X}|} \sum_{x \in \mathcal{X}} p_{x,y} = \frac{A}{|\mathcal{X}|}$$

where $A := \sum_{x \in \mathcal{X}} p_{x,y}$ is a constant given by the sum of the entries in any column of $\mathbb{Q}$, since by the weak-symmetry property all column sums in $\mathbb{Q}$ are identical. Note that $\sum_{y \in \mathcal{Y}} P_Y(y) = 1$ yields that

$$\sum_{y \in \mathcal{Y}} \frac{A}{|\mathcal{X}|} = 1$$

and hence

$$A = \frac{|\mathcal{X}|}{|\mathcal{Y}|}. \tag{4.5.4}$$

Accordingly,

$$P_Y(y) = \frac{A}{|\mathcal{X}|} = \frac{|\mathcal{X}|}{|\mathcal{Y}|} \frac{1}{|\mathcal{X}|} = \frac{1}{|\mathcal{Y}|}$$

for any $y \in \mathcal{Y}$; thus the uniform input distribution induces a uniform output distribution and achieves channel capacity as given by (4.5.3).   □

Observation 4.17 Note that if the weakly symmetric channel has a square (i.e., with $|\mathcal{X}| = |\mathcal{Y}|$) transition matrix Q, then Q is a *doubly stochastic* matrix; i.e., both its row sums and its column sums are equal to 1. Note, however, that having a square transition matrix does not necessarily make a weakly symmetric channel symmetric; e.g., see (4.5.1).

Example 4.18 (Capacity of the BSC) Since the BSC with crossover probability (or bit error rate) ε is symmetric, we directly obtain from Lemma 4.16 that its capacity is achieved by a uniform input distribution and is given by

$$C = \log_2(2) - H(1 - \varepsilon, \varepsilon) = 1 - h_{\mathrm{b}}(\varepsilon), \tag{4.5.5}$$

where $h_{\mathrm{b}}(\cdot)$ is the binary entropy function.

Example 4.19 (Capacity of the q-ary symmetric channel) Similarly, the q-ary symmetric channel with symbol error rate ε described in (4.2.11) is symmetric; hence, by Lemma 4.16, its capacity is given by

$$C = \log_2 q - H\left(1 - \varepsilon, \frac{\varepsilon}{q-1}, \ldots, \frac{\varepsilon}{q-1}\right)$$
$$= \log_2 q + \varepsilon \log_2 \frac{\varepsilon}{q-1} + (1 - \varepsilon) \log_2(1 - \varepsilon).$$

Note that when $q = 2$, the channel capacity is equal to that of the BSC, as expected. Furthermore, when $\varepsilon = 0$, the channel reduces to the identity (noiseless) q-ary channel and its capacity is given by $C = \log_2 q$.

We next note that one can further weaken the weak-symmetry property and define a class of "quasi-symmetric" channels for which the uniform input distribution still achieves capacity and yields a simple closed-form formula for capacity.

Definition 4.20 A DMC with finite input alphabet $\mathcal{X}$, finite output alphabet $\mathcal{Y}$ and channel transition matrix $\mathbb{Q} = [p_{x,y}]$ of size $|\mathcal{X}| \times |\mathcal{Y}|$ is said to be *quasi-symmetric*[14] if $\mathbb{Q}$ can be partitioned along its columns into m weakly symmetric sub-matrices $\mathbb{Q}_1, \mathbb{Q}_2, \dots, \mathbb{Q}_m$ for some integer $m \geq 1$, where each $\mathbb{Q}_i$ sub-matrix has size $|\mathcal{X}| \times |\mathcal{Y}_i|$ for $i = 1, 2, \dots, m$ with $\mathcal{Y}_1 \cup \cdots \cup \mathcal{Y}_m = \mathcal{Y}$ and $\mathcal{Y}_i \cap \mathcal{Y}_j = \emptyset$ $\forall i \neq j$, $i, j = 1, 2, \dots, m$.

Hence, quasi-symmetry is our weakest symmetry notion, since a weakly symmetric channel is clearly quasi-symmetric (just set $m = 1$ in the above definition); we thus have symmetry $\implies$ weak-symmetry $\implies$ quasi-symmetry.

Lemma 4.21 *The capacity of a quasi-symmetric channel $\mathbb{Q}$ as defined above is achieved by a uniform input distribution and is given by*

$$C = \sum_{i=1}^{m} a_i C_i, \qquad (4.5.6)$$

where

$$a_i := \sum_{y \in \mathcal{Y}_i} p_{x,y} = \text{sum of any row in } \mathbb{Q}_i, \quad i = 1, \dots, m,$$

and

$$C_i = \log_2 |\mathcal{Y}_i| - H\left(\text{any row in the matrix } \tfrac{1}{a_i} \mathbb{Q}_i\right), \quad i = 1, \dots, m$$

is the capacity of the ith weakly symmetric "sub-channel" whose transition matrix is obtained by multiplying each entry of $\mathbb{Q}_i$ by $\tfrac{1}{a_i}$ (this normalization renders sub-matrix $\mathbb{Q}_i$ into a stochastic matrix and hence a channel transition matrix).

Proof We first observe that for each $i = 1, \dots, m$, a_i is *independent* of the input value x, since sub-matrix i is weakly symmetric (so any row in $\mathbb{Q}_i$ is a permutation of any other row), and hence, a_i is the sum of *any* row in $\mathbb{Q}_i$.

For each $i = 1, \dots, m$, define

$$P_{Y_i|X}(y|x) := \begin{cases} \frac{p_{x,y}}{a_i} & \text{if } y \in \mathcal{Y}_i \text{ and } x \in \mathcal{X}; \\ 0 & \text{otherwise}, \end{cases}$$

where Y_i is a random variable taking values in $\mathcal{Y}_i$. It can be easily verified that $P_{Y_i|X}(y|x)$ is a legitimate conditional distribution. Thus, $[P_{Y_i|X}(y|x)] = \tfrac{1}{a_i}\mathbb{Q}_i$ is the transition matrix of the weakly symmetric "sub-channel" i with input alphabet $\mathcal{X}$ and output alphabet $\mathcal{Y}_i$. Let $I(X; Y_i)$ denote its mutual information. Since each such sub-channel i is weakly symmetric, we know that its capacity C_i is given by

$$C_i = \max_{P_X} I(X; Y_i) = \log_2 |\mathcal{Y}_i| - H\left(\text{any row in the matrix } \tfrac{1}{a_i}\mathbb{Q}_i\right),$$

---

[14]This notion of "quasi-symmetry" is slightly more general than Gallager's notion [135, p. 94], as we herein allow each sub-matrix to be weakly symmetric (instead of symmetric as in [135]).

where the maximum is achieved by a uniform input distribution.

Now, the mutual information between the input and the output of our original quasi-symmetric channel $\mathbb{Q}$ can be written as

$$
\begin{aligned}
I(X;Y) &= \sum_{y\in\mathcal{Y}}\sum_{x\in\mathcal{X}} P_X(x)\, p_{x,y} \log_2 \frac{p_{x,y}}{\sum_{x'\in\mathcal{X}} P_X(x')p_{x',y}} \\[2mm]
&= \sum_{i=1}^{m}\sum_{y\in\mathcal{Y}_i}\sum_{x\in\mathcal{X}} a_i\, P_X(x)\, \frac{p_{x,y}}{a_i} \log_2 \frac{\frac{p_{x,y}}{a_i}}{\sum_{x'\in\mathcal{X}} P_X(x')\frac{p_{x',y}}{a_i}} \\[2mm]
&= \sum_{i=1}^{m} a_i \sum_{y\in\mathcal{Y}_i}\sum_{x\in\mathcal{X}} P_X(x)P_{Y_i|X}(y|x) \log_2 \frac{P_{Y_i|X}(y|x)}{\sum_{x'\in\mathcal{X}} P_X(x')P_{Y_i|X}(y|x')} \\[2mm]
&= \sum_{i=1}^{m} a_i\, I(X;Y_i).
\end{aligned}
$$

Therefore, the capacity of channel $\mathbb{Q}$ is

$$
\begin{aligned}
C &= \max_{P_X} I(X;Y) \\[2mm]
&= \max_{P_X} \sum_{i=1}^{m} a_i\, I(X;Y_i) \\[2mm]
&= \sum_{i=1}^{m} a_i \max_{P_X} I(X;Y_i) \quad \text{(as the same uniform } P_X \text{ maximizes each } I(X;Y_i)) \\[2mm]
&= \sum_{i=1}^{m} a_i\, C_i.
\end{aligned}
$$

$\square$

Example 4.22 (*Capacity of the BEC*) The BEC with erasure probability α as given in (4.2.6) is quasi-symmetric (but neither weakly symmetric nor symmetric). Indeed, its transition matrix $\mathbb{Q}$ can be partitioned along its columns into two symmetric (hence weakly symmetric) sub-matrices

$$
\mathbb{Q}_1 = \begin{bmatrix} 1-\alpha & 0 \\ 0 & 1-\alpha \end{bmatrix}
$$

and

$$
\mathbb{Q}_2 = \begin{bmatrix} \alpha \\ \alpha \end{bmatrix}.
$$

Thus, applying the capacity formula for quasi-symmetric channels of Lemma 4.21 yields that the capacity of the BEC is given by

$$C = a_1 C_1 + a_2 C_2,$$

where $a_1 = 1 - \alpha$, $a_2 = \alpha$,

$$C_1 = \log_2(2) - H\left(\frac{1-\alpha}{1-\alpha}, \frac{0}{1-\alpha}\right) = 1 - H(1,0) = 1 - 0 = 1,$$

and

$$C_2 = \log_2(1) - H\left(\frac{\alpha}{\alpha}\right) = 0 - 0 = 0.$$

Therefore, the BEC capacity is given by

$$C = (1-\alpha)(1) + (\alpha)(0) = 1 - \alpha. \tag{4.5.7}$$

Example 4.23 (*Capacity of the BSEC*) Similarly, the BSEC with crossover probability ε and erasure probability α as described in (4.2.8) is quasi-symmetric; its transition matrix can be partitioned along its columns into two symmetric sub-matrices

$$\mathbb{Q}_1 = \begin{bmatrix} 1 - \varepsilon - \alpha & \varepsilon \\ \varepsilon & 1 - \varepsilon - \alpha \end{bmatrix}$$

and

$$\mathbb{Q}_2 = \begin{bmatrix} \alpha \\ \alpha \end{bmatrix}.$$

Hence, by Lemma 4.21, the channel capacity is given by $C = a_1 C_1 + a_2 C_2$ where $a_1 = 1 - \alpha$, $a_2 = \alpha$,

$$C_1 = \log_2(2) - H\left(\frac{1-\varepsilon-\alpha}{1-\alpha}, \frac{\varepsilon}{1-\alpha}\right) = 1 - h_b\left(\frac{1-\varepsilon-\alpha}{1-\alpha}\right),$$

and

$$C_2 = \log_2(1) - H\left(\frac{\alpha}{\alpha}\right) = 0.$$

We thus obtain that

$$C = (1-\alpha)\left[1 - h_b\left(\frac{1-\varepsilon-\alpha}{1-\alpha}\right)\right] + (\alpha)(0)$$

$$= (1-\alpha)\left[1 - h_b\left(\frac{1-\varepsilon-\alpha}{1-\alpha}\right)\right]. \tag{4.5.8}$$

As already noted, the BSEC is a combination of the BSC with bit error rate ε and the BEC with erasure probability α. Indeed, setting $\alpha = 0$ in (4.5.8) yields that $C = 1 - h_b(1-\varepsilon) = 1 - h_b(\varepsilon)$ which is the BSC capacity. Furthermore, setting $\varepsilon = 0$ results in $C = 1 - \alpha$, the BEC capacity.

4.5.2 Karush–Kuhn–Tucker Conditions for Channel Capacity

When the channel does not satisfy any symmetry property, the following necessary and sufficient Karush–Kuhn–Tucker (KKT) conditions (e.g., cf. Appendix B.8, [135, pp. 87–91] or [46, 56]) for calculating channel capacity can be quite useful.

Definition 4.24 *(Mutual information for a specific input symbol)* The mutual information for a specific input symbol is defined as

$$I(x; Y) := \sum_{y \in \mathcal{Y}} P_{Y|X}(y|x) \log_2 \frac{P_{Y|X}(y|x)}{P_Y(y)}.$$

From the above definition, the mutual information becomes

$$I(X; Y) = \sum_{x \in \mathcal{X}} P_X(x) \sum_{y \in \mathcal{Y}} P_{Y|X}(y|x) \log_2 \frac{P_{Y|X}(y|x)}{P_Y(y)}$$
$$= \sum_{x \in \mathcal{X}} P_X(x) I(x; Y).$$

Lemma 4.25 (KKT conditions for channel capacity) *For a given DMC, an input distribution P_X achieves its channel capacity iff there exists a constant C such that*

$$\begin{cases} I(x; Y) = C \quad \forall x \in \mathcal{X} \text{ with } P_X(x) > 0; \\ I(x; Y) \leq C \quad \forall x \in \mathcal{X} \text{ with } P_X(x) = 0. \end{cases} \qquad (4.5.9)$$

Furthermore, the constant C is the channel capacity (justifying the choice of notation).

Proof The forward (if) part holds directly; hence, we only prove the converse (only-if) part. Without loss of generality, we assume that $P_X(x) < 1$ for all $x \in \mathcal{X}$, since $P_X(x) = 1$ for some x implies that $I(X; Y) = 0$. The problem of calculating the channel capacity is to maximize

$$I(X; Y) = \sum_{x \in \mathcal{X}} \sum_{y \in \mathcal{Y}} P_X(x) P_{Y|X}(y|x) \log_2 \frac{P_{Y|X}(y|x)}{\sum_{x' \in \mathcal{X}} P_X(x') P_{Y|X}(y|x')}, \qquad (4.5.10)$$

subject to the condition

$$\sum_{x \in \mathcal{X}} P_X(x) = 1 \qquad (4.5.11)$$

for a given channel distribution $P_{Y|X}$. By using the Lagrange multipliers method (e.g., see Appendix B.8 or [46]), maximizing (4.5.10) subject to (4.5.11) is equivalent to maximize:

$$f(P_X) := \sum_{\substack{x \in \mathcal{X} \\ y \in \mathcal{Y}}} P_X(x) P_{Y|X}(y|x) \log_2 \frac{P_{Y|X}(y|x)}{\sum_{x' \in \mathcal{X}} P_X(x') P_{Y|X}(y|x')} + \lambda \left(\sum_{x \in \mathcal{X}} P_X(x) - 1 \right).$$

We then take the derivative of the above quantity with respect to $P_X(x'')$, and obtain that[15]

$$\frac{\partial f(P_X)}{\partial P_X(x'')} = I(x''; Y) - \log_2(e) + \lambda.$$

By Property 2 of Lemma 2.46, $I(X; Y) = I(P_X, P_{Y|X})$ is a concave function in P_X (for a fixed $P_{Y|X}$). Therefore, the maximum of $I(P_X, P_{Y|X})$ occurs for a zero derivative when $P_X(x)$ does not lie on the boundary, namely, $1 > P_X(x) > 0$. For those $P_X(x)$ lying on the boundary, i.e., $P_X(x) = 0$, the maximum occurs iff a displacement from the boundary to the interior decreases the quantity, which implies a nonpositive derivative, namely,

$$I(x; Y) \le -\lambda + \log_2(e), \quad \text{for those } x \text{ with } P_X(x) = 0.$$

To summarize, if an input distribution P_X achieves the channel capacity, then

$$\begin{cases} I(x''; Y) = -\lambda + \log_2(e), \text{ for } P_X(x'') > 0; \\ I(x''; Y) \le -\lambda + \log_2(e), \text{ for } P_X(x'') = 0 \end{cases}$$

---

[15]The details for taking the derivative are as follows:

$$\frac{\partial}{\partial P_X(x'')} \left\{ \sum_{x \in \mathcal{X}} \sum_{y \in \mathcal{Y}} P_X(x) P_{Y|X}(y|x) \log_2 P_{Y|X}(y|x) \right.$$

$$\left. - \sum_{x \in \mathcal{X}} \sum_{y \in \mathcal{Y}} P_X(x) P_{Y|X}(y|x) \log_2 \left[\sum_{x' \in \mathcal{X}} P_X(x') P_{Y|X}(y|x') \right] + \lambda \left(\sum_{x \in \mathcal{X}} P_X(x) - 1 \right) \right\}$$

$$= \sum_{y \in \mathcal{Y}} P_{Y|X}(y|x'') \log_2 P_{Y|X}(y|x'') - \left(\sum_{y \in \mathcal{Y}} P_{Y|X}(y|x'') \log_2 \left[\sum_{x' \in \mathcal{X}} P_X(x') P_{Y|X}(y|x') \right] \right.$$

$$+ \log_2(e) \sum_{x \in \mathcal{X}} \sum_{y \in \mathcal{Y}} P_X(x) P_{Y|X}(y|x) \frac{P_{Y|X}(y|x'')}{\sum_{x' \in \mathcal{X}} P_X(x') P_{Y|X}(y|x')} \right) + \lambda$$

$$= I(x''; Y) - \log_2(e) \sum_{y \in \mathcal{Y}} \left[\sum_{x \in \mathcal{X}} P_X(x) P_{Y|X}(y|x) \right] \frac{P_{Y|X}(y|x'')}{\sum_{x' \in \mathcal{X}} P_X(x') P_{Y|X}(y|x')} + \lambda$$

$$= I(x''; Y) - \log_2(e) \sum_{y \in \mathcal{Y}} P_{Y|X}(y|x'') + \lambda$$

$$= I(x''; Y) - \log_2(e) + \lambda.$$

for some λ. With the above result, setting $C = -\lambda + \log_2(e)$ yields (4.5.9). Finally, multiplying both sides of each equation in (4.5.9) by $P_X(x)$ and summing over x yields that $\max_{P_X} I(X; Y)$ on the left and the constant C on the right, thus proving that the constant C is indeed the channel's capacity. $\square$

Example 4.26 (Quasi-symmetric channels) For a quasi-symmetric channel, one can directly verify that the uniform input distribution satisfies the KKT conditions of Lemma 4.25 and yields that the channel capacity is given by (4.5.6); this is left as an exercise. As we already saw, the BSC, the q-ary symmetric channel, the BEC and the BSEC are all quasi-symmetric.

Example 4.27 Consider a DMC with a ternary input alphabet $\mathcal{X} = \{0, 1, 2\}$, binary output alphabet $\mathcal{Y} = \{0, 1\}$ and the following transition matrix:

$$\mathbb{Q} = \begin{bmatrix} 1 & 0 \\ \frac{1}{2} & \frac{1}{2} \\ 0 & 1 \end{bmatrix}.$$

This channel is not quasi-symmetric. However, one may guess that the capacity of this channel is achieved by the input distribution $(P_X(0), P_X(1), P_X(2)) = (\frac{1}{2}, 0, \frac{1}{2})$ since the input $x = 1$ has an equal conditional probability of being received as 0 or 1 at the output. Under this input distribution, we obtain that $I(x = 0; Y) = I(x = 2; Y) = 1$ and that $I(x = 1; Y) = 0$. Thus, the KKT conditions of (4.5.9) are satisfied; hence confirming that the above input distribution achieves channel capacity and that channel capacity is equal to 1 bit.

Observation 4.28 (Capacity achieved by a uniform input distribution) We close this section by noting that there is a class of DMCs that is larger than that of quasi-symmetric channels for which the uniform input distribution achieves capacity. It concerns the class of so-called "T-symmetric" channels [319, Sect. 5, Definition 1] for which

$$T(x) := I(x; Y) - \log_2 |\mathcal{X}| = \sum_{y \in \mathcal{Y}} P_{Y|X}(y|x) \log_2 \frac{P_{Y|X}(y|x)}{\sum_{x' \in \mathcal{X}} P_{Y|X}(y|x')}$$

is a constant function of x (i.e., independent of x), where $I(x; Y)$ is the mutual information for input x under a uniform input distribution. Indeed, the T-symmetry condition is equivalent to the property of having the uniform input distribution achieve capacity. This directly follows from the KKT conditions of Lemma 4.25. An example of a T-symmetric channel that is not quasi-symmetric is the binary-input ternary-output channel with the following transition matrix:

$$\mathbb{Q} = \begin{bmatrix} \frac{1}{3} & \frac{1}{3} & \frac{1}{3} \\ \frac{1}{6} & \frac{1}{6} & \frac{2}{3} \end{bmatrix}.$$

Hence, its capacity is achieved by the uniform input distribution. See [319, Fig. 2] for (infinitely many) other examples of T-symmetric channels. However, unlike quasi-symmetric channels, T-symmetric channels do not admit in general a simple closed-form expression for their capacity [such as the one given in (4.5.6)].

4.6 Lossless Joint Source-Channel Coding and Shannon's Separation Principle

We next establish Shannon's *lossless joint source-channel coding theorem*[16] which provides explicit (and directly verifiable) conditions for any communication system in terms of its source and channel information-theoretic quantities under which the source can be reliably transmitted (i.e., with asymptotically vanishing error probability). More specifically, this theorem consists of two parts: (i) a forward part which reveals that if the minimal achievable compression (or source coding) rate of a source is strictly smaller than the capacity of a channel, then the source can be reliably sent over the channel via rate-one source-channel block codes; (ii) a converse part which states that if the source's minimal achievable compression rate is strictly larger than the channel capacity, then the source cannot be reliably sent over the channel via rate-one source-channel block codes. The theorem (under minor modifications) has also a more general version in terms of reliable transmissibility of the source over the channel via source-channel block codes of arbitrary rate (not necessarily equal to one).

This key theorem is usually referred to as Shannon's *source-channel separation theorem or principle*; this renaming is explained in the following. First, the theorem's necessary and sufficient conditions for reliable transmissibility are a function of entirely "separable" or "disentangled" information quantities, the source's minimal compression rate and the channel's capacity with no quantities that depends on both the source and the channel; this can be seen as a "functional separation" property or condition. Second, the proof of the forward part, which (as we will see) consists of properly combining Shannon's source coding (Theorems 3.6 or 3.15) and channel coding (Theorem 4.11) theorems, shows that reliable transmissibility can be realized by separating (or decomposing) the source-channel coding function into two distinct and independently conceived source and channel coding operations applied in tandem, where the source code depends only on the source statistics and, similarly, the channel code is a sole function of the channel statistics. In other words, we have "operational separation" in that a separate (tandem or two-stage) source and channel coding scheme as depicted in Fig. 4.8 is as good (in terms of asymptotic reliable transmissibility) as the more general joint source-channel coding scheme shown in Fig. 4.9 in which the coding operation can include a combined (one-stage) code designed with respect to both the source and the channel or jointly coordinated source and channel codes. Now, gathering the above two facts with the theorem's converse part—with the exception of the unresolved case where the source's minimal achievable compression rate is

---

[16]This theorem is sometimes referred to as the *lossless information transmission theorem*.

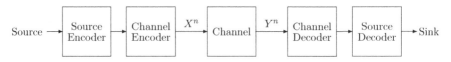

Fig. 4.8 A separate (tandem) source-channel coding scheme

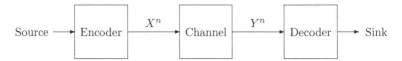

Fig. 4.9 A joint source-channel coding scheme

exactly equal to the channel capacity—implies that either reliable transmissibility of the source over the channel is achievable via separate source and channel coding (under the transmissibility condition) or it is not at all achievable, hence justifying calling the theorem by the separation principle.

We will prove the theorem by assuming that the source is stationary ergodic[17] in the forward part and just stationary in the converse part and that the channel is a DMC; note that the theorem can be extended to more general sources and channels with memory (see [75, 96, 394]).

Definition 4.29 *(Source-channel block code)* Given a discrete source $\{V_i\}_{i=1}^{\infty}$ with finite alphabet $\mathcal{V}$ and a discrete channel $\{P_{Y^n|X^n}\}_{n=1}^{\infty}$ with finite input and output alphabets $\mathcal{X}$ and $\mathcal{Y}$, respectively, an m-to-n source-channel block code $\mathcal{C}_{m,n}$ with rate $\frac{m}{n}$ source symbol/channel symbol is a pair of mappings $(f^{(sc)}, g^{(sc)})$, where[18]

$$f^{(sc)}: \mathcal{V}^m \to \mathcal{X}^n$$

and

$$g^{(sc)}: \mathcal{Y}^n \to \mathcal{V}^m.$$

The code's operation is illustrated in Fig. 4.10. The source m-tuple V^m is encoded via the source-channel encoding function $f^{(sc)}$, yielding the codeword $X^n = f^{(sc)}(V^m)$ as the channel input. The channel output Y^n, which is dependent on V^m only via X^n (i.e., we have the Markov chain $V^m \to X^n \to Y^n$), is decoded via $g^{(sc)}$ to obtain the source tuple estimate $\hat{V}^m = g^{(sc)}(Y^n)$.

An error is made by the decoder if $V^m \neq \hat{V}^m$, and the code's error probability is given by

---

[17]The minimal achievable compression rate of such sources is given by the entropy rate, see Theorem 3.15.

[18]Note that $n = n_m$; that is, the channel blocklength n is in general a function of the source blocklength m. Similarly, $f^{(sc)} = f_m^{(sc)}$ and $g^{(sc)} = g_m^{(sc)}$; i.e., the encoding and decoding functions are implicitly dependent on m.

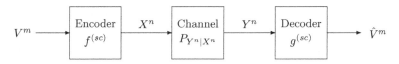

Fig. 4.10 An m-to-n block source-channel coding system

$$P_e(\mathcal{C}_{m,n}) := \Pr[V^m \neq \hat{V}^m]$$
$$= \sum_{v^m \in \mathcal{V}^m} \sum_{y^n \in \mathcal{Y}^n : g^{(sc)}(y^n) \neq v^m} P_{V^m}(v^m) P_{Y^n|X^n}(y^n | f^{(sc)}(v^m))$$

where P_{V^m} and $P_{Y^n|X^n}$ are the source and channel distributions, respectively.

We next prove Shannon's lossless joint source-channel coding theorem when source m-tuples are transmitted via m-tuple codewords or m uses of the channel (i.e., when $n = m$ or for rate-one source-channel block codes). The source is assumed to have memory (as indicated below), while the channel is memoryless.

Theorem 4.30 (Lossless joint source-channel coding theorem for rate-one block codes) *Consider a discrete source $\{V_i\}_{i=1}^{\infty}$ with finite alphabet $\mathcal{V}$ and entropy rate[19] $H(\mathcal{V})$ and a DMC with input alphabet $\mathcal{X}$, output alphabet $\mathcal{Y}$ and capacity C, where both $H(\mathcal{V})$ and C are measured in the same units (i.e., they both use the same base of the logarithm). Then, the following hold:*

- *Forward part (achievability): For any $0 < \epsilon < 1$ and given that the source is stationary ergodic, if*

$$H(\mathcal{V}) < C,$$

then there exists a sequence of rate-one source-channel codes $\{\mathcal{C}_{m,m}\}_{m=1}^{\infty}$ such that

$$P_e(\mathcal{C}_{m,m}) < \epsilon \quad \text{for sufficiently large } m,$$

where $P_e(\mathcal{C}_{m,m})$ is the error probability of the source-channel code $\mathcal{C}_{m,m}$.
- *Converse part: For any $0 < \epsilon < 1$ and given that the source is stationary, if*

$$H(\mathcal{V}) > C,$$

then any sequence of rate-one source-channel codes $\{\mathcal{C}_{m,m}\}_{m=1}^{\infty}$ satisfies

$$P_e(\mathcal{C}_{m,m}) > (1 - \epsilon)\mu \quad \text{for sufficiently large } m, \quad (4.6.1)$$

where $\mu = H_D(\mathcal{V}) - C_D$ with $D = |\mathcal{V}|$, and $H_D(\mathcal{V})$ and C_D are entropy rate and channel capacity measured in D-ary digits, i.e., the codes' error probability is bounded away from zero and it is not possible to transmit the source over

---

[19]We assume the source entropy rate exists as specified below.

the channel via rate-one source-channel block codes with arbitrarily low error probability.[20]

Proof of the forward part: Without loss of generality, we assume throughout this proof that both the source entropy rate $H(\mathcal{V})$ and the channel capacity C are measured in nats (i.e., they are both expressed using the natural logarithm).

We will show the existence of the desired rate-one source-channel codes $\mathcal{C}_{m,m}$ via a separate (tandem or two-stage) source and channel coding scheme as the one depicted in Fig. 4.8.

Let $\gamma := C - H(\mathcal{V}) > 0$. Now, given any $0 < \epsilon < 1$, by the lossless source coding theorem for stationary ergodic sources (Theorem 3.15), there exists a sequence of source codes of blocklength m and size M_m with encoder

$$f_s : \mathcal{V}^m \to \{1, 2, \ldots, M_m\}$$

and decoder

$$g_s : \{1, 2, \ldots, M_m\} \to \mathcal{V}^m$$

such that

$$\frac{1}{m} \log M_m < H(\mathcal{V}) + \gamma/2 \qquad (4.6.2)$$

and

$$\Pr\left[g_s(f_s(V^m)) \neq V^m \right] < \epsilon/2$$

for m sufficiently large.[21]

Furthermore, by the channel coding theorem under the maximal probability of error criterion (see Observation 4.6 and Theorem 4.11), there exists a sequence of channel codes of blocklength m and size $\bar{M}_m$ with encoder

$$f_c : \{1, 2, \ldots, \bar{M}_m\} \to \mathcal{X}^m$$

---

[20]Note that (4.6.1) actually implies that $\liminf_{m\to\infty} P_e(\mathcal{C}_{m,m}) \geq \lim_{\epsilon\downarrow 0}(1-\epsilon)\mu = \mu$, where the error probability lower bound has nothing to do with ϵ. Here, we state the converse of Theorem 4.30 in a form in parallel to the converse statements in Theorems 3.6, 3.15 and 4.11.

[21]Theorem 3.15 indicates that for any $0 < \varepsilon' := \min\{\varepsilon/2, \gamma/(2\log(2))\} < 1$, there exists δ with $0 < \delta < \varepsilon'$ and a sequence of *binary* block codes $\{\mathcal{C}_m = (m, M_m)\}_{m=1}^\infty$ with

$$\limsup_{m\to\infty} \frac{1}{m} \log_2 M_m < H_2(\mathcal{V}) + \delta, \qquad (4.6.3)$$

and probability of decoding error satisfying $P_e(\mathcal{C}_m) < \varepsilon'$ $(\leq \varepsilon/2)$ for sufficiently large m, where $H_2(\mathcal{V})$ is the entropy rate measured in bits. Here, (4.6.3) implies that $\frac{1}{m} \log_2 M_m < H_2(\mathcal{V}) + \delta$ for sufficiently large m. Hence,

$$\frac{1}{m} \log M_m < H(\mathcal{V}) + \delta \log(2) < H(\mathcal{V}) + \varepsilon' \log(2) \leq H(\mathcal{V}) + \gamma/2$$

for sufficiently large m.

and decoder

$$g_c: \mathcal{Y}^m \rightarrow \{1, 2, \ldots, \bar{M}_m\}$$

such that[22]

$$\frac{1}{m} \log \bar{M}_m > C - \gamma/2 \left(= H(V) + \gamma/2 > \frac{1}{m} \log M_m \right) \qquad (4.6.5)$$

and

$$\lambda := \max_{w \in \{1, \ldots, \bar{M}_m\}} \Pr \left[g_c(Y^m) \neq w | X^m = f_c(w) \right] < \epsilon/2$$

for m sufficiently large.

Now we form our source-channel code by concatenating in tandem the above source and channel codes. Specifically, the m-to-m source-channel code $\mathcal{C}_{m,m}$ has the following encoder–decoder pair $(f^{(sc)}, g^{(sc)})$:

$$f^{(sc)}: \mathcal{V}^m \rightarrow \mathcal{X}^m \quad \text{with} \quad f^{(sc)}(v^m) = f_c(f_s(v^m)) \quad \forall v^m \in \mathcal{V}^m$$

and

$$g^{(sc)}: \mathcal{Y}^m \rightarrow \mathcal{V}^m$$

with

$$g^{(sc)}(y^m) = \begin{cases} g_s(g_c(y^m)), & \text{if } g_c(y^m) \in \{1, 2, \ldots, M_m\} \\ \text{arbitrary}, & \text{otherwise} \end{cases} \quad \forall y^m \in \mathcal{Y}^m.$$

The above construction is possible since $\{1, 2, \ldots, M_m\}$ is a subset of $\{1, 2, \ldots, \bar{M}_m\}$. The source-channel code's probability of error can be analyzed by considering the

---

[22]Theorem 4.11 and its proof of forward part indicate that for any $0 < \varepsilon' := \min\{\varepsilon/4, \gamma/(16 \log(2))\} < 1$, there exist $0 < \gamma' < \min\{4\varepsilon', C_2\} = \min\{\varepsilon, \gamma/(4 \log(2)), C_2\}$ and a sequence of data transmission block codes $\{\mathcal{C}_m = (m, \bar{M}_m')\}_{m=1}^\infty$ satisfying

$$C_2 - \gamma' < \frac{1}{m} \log_2 \bar{M}_m' \leq C_2 - \frac{\gamma'}{2} \qquad (4.6.4)$$

and

$$P_e(\mathcal{C}_m) < \varepsilon' \quad \text{for sufficiently large } m,$$

provided that $C_2 > 0$, where C_2 is the channel capacity measured in bits.

Observation 4.6 indicates that by throwing away from $\mathcal{C}_m$ half of its codewords with largest conditional probability of error, a new code $\mathcal{C}_m = (m, \bar{M}_m) = (m, \bar{M}_m'/2)$ is obtained, which satisfies $\lambda(\mathcal{C}_m) \leq 2P_e(\mathcal{C}_m) < 2\varepsilon' \leq \varepsilon/2$.

Equation (4.6.4) then implies that for $m > 1/\gamma'$ sufficiently large,

$$\frac{1}{m} \log \bar{M}_m = \frac{1}{m} \log \bar{M}_m' - \frac{1}{m} \log(2) > C - \gamma' \log(2) - \frac{1}{m} \log(2) > C - 2\gamma' \log(2) > C - \gamma/2.$$

cases of whether or not a channel decoding error occurs as follows:

$$
\begin{aligned}
P_e(\mathcal{C}_{m,m}) &= \Pr[g^{(sc)}(Y^m) \neq V^m] \\
&= \Pr[g^{(sc)}(Y^m) \neq V^m, g_c(Y^m) = f_s(V^m)] \\
&\qquad + \Pr[g^{(sc)}(Y^m) \neq V^m, g_c(Y^m) \neq f_s(V^m)] \\
&= \Pr[g_s(g_c(Y^m)) \neq V^m, g_c(Y^m) = f_s(V^m)] \\
&\qquad + \Pr[g^{(sc)}(Y^m) \neq V^m, g_c(Y^m) \neq f_s(V^m)] \\
&\leq \Pr[g_s(f_s(V^m)) \neq V^m] + \Pr[g_c(Y^m) \neq f_s(V^m)] \\
&= \Pr[g_s(f_s(V^m)) \neq V^m] \\
&\qquad + \sum_{w \in \{1,2,\dots,M_m\}} \Pr[f_s(V^m) = w] \Pr[g_c(Y^m) \neq w | f_s(V^m) = w] \\
&= \Pr[g_s(f_s(V^m)) \neq V^m] \\
&\qquad + \sum_{w \in \{1,2,\dots,M_m\}} \Pr[X^m = f_c(w)] \Pr[g_c(Y^m) \neq w | X^m = f_c(w)] \\
&\leq \Pr[g_s(f_s(V^m)) \neq V^m] + \lambda \\
&< \epsilon/2 + \epsilon/2 = \epsilon
\end{aligned}
$$

for m sufficiently large. Thus, the source can be reliably sent over the channel via rate-one block source-channel codes as long as $H(V) < C$.   □
Proof of the converse part: For simplicity, we assume in this proof that $H(V)$ and C are measured in bits.

For any m-to-m source-channel code $\mathcal{C}_{m,m}$, we can write

$$
H(V) \leq \frac{1}{m} H(V^m) \tag{4.6.6}
$$

$$
= \frac{1}{m} H(V^m | \hat{V}^m) + \frac{1}{m} I(V^m; \hat{V}^m)
$$

$$
\leq \frac{1}{m} \left[P_e(\mathcal{C}_{m,m}) \log_2(|\mathcal{V}|^m) + 1 \right] + \frac{1}{m} I(V^m; \hat{V}^m) \tag{4.6.7}
$$

$$
\leq P_e(\mathcal{C}_{m,m}) \log_2 |\mathcal{V}| + \frac{1}{m} + \frac{1}{m} I(X^m; Y^m) \tag{4.6.8}
$$

$$
\leq P_e(\mathcal{C}_{m,m}) \log_2 |\mathcal{V}| + \frac{1}{m} + C, \tag{4.6.9}
$$

where

- Equation (4.6.6) is due to the fact that $(1/m)H(V^m)$ is nonincreasing in m and converges to $H(V)$ as $m \to \infty$ since the source is stationary (see Observation 3.12),
- Equation (4.6.7) follows from Fano's inequality,

$$
H(V^m | \hat{V}^m) \leq P_e(\mathcal{C}_{m,m}) \log_2(|\mathcal{V}|^m) + h_b(P_e(\mathcal{C}_{m,m})) \leq P_e(\mathcal{C}_{m,m}) \log_2(|\mathcal{V}|^m) + 1,
$$

- Equation (4.6.8) is due to the data processing inequality since $V^m \to X^m \to Y^m \to \hat{V}^m$ form a Markov chain, and
- Equation (4.6.9) holds by (4.3.8) since the channel is a DMC.

Note that in the above derivation, the information measures are all measured in bits. This implies that for $m \geq \log_D(2)/(\varepsilon\mu)$,

$$P_e(\mathcal{C}_{m,m}) \geq \frac{H(V) - C}{\log_2(|V|)} - \frac{1}{m\log_2(|V|)} = H_D(V) - C_D - \frac{\log_D(2)}{m} \geq (1 - \varepsilon)\mu.$$

□

Observation 4.31 We make the following remarks regarding the above joint source-channel coding theorem:

- In general, it is not known whether the source can be (asymptotically) reliably transmitted over the DMC when

$$H(V) = C$$

even if the source is a DMS. This is because separate source and channel codings are used to prove the forward part of the theorem and the facts that the source coding rate approaches the source entropy rate from above [cf. (4.6.2)] while the channel coding rate approaches channel capacity from below [cf. (4.6.5)].
- The above theorem directly holds for DMSs since any DMS is stationary and ergodic.
- We can expand the forward part of the theorem above by replacing the requirement that the source be stationary ergodic with the more general condition that the source be *information stable*.[23] Note that time-invariant irreducible Markov sources (that are not necessarily stationary) are information stable.

The above lossless joint source-channel coding theorem can be readily generalized for m-to-n source-channel codes—i.e., codes with rate not necessarily equal to one— as follows (its proof, which is similar to the previous theorem, is left as an exercise).

Theorem 4.32 (Lossless joint source-channel coding theorem for general rate block codes) *Consider a discrete source $\{V_i\}_{i=1}^{\infty}$ with finite alphabet V and entropy rate $H(V)$ and a DMC with input alphabet $\mathcal{X}$, output alphabet $\mathcal{Y}$ and capacity C, where both $H(V)$ and C are measured in the same units. Then, the following holds:*

- *Forward part (achievability): For any $0 < \epsilon < 1$ and given that the source is stationary ergodic, there exists a sequence of m-to-n_m source-channel codes $\{\mathcal{C}_{m,n_m}\}_{m=1}^{\infty}$ such that*

$$P_e(\mathcal{C}_{m,n_m}) < \epsilon \quad \text{for sufficiently large } m$$

---

[23] See [75, 96, 303, 394] for a definition of information stable sources, whose property is slightly more general than the Generalized AEP property given in Theorem 3.14.

if

$$\limsup_{m\to\infty} \frac{m}{n_m} < \frac{C}{H(\mathcal{V})}.$$

- *Converse part: For any $0 < \epsilon < 1$ and given that the source is stationary, any sequence of m-to-n_m source-channel codes $\{\mathscr{C}_{m,n_m}\}_{m=1}^{\infty}$ with*

$$\liminf_{m\to\infty} \frac{m}{n_m} > \frac{C}{H(\mathcal{V})},$$

satisfies

$$P_e(\mathscr{C}_{m,n_m}) > (1-\epsilon)\mu \qquad \text{for sufficiently large } m,$$

for some positive constant μ that depends on $\liminf_{m\to\infty}(m/n_m)$, $H(\mathcal{V})$ and C, i.e., the codes' error probability is bounded away from zero and it is not possible to transmit the source over the channel via m-to-n_m source-channel block codes with arbitrarily low error probability.

Discussion: separate versus joint source-channel coding

Shannon's separation principle has provided the linchpin for most modern communication systems where source coding and channel coding schemes are separately constructed (with the source (respectively, channel) code designed by only taking into account the source (respectively, channel) characteristics) and applied in tandem without the risk of sacrificing optimality in terms of reliable transmissibility under unlimited coding delay and complexity. This result is the raison d'être for separately studying the practices of source coding or data compression (e.g., see [42, 142, 158, 290, 326, 330]) and channel coding (e.g., see [208, 248, 254, 321, 407]). Furthermore, by disentangling the source and channel coding operations, separate coding offers appealing properties such as system modularity and flexibility. For example, if one needs to send different sources over the same channel, using the separate coding approach, one only needs to modify the source code while keeping the channel code unchanged (analogously, if a single source is to be communicated over different channels, one only has to adapt the channel code).

However, in practical implementations, there is a price to pay in delay and complexity for extremely long coding blocklengths (particularly when delay and complexity constraints are quite stringent such as in wireless communications systems). To begin, note that joint source-channel coding might be expected to offer improvements for the combination of a source with substantial redundancy and a channel with significant noise, since, for such a system, separate coding would involve source coding to remove redundancy followed by channel coding to insert redundancy. It is a natural conjecture that this is not the most efficient approach even if the blocklength is allowed to grow without bound. Indeed, Shannon [340] made this point as follows:

> ... However, any redundancy in the source will usually help if it is utilized at the receiving point. In particular, if the source already has a certain redundancy and no attempt is made to eliminate it in matching to the channel, this redundancy will help

combat noise. For example, in a noiseless telegraph channel one could save about 50% in time by proper encoding of the messages. This is not done and most of the redundancy of English remains in the channel symbols. This has the advantage, however, of allowing considerable noise in the channel. A sizable fraction of the letters can be received incorrectly and still reconstructed by the context. In fact this is probably not a bad approximation to the ideal in many cases . . .

We make the following observations regarding the merits of joint versus separate source-channel coding:

- Under finite coding blocklengths and/or complexity, many studies have demonstrated that joint source-channel coding can provide better performance than separate coding (e.g., see [13, 14, 37, 100, 127, 200, 247, 410, 427] and the references therein).
- Even in the infinite blocklength regime where separate coding is optimal in terms of reliable transmissibility, it can be shown that for a large class of systems, joint source-channel coding can achieve an *error exponent*[24] that is as large as *double* the error exponent resulting from separate coding [422–424]. This indicates that one can realize via joint source-channel coding the same performance as separate coding, while reducing the coding delay by *half* (this result translates into notable power savings of more than 2 dB when sending binary sources over channels with Gaussian noise, fading an output quantization [422]). These findings provide an information-theoretic rationale for adopting joint source-channel coding over separate coding.
- Finally, it is important to point out that, with the exception of certain network topologies [173, 383, 425] where separation is optimal, the separation theorem does *not* in general hold for *multiuser (multiterminal)* systems (cf., [81, 83, 106, 174]),
  and thus, in such systems, it is more beneficial to perform joint source-channel coding.

The study of joint source-channel coding dates back to as early as the 1960s. Over the years, many works have introduced joint source-channel coding techniques and illustrated (analytically or numerically) their benefits (in terms of both performance improvement and increased robustness to variations in channel noise) over separate coding for given source and channel conditions and fixed complexity and/or delay constraints. In joint source-channel coding systems, the designs of the source and channel codes are either well coordinated or combined into a single step. Examples of

---

[24]The *error exponent or reliability function* of a coding system is the largest rate of exponential decay of its decoding error probability as the coding blocklength grows without bound [51, 87, 95, 107, 114, 135, 177, 178, 205, 347, 348]. Roughly speaking, the error exponent is a number E with the property that the decoding error probability of a good code is approximately e^{-nE} for large coding blocklength n. In addition to revealing the fundamental trade-off between the error probability of optimal codes and their blocklength for a given coding rate and providing insight on the behavior of optimal codes, such a function provides a powerful tool for proving the achievability part of coding theorems (e.g., [135]), for comparing the performance of competing coding schemes (e.g., weighing joint against separate coding [422]) and for communications system design [194].

•

(both constructive and theoretical) previous lossless and lossy joint source-channel coding investigations for single-user[25] systems include the following:

(a) *Fundamental limits*: joint source-channel coding theorems and the separation principle [21, 34, 75, 96, 103, 135, 161, 164, 172, 187, 231, 271, 273, 351, 365, 373, 386, 394, 399], and joint source-channel coding exponents [69, 70, 84, 85, 135, 220, 422–424].

(b) *Channel-optimized source codes* (i.e., source codes that are robust against channel noise) [15, 32, 33, 39, 102, 115–117, 121, 126, 131, 143, 155, 167, 218, 238–240, 247, 272, 293, 295, 296, 354–356, 369, 375, 392, 419].

(c) *Source-optimized channel codes* (i.e., channel codes that exploit the source's redundancy) [14, 19, 62, 91, 93, 100, 118, 122, 127, 139, 169, 198, 234, 263, 331, 336, 410, 427, 428], *uncoded source-channel matching with joint decoding* [13, 92, 140, 230, 285, 294, 334, 335, 366, 406] and *source-matched channel signaling* [109, 229, 276, 368].

(d) *Jointly coordinated source and channel codes* [61, 101, 124, 132, 149, 150, 152, 166, 168, 171, 183, 184, 189, 190, 204, 217, 241, 268, 275, 282, 283, 286, 288, 332, 381, 402, 416, 417].

(e) *Hybrid digital-analog source-channel coding and analog mapping* [8, 57, 64, 71, 77, 79, 112, 130, 138, 147, 185, 193, 219, 221, 232, 244, 245, 274, 314, 320, 324, 335, 341, 357, 358, 367, 382, 391, 401, 405, 409, 429].

The above references while numerous are not exhaustive as the field of joint source-channel coding has been quite active, particularly over the last decades.

Problems

1. Prove the Shannon–McMillan–Breiman theorem for pairs (Theorem 4.9).
2. The proof of Shannon's channel coding theorem is based on the random coding technique. What is the codeword-selecting distribution of the random codebook? What is the decoding rule in the proof?
3. Show that processing the output of a DMC (via a given function) does not strictly increase its capacity.
4. Consider the system shown in the block diagram below. Can the channel capacity between channel input X and channel output Z be strictly larger than the channel capacity between channel input X and channel output Y? Which lemma or theorem is your answer based on?

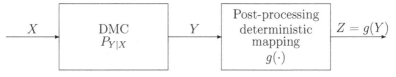

---

[25]We underscore that, even though not listed here, the literature on joint source-channel coding for multiuser systems is also quite extensive and ongoing.

5. Consider a DMC with input X and output Y. Assume that the input alphabet is $\mathcal{X} = \{1, 2\}$, the output alphabet is $\mathcal{Y} = \{0, 1, 2, 3\}$, and the transition probability is given by

$$P_{Y|X}(y|x) = \begin{cases} 1 - 2\varepsilon, & \text{if } x = y; \\ \varepsilon, & \text{if } |x - y| = 1; \\ 0, & \text{otherwise,} \end{cases}$$

where $0 < \epsilon < 1/2$.

(a) Determine the channel probability transition matrix $\mathbb{Q} := [P_{Y|X}(y, x)]$.
(b) Compute the capacity of this channel. What is the maximizing input distribution that achieves capacity?

6. *Binary-input additive discrete-noise channel*: Find the capacity of a DMC whose output Y is given by $Y = X + Z$, where X and Z are the channel input and noise, respectively. Assume that the noise is independent of the input and that it has alphabet $\mathcal{Z} = \{-b, 0, b\}$ such that $P_Z(-b) = P_Z(0) = P_Z(b) = 1/3$, where $b > 0$ is a fixed real number. Also assume that the input alphabet is given by $\mathcal{X} = \{-a, a\}$ for some given real number $a > 0$. Discuss the dependence of the channel capacity on the values of a and b.

7. *The Z-channel*: Find the capacity of DMC called the Z-channel and described by the following transition diagram (where $0 \le \beta \le 1/2$).

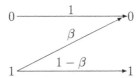

8. *Functional representation of the Z-channel*: Consider the DMC of Problem 4.7 above.

(a) Give a functional representation of the channel by explicitly expressing, at any time instant i, the channel output Y_i in terms of the input X_i and a binary noise random variable Z_i, which is independent of the input and is generated from a memoryless process $\{Z_i\}$ with $\Pr[Z_i = 0] = \beta$.
(b) Show that the channel's input–output mutual information satisfies

$$I(X; Y) \ge H(Y) - H(Z).$$

(c) Show that the capacity of the Z-channel is no smaller than that of a BSC with crossover probability $1 - \beta$ (i.e., a binary modulo-2 additive noise channel with $\{Z_i\}$ as its noise process):

$$C \ge 1 - h_b(\beta)$$

where $h_b(\cdot)$ is the binary entropy function.

9. A DMC has identical input and output alphabets given by $\{0, 1, 2, 3, 4\}$. Let X be the channel input, and Y be the channel output. Suppose that

$$P_{Y|X}(i|i) = \frac{1}{2} \quad \forall\, i \in \{0, 1, 2, 3, 4\}.$$

   (a) Find the channel transition matrix that maximizes $H(Y|X)$.
   (b) Using the channel transition matrix obtained in (a), evaluate the channel capacity.

10. *Binary channel*: Consider a binary memoryless channel with the following probability transition matrix:

$$\mathbb{Q} = \begin{bmatrix} 1 - \alpha & \alpha \\ \beta & 1 - \beta \end{bmatrix},$$

   where $\alpha > 0$, $\beta > 0$ and $\alpha + \beta < 1$.

   (a) Determine the capacity C of this channel in terms of α and β.
   (b) What does the expression of C reduce to if $\alpha = \beta$?

11. Find the capacity of the asymmetric binary channel with errors and erasures described in (4.2.10). Verify that the channel capacity reduces to that of the BSEC when setting $\varepsilon' = \varepsilon$ and $\alpha' = \alpha$.

12. Find the capacity of the binary-input quaternary-output DMC given in (4.5.2). For what values of ε is capacity maximized, and for what values of ε is capacity minimized?

13. *Nonbinary erasure channel*: Find the capacity of the q-ary erasure channel described in (4.2.12) and compare the result with the capacity of the BEC.

14. *Product of two channels:* Consider two DMCs

$$(\mathcal{X}_1, P_{Y_1|X_1}, \mathcal{Y}_1) \quad \text{and} \quad (\mathcal{X}_2, P_{Y_2|X_2}, \mathcal{Y}_2)$$

   with capacity C_1 and C_2, respectively. A new channel $(\mathcal{X}_1 \times \mathcal{X}_2, P_{Y_1|X_1} \times P_{Y_2|X_2}, \mathcal{Y}_1 \times \mathcal{Y}_2)$ is formed in which $x_1 \in \mathcal{X}_1$ and $x_2 \in \mathcal{X}_2$ are simultaneously sent, resulting in Y_1, Y_2. Find the capacity of this channel, which was first introduced by Shannon [344].

15. *The sum channel*: This channel, originally due to Shannon [344], operates by signaling over two DMCs with disjoint input and output alphabets, as described below.

   (a) Let $(\mathcal{X}_1, P_{Y_1|X_1}, \mathcal{Y}_1)$ be a DMC with finite input alphabet $\mathcal{X}_1$, finite output alphabet $\mathcal{Y}_1$, transition distribution $P_{Y_1|X_1}(y|x)$ and capacity C_1. Similarly, let $(\mathcal{X}_2, P_{Y_2|X_2}, \mathcal{Y}_2)$ be another DMC with capacity C_2. Assume that $\mathcal{X}_1 \cap \mathcal{X}_2 = \emptyset$ and that $\mathcal{Y}_1 \cap \mathcal{Y}_2 = \emptyset$.
      Now let $(\mathcal{X}, P_{Y|X}, \mathcal{Y})$ be the *sum* of these two channels where $\mathcal{X} = \mathcal{X}_1 \cup \mathcal{X}_2$, $\mathcal{Y} = \mathcal{Y}_1 \cup \mathcal{Y}_2$ and

$$P_{Y|X}(y|x) = \begin{cases} P_{Y_1|X_1}(y|x) & \text{if } x \in \mathcal{X}_1, y \in \mathcal{Y}_1 \\ P_{Y_2|X_2}(y|x) & \text{if } x \in \mathcal{X}_2, y \in \mathcal{Y}_2 \\ 0 & \text{otherwise.} \end{cases}$$

Show that the capacity of the sum channel is given by

$$C_{\text{sum}} = \log_2 \left(2^{C_1} + 2^{C_2} \right) \qquad \text{bits/channel use.}$$

Hint: Introduce a Bernoulli random variable Z with $\Pr[Z = 1] = \alpha$ such that $Z = 1$ if $X \in \mathcal{X}_1$ (when the first channel is used), and $Z = 2$ if $X \in \mathcal{X}_2$ (when the second channel is used). Then show that

$$I(X;Y) = I(X, Z; Y)$$
$$= h_b(\alpha) + \alpha I(X_1; Y_1) + (1 - \alpha)I(X_2; Y_2),$$

where $h_b(\cdot)$ is the binary entropy function, and $I(X_i; Y_i)$ is the mutual information for channel $P_{Y_i|X_i}(y|x)$, $i = 1, 2$. Then maximize (jointly) over the input distribution and α.

(b) Compute C_{sum} above if the first channel is a BSC with crossover probability 0.11, and the second channel is a BEC with erasure probability 0.5.

16. Prove that the quasi-symmetric channel satisfies the KKT conditions of Lemma 4.25 and yields the channel capacity given by (4.5.6).
17. Let the channel transition probability $P_{Y|X}$ of a DMC be defined as the following figure, where $0 < \epsilon < 0.5$.

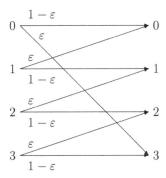

(a) Is the channel *weakly symmetric*? Is the channel *symmetric*?
(b) Determine the channel capacity of this channel. Also, indicate the input distribution that achieves the channel capacity.

18. Let the relation between the channel input $\{X_n\}_{n=1}^{\infty}$ and channel output $\{Y_n\}_{n=1}^{\infty}$ be given by

$$Y_n = (\alpha_n \times X_n) \oplus N_n \text{ for each } n,$$

where α_n, X_n, Y_n, and N_n all take values from $\{0, 1\}$, and "$\oplus$" represents the modulo-2 addition operation. Assume that the attenuation $\{\alpha_n\}_{n=1}^{\infty}$, channel input $\{X_n\}_{n=1}^{\infty}$ and noise $\{N_n\}_{n=1}^{\infty}$ processes are independent of each other. Also, $\{\alpha_n\}_{n=1}^{\infty}$ and $\{N_n\}_{n=1}^{\infty}$ are i.i.d. with

$$\Pr[\alpha_n = 1] = \Pr[\alpha_n = 0] = \frac{1}{2}$$

and

$$\Pr[N_n = 1] = 1 - \Pr[N_n = 0] = \varepsilon \in (0, 1/2).$$

(a) Show that the channel is a DMC and derive its transition probability matrix

$$\begin{bmatrix} P_{Y_j|X_j}(0|0) & P_{Y_j|X_j}(1|0) \\ P_{Y_j|X_j}(0|1) & P_{Y_j|X_j}(1|1) \end{bmatrix}.$$

(b) Determine the channel capacity C.
   Hint: Use the KKT conditions for channel capacity (Lemma 4.25).

(c) Suppose that α^n is known and consists of k 1's. Find the maximum $I(X^n; Y^n)$ for the same channel with known α^n.
   Hint: For known α^n, $\{(X_j, Y_j)\}_{j=1}^{n}$ are independent. Recall $I(X^n; Y^n) \leq \sum_{j=1}^{n} I(X_j; Y_j)$.

(d) Some researchers attempt to derive the capacity of the channel in (b) in terms of the following steps:
   • Derive the maximum mutual information between channel input X^n and output Y^n for a given α^n [namely, the solution in (c)].
   • Calculate the expectation value of the maximum mutual information obtained from the previous step according to the statistics of α^n.
   • Then, the capacity of the channel is equal to this "expected value" divided by n.
   Does this "expected capacity" $\bar{C}$ coincide with that in (b)?

19. *Maximum likelihood vs. minimum Hamming distance decoding:* Given a channel with finite input and output alphabets $\mathcal{X}$ and $\mathcal{Y}$, respectively, and given codebook $\mathcal{C} = \{c_1, \ldots, c_M\}$ of size M and blocklength n with $c_i = (c_{i,1}, \ldots, c_{i,n}) \in \mathcal{X}^n$, if an n-tuple $y^n \in \mathcal{Y}^n$ is received at the channel output, then under maximum likelihood (ML) decoding, y^n is decoded into the codeword $c^* \in \mathcal{C}$ that maximizes $P(Y^n = y^n | X^n = c)$ among all codewords $c \in \mathcal{C}$. It can be shown that ML decoding minimizes the probability of decoding error when the channel input n-tuple is chosen uniformly among all the codewords.
   Recall that the Hamming distance $d_H(x^n, y^n)$ between two n-tuples x^n and y^n taking values in $\mathcal{X}^n$ is defined as the number of positions where x^n and y^n differ. For a BSC using a binary codebook $\mathcal{C} = \{c_1, \ldots, c_M\} \subseteq \{0, 1\}^n$ of size M and blocklength n, if a received n-tuple y^n is received at the channel output, then

under minimum Hamming distance decoding, y^n is decoded into the codeword $c \in C$ that minimizes $d_H(c, y^n)$ among all codewords $c \in C$.

Prove that minimum Hamming distance decoding is equivalent to ML decoding for the BSC if its crossover probability ϵ satisfies: $\varepsilon \leq 1/2$.

Note: Note that ML decoding is not necessarily optimal if the codewords are not selected via a uniform probability distribution. In that more general case, which often occurs in practical systems (e.g., see [14, 169]), the optimal decoding rule is the so-called *maximum a posteriori* (MAP) rule, which selects the codeword $c^* \in C$ that maximizes $P(X^n = c|Y^n = y^n)$ among all codewords $c \in C$. It can readily be seen using Bayes' rule that MAP decoding reduces to ML decoding when the codewords are governed by a uniform distribution.

20. Suppose that blocklength $n = 2$ and code size $M = 2$. Assume each code bit is either 0 or 1.

    (a) What is the number of all possible codebook designs? (Note: This number includes those lousy code designs, such as $\{00, 00\}$.)

    (b) Suppose that one randomly draws one of these possible code designs according to a uniform distribution and applies the selected code to BSC with crossover probability ε. Then what is the expected error probability, if the decoder simply selects the codeword whose Hamming distance to the received vector is the smallest? (When both codewords have the same Hamming distance to the received vector, the decoder chooses one of them at random as the transmitted codeword.)

    (c) Explain why the error in (b) does not vanish as $\varepsilon \downarrow 0$.
        Hint: The error of random (n, M) code is lower bounded by the error of random $(n, 2)$ code for $M \geq 2$.

21. *Fano's inequality:* Assume that the alphabets for random variables X and Y are both given by

$$\mathcal{X} = \mathcal{Y} = \{1, 2, 3, 4, 5\}.$$

Let

$$\hat{x} = g(y)$$

be an estimate of x from observing y. Define the probability of estimation error as

$$P_e = \Pr\{g(Y) \neq X\}.$$

Then, Fano's inequality gives a lower bound for P_e as

$$h_b(P_e) + 2P_e \geq H(X|Y),$$

where $h_b(p) = p \log_2 \frac{1}{p} + (1 - p) \log_2 \frac{1}{1-p}$ is the binary entropy function. The curve for

$$h_b(P_e) + 2P_e = H(X|Y)$$

in terms of $H(X|Y)$ versus P_e is plotted in the figure below.

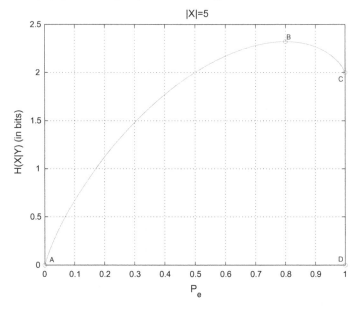

(a) Point A on the above figure shows that if $H(X|Y) = 0$, zero estimation error, namely, $P_e = 0$, can be achieved. In this case, characterize the distribution $P_{X|Y}$. Also, give an estimator $g(\cdot)$ that achieves $P_e = 0$. *Hint:* Think what kind of relation between X and Y can render $H(X|Y) = 0$.

(b) Point B on the above figure indicates that when $H(X|Y) = \log_2(5)$, the estimation error can only be equal to 0.8. In this case, characterize the distributions $P_{X|Y}$ and P_X. Prove that at $H(X|Y) = \log_2(5)$, all estimators yield $P_e = 0.8$.
Hint: Think what kind of relation between X and Y can result in $H(X|Y) = \log_2(5)$.

(c) Point C on the above figure hints that when $H(X|Y) = 2$, the estimation error can be as worse as 1. Give an estimator $g(\cdot)$ that leads to $P_e = 1$, if $P_{X|Y}(x|y) = 1/4$ for $x \neq y$, and $P_{X|Y}(x|y) = 0$ for $x = y$.

(d) Similarly, point D on the above figure hints that when $H(X|Y) = 0$, the estimation error can be as worse as 1. Give an estimator $g(\cdot)$ that leads to $P_e = 1$ at $H(X|Y) = 0$.

22. Decide whether the following statement is *true* or *false*. Consider a discrete memoryless channel with input alphabet $\mathcal{X}$, output alphabet $\mathcal{Y}$ and transition distribution $P_{Y|X}(y|x) := \Pr\{Y = y|X = x\}$. Let $P_{X_1}(\cdot)$ and $P_{X_2}(\cdot)$ be two possible input distributions, and $P_{Y_1}(\cdot)$ and $P_{Y_2}(\cdot)$ be the corresponding output distributions; i.e., $\forall y \in \mathcal{Y}$, $P_{Y_i}(y) = \sum_{x \in \mathcal{X}} P_{Y|X}(y|x)P_{X_i}(x)$, $i = 1, 2$. Then,

$$D(P_{X_1}||P_{X_2}) \geq D(P_{Y_1}||P_{Y_2}).$$

23. Consider a system consisting of two (parallel) discrete memoryless channels with transition probability matrices $Q_1 = [p_1(y_1|x)]$ and $Q_2 = [p_2(y_2|x)]$. These channels have a common input alphabet $\mathcal{X}$ and output alphabets $\mathcal{Y}_1$ and $\mathcal{Y}_2$, respectively, where $\mathcal{Y}_1$ and $\mathcal{Y}_2$ are disjoint. Let X denote the common input to the two channels, and let Y_1 and Y_2 be the corresponding outputs in channels Q_1 and Q_2, respectively. Now let Y be an overall output of the system which switches between Y_1 and Y_2 according to the values of a binary random variable $Z \in \{1, 2\}$ as follows:

$$Y = \begin{cases} Y_1 & \text{if } Z = 1; \\ Y_2 & \text{if } Z = 2; \end{cases}$$

where Z is independent of the input X and has distribution $P(Z = 1) = \lambda$.

(a) Express the system's mutual information $I(X; Y)$ in terms of λ, $I(X; Y_1)$ and $I(X; Y_2)$.

(b) Find an upper bound on the system's capacity $C = \max_{P_X} I(X; Y)$ in terms of λ, C_1 and C_2, where C_i is the capacity of channel Q_i, $i = 1, 2$.

(c) If both C_1 and C_2 can be achieved by the same input distribution, show that the upper bound on C in (b) is exact.

24. *Cascade channel*:   Consider a channel with input alphabet $\mathcal{X}$, output alphabet $\mathcal{Y}$ and transition probability matrix $Q_1 = [p_1(y|x)]$. Consider another channel with input alphabet $\mathcal{Y}$, output alphabet $\mathcal{Z}$ and transition probability matrix $Q_2 = [p_2(z|y)]$. Let C_1 denote the capacity of channel Q_1, and let C_2 denote the capacity of channel Q_2.

Define a new *cascade* channel with input alphabet $\mathcal{X}$, output alphabet $\mathcal{Z}$ and transition probability matrix $Q = [p(z|x)]$ obtained by feeding the output of channel Q_1 into the input of channel Q_2. Let C denote the capacity of channel Q.

(a) Show that $p(z|x) = \sum_{y \in \mathcal{Y}} p_2(z|y) p_1(y|x)$ and that $C \le \min\{C_1, C_2\}$.

(b) If $\mathcal{X} = \{0, 1\}$, $\mathcal{Y} = \mathcal{Z} = \{a, b, c\}$, Q_1 is described by

$$Q_1 = [p_1(y|x)] = \begin{bmatrix} 1 - \alpha & \alpha & 0 \\ 0 & \alpha & 1 - \alpha \end{bmatrix}, \quad 0 \le \alpha \le 1,$$

and Q_2 is described by

$$Q_2 = [p_2(z|y)] = \begin{bmatrix} 1 - \epsilon & \epsilon/2 & \epsilon/2 \\ \epsilon/2 & 1 - \epsilon & \epsilon/2 \\ \epsilon/2 & \epsilon/2 & 1 - \epsilon \end{bmatrix}, \quad 0 \le \epsilon \le 1,$$

find the capacities C_1 and C_2.

(c) Given the channels Q_1 and Q_2 described in part (b), find C in terms of α and ϵ.

(d) Compute the value of C obtained in part (c) if $\epsilon = 2/3$. Explain qualitatively.

25. Let X be a binary random variable with alphabet $\mathcal{X} = \{0, 1\}$. Let Z denote another random variable that is independent of X and taking values in $\mathcal{Z} = \{0, 1, 2, 3\}$ such that $\Pr[Z = 0] = \Pr[Z = 1] = \Pr[Z = 2] = \epsilon$, where $0 < \epsilon \leq 1/3$. Consider a DMC with input X, noise Z, and output Y described by the equation

$$Y = 3X + (-1)^X Z,$$

where X and Z are as defined above.

(a) Determine the channel transition probability matrix $Q = [p(y|x)]$.
(b) Compute the capacity C of this channel in terms of ϵ. What is the maximizing input distribution that achieves capacity?
(c) For what value of ϵ is the noise entropy $H(Z)$ maximized? What is the value of C for this choice of ϵ? Comment on the result.

26. *A channel with skewed errors:* This problem presents an additive noise channel in which the nonzero noise values can be partitioned into two distinct sets: the set of "common" errors $\mathcal{A}$ and the set of "uncommon" errors $\mathcal{B}$.

Consider a DMC with identical input and output alphabets given by $\mathcal{X} = \mathcal{Y} = \{0, 1, \ldots, q - 1\}$ where q is a fixed positive integer. The channel is a modulo-q additive noise channel, whose output Y is given by

$$Y = X \oplus_q Z,$$

where $\oplus_q$ denotes addition modulo-q, X is the channel input, and Z is the channel noise which is independent of X and has alphabet $\mathcal{Z} = \mathcal{X} = \mathcal{Y}$. Given a partition of $\mathcal{Z}$ via sets

$$\mathcal{A} = \{1, 2, \ldots, r\},$$

and

$$\mathcal{B} = \{r + 1, r + 2, \ldots, q - 1\},$$

for a fixed integer $0 < r < q - 1$, the distribution of Z is described as follows:

- $P(Z \neq 0) = \epsilon$.
- $P(Z \in \mathcal{A}|Z \neq 0) = \gamma$.
- $P(Z = i)$ is constant for all $i \in \mathcal{A}$.
- $P(Z = j)$ is constant for all $j \in \mathcal{B}$.

(a) Determine $P(Z = z)$ for all $z \in \mathcal{Z}$.
(b) Find the capacity of the channel in terms of q, r, ϵ and γ.
(c) Find the values of ϵ and γ that minimize capacity and determine the corresponding minimal capacity. Interpret the results qualitatively.

Note: This channel was introduced in [129] to model nonbinary data transmission and storage channels in which some types of errors (designated as "common errors") occur much more frequently than others. A family of codes, called *focused error control codes*, was developed in [129] to provide a certain level of protection against the common errors of the channel while guaranteeing another lower level of protection against uncommon errors; hence the levels of protection are determined based not only on the numbers of errors but on the kind of errors as well (unlike traditional channel codes). The performance of these codes was assessed in [10].

27. *Effect of memory on capacity:* This problem illustrates the adage "memory increases (operational) capacity." Given an integer $q \geq 2$, consider a q-ary additive noise channel described by

$$Y_i = X_i \oplus_q Z_i, \quad i = 1, 2 \ldots,$$

where $\oplus_q$ denotes addition modulo-q and Y_i, X_i and Z_i are the channel output, input and noise at time instant i, all with identical alphabet $\mathcal{Y} = \mathcal{X} = \mathcal{Z} = \{0, 1, \ldots, q-1\}$. We assume that the input and noise processes are independent of each other and that the noise process $\{Z_i\}_{i=1}^{\infty}$ is stationary ergodic. It can be shown via an extended version of Theorem 4.11 that the operational capacity of this channel with memory is given by [96, 191]:

$$C_{op} = \lim_{n \to \infty} \max_{p(x^n)} \frac{1}{n} I(X^n; Y^n).$$

Now consider an "equivalent" memoryless channel in the sense that it has a memoryless additive noise $\{\tilde{Z}_i\}_{i=1}^{\infty}$ with identical marginal distribution as noise $\{Z_i\}_{i=1}^{\infty}$: $P_{\tilde{Z}_i}(z) = P_{Z_i}(z)$ for all i and $z \in \mathcal{Z}$. Letting $\tilde{C}$ denote the (operational) capacity of the equivalent memoryless channel, show that

$$C_{op} \geq \tilde{C}.$$

Note: The adage "memory increases (operational) capacity" *does not* hold for arbitrary channels. It is only valid for well-behaved channels with memory [97], such as the above additive noise channel with stationary ergodic noise or more generally for *information stable*[26] channels [96, 191, 303] whose capacity is given by[27]

$$C_{op} = \liminf_{n \to \infty} \max_{p(x^n)} \frac{1}{n} I(X^n; Y^n). \tag{4.7.1}$$

---

[26]Loosely speaking, a channel is information stable if the input process which maximizes the channel's block mutual information yields a joint input–output process that behaves ergodically and satisfies the joint AEP (see [75, 96, 191, 303, 394] for a precise definition).

[27]Note that a formula of the capacity of more general (not necessarily information stable) channels with memory does exist in terms of a generalized (spectral) mutual information rate, see [172, 396].

However, one can find counterexamples to this adage, such as in [3] regarding non-ergodic "averaged" channels [4, 199]. Examples of such averaged channels include additive noise channels with stationary but non-ergodic noise, in particular, the Polya contagion channel [12] whose noise process is described in Example 3.16.

28. *Feedback capacity.* Consider a (not necessarily memoryless) discrete channel with input alphabet $\mathcal{X}$, output alphabet $\mathcal{Y}$ and n-fold transition distributions $P_{Y^n|X^n}, n = 1, 2, \ldots$. The channel is to be used with *feedback* as shown in the figure below.

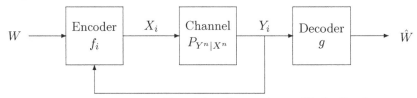

Channel coding system with feedback.

More specifically, there is a noiseless feedback link from the channel output to the transmitter with one time unit of delay. As a result, at each time instance i, the channel input X_i is a function of both the message W and all past channel outputs $Y^{i-1} = (Y_1, \ldots, Y_{i-1})$. More formally, an (n, M_n) feedback channel code consists of a sequence of encoding functions

$$f_i : \{1, 2, \ldots, M_n\} \times \mathcal{Y}^{i-1} \to \mathcal{X}$$

for $i = 1, \ldots, n$ and a decoding function

$$g : \mathcal{Y}^n \to \{1, 2, \ldots, M_n\}.$$

To send message W, assumed to be uniformly distributed over the message set $\{1, 2, \ldots, M_n\}$, the transmitter sends the channel codeword $X^n = (X_i, \ldots, X_n)$, where $X_i = f_i(W, Y^{i-1})$, $i = 1, \ldots, n$ (for $i = 1$, $X_1 = f_1(W)$), which is received as Y^n at the channel output. The decoder then provides the message estimate via $\hat{W} = g(Y^n)$ and the resulting average error probability is $P_e = \Pr[\hat{W} \neq W]$.

Causality channel condition: By the nature in which the channel is operated with or without feedback, we assume that a causal condition holds in the form of the following Markov chain property :

$$W \to (X^i, Y^{i-1}) \to Y_i, \text{ for } i = 1, 2, \ldots, \qquad (4.7.2)$$

where $Y^{i-1} = \emptyset$ for $i = 1$.

We say that a rate R is *achievable with feedback* if there exists a sequence of (n, M_n) feedback channel codes with

$$\liminf_{n \to \infty} \frac{1}{n} \log_2 M_n \geq R \quad \text{and} \quad \lim_{n \to \infty} P_e = 0.$$

The *feedback operational capacity* $C_{op,FB}$ of the channel is defined as the supremum of all achievable rates with feedback:

$$C_{op,FB} = \sup\{R : R \text{ is achievable with feedback}\}.$$

Comparing this definition of feedback operational capacity with the one when no feedback exists given in Definition 4.10 and studied in Theorem 4.11, we readily observe that, in general,

$$C_{op,FB} \geq C_{op}$$

since non-feedback codes belong to the class of feedback codes. This inequality is intuitively not surprising as in the presence of feedback, the transmitter can use the previously received output symbols to better understand the channel behavior and hence send codewords that are more robust to channel noise, potentially increasing the rate at which information can be transferred reliably over the channel.

(a) Show that for DMCs, feedback does not increase operational capacity:

$$C_{op,FB} = C_{op} = \max_{P_X} I(X; Y).$$

Note that for a DMC with feedback, property (4.2.1) does not hold since current inputs depend on past outputs. However, by the memorylessness nature of the channel, we assume the following causality Markov chain condition:

$$(W, X^{i-1}, Y^{i-1}) \to X_i \to Y_i \qquad (4.7.3)$$

for the channel, which is a simplified version of (4.7.2), see also [415, Definition 7.4]. Condition (4.7.3) can be seen as a generalized definition of a DMC used with or without feedback coding.

(b) Consider the q-ary channel of Problem 4.27 with stationary ergodic additive noise. Assume that the noise process is independent of the message W.[28] Show that although this channel has memory, feedback does not increase its operational capacity:

$$C_{op,FB} = C_{op} = \log_2 q - \lim_{n \to \infty} \frac{1}{n} H(Z^n).$$

We point out that the classical Gilbert–Elliott burst noise channel [108, 145, 277] is a special instance of this channel.

---

[28] This intrinsically natural assumption, which is equivalent to requiring that the channel input and noise processes are independent of each other when no feedback is present, ensures that (4.7.2) holds for this channel.

Note: Result (a) is due to Shannon [343]. Even though feedback does not help increase capacity for a DMC, it can have several benefits such a simplifying the coding scheme and speeding the rate at which the error probability of good codes decays to zero (e.g., see [87, 284]). Result (b), which was shown in [9] for arbitrary additive noise processes with memory, stems from the fact that the channel has a symmetry property in the sense that a uniform input maximizes the mutual information between channel input and output tuples. Similar results for channels with memory satisfying various symmetry properties have appeared in [11, 292, 333, 361]. However, it can be shown that for channels with memory and asymmetric structures,

$$C_{op,FB} > C_{op},$$

see, for example, [415, Problem 7.12] and [16] where average input costs are imposed.

We further point out that for information stable channels, the feedback operational capacity is given by [215, 291, 378]

$$C_{op,FB} = \liminf_{n \to \infty} \max_{P_{X^n \| Y^{n-1}}} \frac{1}{n} I(X^n \to Y^n), \qquad (4.7.4)$$

where for $x^n \in \mathcal{X}^n$ and $y^{n-1} \in \mathcal{Y}^{n-1}$,

$$P_{X^n \| Y^{n-1}}(x^n \| y^{n-1}) := \prod_{i=1}^n P_{X_i | X^{i-1} Y^{i-1}}(x_i | x^{i-1}, y^{i-1})$$

is a causal conditional probability that represents feedback strategies and

$$I(X^n \to Y^n) := \sum_{i=1}^n I(Y_i; X^i | Y^{i-1})$$

is a causal version of the mutual information between tuples, known as *directed information* [233, 261, 265]. It can be verified that (4.7.4) reduces to (4.7.1) in the absence of feedback.[29] Finally, an alternative (albeit more complicated) expression to (4.7.4), that uses the standard mutual information, is given by [72]

$$C_{op,FB} = \liminf_{n \to \infty} \max_{f^n} \frac{1}{n} I(W; Y^n), \qquad (4.7.5)$$

where W is uniformly distributed over $\{1, 2 \ldots, M_n\}$ and the maximization is taken over all feedback encoding functions $f^n = (f_1, f_2, \ldots, f_n)$, where

---

[29]For arbitrary channels with memory, a generalized expression for $C_{op,FB}$ is established in [376, 378] in terms of a generalized (spectral) directed information rate.

$f_i : \{1, 2 \ldots, M_n\} \times \mathcal{Y}^{i-1} \to \mathcal{X}$ for $i = 1, 2, \ldots, n$ (note that the optimization over f^n in (4.7.5) necessitates optimizing over M_n).[30]

29. Suppose you wish to encode a binary DMS with $P_X(0) = 3/4$ using a rate-1 source-channel block code for transmission over a BEC with erasure probability α. For what values of α, can the source be recovered reliably (i.e., with arbitrarily low error probability) at the receiver?

30. Consider the binary Polya contagion Markov source of memory two treated in Problem 3.10; see also Example 3.17 with $M = 2$. We are interested in sending this source over the BSC with crossover probability ε using rate-R_{sc} block source-channel codes.

   (a) Write down the sufficient condition for reliable transmissibility of the source over the BSC via rate-R_{sc} source-channel codes in terms of ε, R_{sc} and the source parameters $\rho := R/T$ and $\delta := \Delta/T$.
   (b) If $\rho = \delta = 1/2$ and $\varepsilon = 1/4$, determine the permissible range of rates R_{sc} for reliably communicating the source over the channel.

31. Consider a DMC with input alphabet $\mathcal{X} = \{0, 1, 2, 3, 4\}$, output alphabet $\mathcal{Y} = \{0, 1, 2, 3, 4, 5\}$ and the following transition matrix

$$Q = \begin{bmatrix} 1 - 2\alpha & \alpha & \alpha & 0 & 0 & 0 \\ \alpha & \alpha & 1 - 2\alpha & 0 & 0 & 0 \\ 0 & 0 & 0 & 1 - \beta & \beta/2 & \beta/2 \\ 0 & 0 & 0 & \beta/2 & 1 - \beta & \beta/2 \\ 0 & 0 & 0 & \beta/2 & \beta/2 & 1 - \beta \end{bmatrix},$$

where $0 < \alpha < 1/2$ and $0 < \beta < 1$.

   (a) Determine the capacity C of this channel in terms of α and β. What is the maximizing input distribution that achieves capacity?
   (b) Find the values of α and β that will yield the smallest possible value of C.
   (c) Show that any (not necessarily memoryless) binary source $\{U_i\}_{i=1}^{\infty}$ with arbitrary distribution can be sent without any loss via a rate-one source-channel code over the channel with the parameters α and β obtained in part (b).

---

[30]This result was actually shown in [72] for general channels with memory in terms of a generalized (spectral) mutual information rate.

Chapter 5
Differential Entropy and Gaussian Channels

We have so far examined information measures and their operational characterization for discrete-time discrete-alphabet systems. In this chapter, we turn our focus to continuous-alphabet (real-valued) systems. Except for a brief interlude with the continuous-time (waveform) Gaussian channel, we consider discrete-time systems, as treated throughout the book.

We first recall that a real-valued (continuous) random variable X is described by its cumulative distribution function (cdf)

$$F_X(x) := \Pr[X \leq x]$$

for $x \in \mathbb{R}$, the set of real numbers. The distribution of X is called absolutely continuous (with respect to the Lebesgue measure) if a probability density function (pdf) $f_X(\cdot)$ exists such that

$$F_X(x) = \int_{-\infty}^{x} f_X(t)dt,$$

where $f_X(t) \geq 0 \ \forall t$ and $\int_{-\infty}^{+\infty} f_X(t)dt = 1$. If $F_X(\cdot)$ is differentiable everywhere, then the pdf $f_X(\cdot)$ exists and is given by the derivative of $F_X(\cdot)$: $f_X(t) = \frac{dF_X(t)}{dt}$. The *support* of a random variable X with pdf $f_X(\cdot)$ is denoted by S_X and can be conveniently given as

$$S_X = \{x \in \mathbb{R} \colon f_X(x) > 0\}.$$

We will deal with random variables that admit a pdf.[1]

---

[1] A rigorous (measure-theoretic) study for general continuous systems, initiated by Kolmogorov [222], can be found in [196, 303].

© Springer Nature Singapore Pte Ltd. 2018   165
F. Alajaji and P.-N. Chen, *An Introduction to Single-User Information Theory*,
Springer Undergraduate Texts in Mathematics and Technology,
https://doi.org/10.1007/978-981-10-8001-2_5

5.1 Differential Entropy

Recall that the definition of entropy for a discrete random variable X representing a DMS is

$$H(X) := \sum_{x \in \mathcal{X}} -P_X(x) \log_2 P_X(x) \quad \text{(in bits)}.$$

As already seen in Shannon's source coding theorem, this quantity is the minimum average code rate achievable for the lossless compression of the DMS. But if the random variable takes on values in a continuum, the minimum number of bits per symbol needed to losslessly describe it must be infinite. This is illustrated in the following example, where we take a discrete approximation (quantization) of a random variable uniformly distributed on the unit interval and study the entropy of the quantized random variable as the quantization becomes finer and finer.

Example 5.1 Consider a real-valued random variable X that is uniformly distributed on the unit interval, i.e., with pdf given by

$$f_X(x) = \begin{cases} 1 & \text{if } x \in [0, 1); \\ 0 & \text{otherwise}. \end{cases}$$

Given a positive integer m, we can discretize X by uniformly quantizing it into m levels by partitioning the support of X into equal-length segments of size $\Delta = \frac{1}{m}$ (Δ is called the quantization step-size) such that

$$q_m(X) = \frac{i}{m}, \quad \text{if } \frac{i-1}{m} \leq X < \frac{i}{m},$$

for $1 \leq i \leq m$. Then, the entropy of the quantized random variable $q_m(X)$ is given by

$$H(q_m(X)) = -\sum_{i=1}^{m} \frac{1}{m} \log_2 \left(\frac{1}{m} \right) = \log_2 m \quad \text{(in bits)}.$$

Since the entropy $H(q_m(X))$ of the quantized version of X is a lower bound to the entropy of X (as $q_m(X)$ is a function of X) and satisfies in the limit

$$\lim_{m \to \infty} H(q_m(X)) = \lim_{m \to \infty} \log_2 m = \infty,$$

we obtain that the entropy of X is infinite.

The above example indicates that to compress a continuous source without incurring any loss or distortion indeed requires an infinite number of bits. Thus when studying continuous sources, the entropy measure is limited in its effectiveness and the introduction of a new measure is necessary. Such new measure is indeed obtained upon close examination of the entropy of a uniformly quantized real-valued random variable minus the quantization accuracy as the accuracy increases without bound.

Lemma 5.2 *Consider a real-valued random variable X with support $[a, b)$ and pdf f_X such that $-f_X \log_2 f_X$ is integrable[2] (where $-\int_a^b f_X(x) \log_2 f_X(x)dx$ is finite). Then a uniform quantization of X with an n-bit accuracy (i.e., with a quantization step-size of $\Delta = 2^{-n}$) yields an entropy approximately equal to $-\int_a^b f_X(x) \log_2 f_X(x)dx + n$ bits for n sufficiently large. In other words,*

$$\lim_{n \to \infty} [H(q_n(X)) - n] = -\int_a^b f_X(x) \log_2 f_X(x)dx,$$

where $q_n(X)$ is the uniformly quantized version of X with quantization step-size $\Delta = 2^{-n}$.

Proof

Step 1: Mean value theorem. Let $\Delta = 2^{-n}$ be the quantization step-size, and let

$$t_i := \begin{cases} a + i\Delta, & i = 0, 1, \ldots, j - 1 \\ b, & i = j, \end{cases}$$

where $j = \lceil (b - a)2^n \rceil$. From the mean value theorem (e.g., cf. [262]), we can choose $x_i \in [t_{i-1}, t_i]$ for $1 \le i \le j$ such that

$$p_i := \int_{t_{i-1}}^{t_i} f_X(x)dx = f_X(x_i)(t_i - t_{i-1}) = \Delta \cdot f_X(x_i).$$

Step 2: Definition of $h^{(n)}(X)$. Let

$$h^{(n)}(X) := -\sum_{i=1}^{j} [f_X(x_i) \log_2 f_X(x_i)]2^{-n}.$$

Since $-f_X(x) \log_2 f_X(x)$ is integrable,

$$h^{(n)}(X) \to -\int_a^b f_X(x) \log_2 f_X(x)dx \quad \text{as } n \to \infty.$$

Therefore, given any $\varepsilon > 0$, there exists N such that for all $n > N$,

$$\left| -\int_a^b f_X(x) \log_2 f_X(x)dx - h^{(n)}(X) \right| < \varepsilon.$$

---

[2]By integrability, we mean the usual Riemann integrability (e.g., see [323]).

Step 3: Computation of $H(q_n(X))$. The entropy of the (uniformly) quantized version of X, $q_n(X)$, is given by

$$H(q_n(X)) = -\sum_{i=1}^{j} p_i \log_2 p_i$$

$$= -\sum_{i=1}^{j} (f_X(x_i)\Delta) \log_2(f_X(x_i)\Delta)$$

$$= -\sum_{i=1}^{j} (f_X(x_i)2^{-n}) \log_2(f_X(x_i)2^{-n}),$$

where the p_i's are the probabilities of the different values of $q_n(X)$.

Step 4: $H(q_n(X)) - h^{(n)}(X)$.

From Steps 2 and 3,

$$H(q_n(X)) - h^{(n)}(X) = -\sum_{i=1}^{j} [f_X(x_i)2^{-n}] \log_2(2^{-n})$$

$$= n \sum_{i=1}^{j} \int_{t_{i-1}}^{t_i} f_X(x)dx$$

$$= n \int_a^b f_X(x)dx = n.$$

Hence, we have that for $n > N$,

$$\left[-\int_a^b f_X(x) \log_2 f_X(x)dx + n \right] - \varepsilon < H(q_n(X))$$

$$= h^{(n)}(X) + n$$

$$< \left[-\int_a^b f_X(x) \log_2 f_X(x)dx + n \right] + \varepsilon,$$

yielding that

$$\lim_{n \to \infty} [H(q_n(X)) - n] = -\int_a^b f_X(x) \log_2 f_X(x)dx.$$

$\square$

More generally, the following result due to Rényi [316] can be shown for (absolutely continuous) random variables with arbitrary support.

Theorem 5.3 [316, Theorem 1] *For any real-valued random variable with pdf* f_X, *if* $-\sum_i p_i \log_2 p_i$ *is finite, where the (possibly countably many) p_i's are the* *probabilities of the different values of the uniformly quantized $q_n(X)$ over support S_X, then*

$$\lim_{n \to \infty} [H(q_n(X)) - n] = - \int_{S_X} f_X(x) \log_2 f_X(x) dx$$

provided the integral on the right-hand side exists.

In light of the above results, we can define the following information measure [340]:

Definition 5.4 (*Differential entropy*) The differential entropy (in bits) of a continuous random variable X with pdf f_X and support S_X is defined as

$$h(X) := - \int_{S_X} f_X(x) \cdot \log_2 f_X(x) dx = E[- \log_2 f_X(X)],$$

when the integral exists.

Thus, the differential entropy $h(X)$ of a real-valued random variable X has an *operational meaning* in the following sense. Since $H(q_n(X))$ is the minimum average number of bits needed to losslessly describe $q_n(X)$, we thus obtain that $h(X) + n$ is approximately needed to describe X when uniformly quantizing it with an n-bit accuracy. Therefore, we may conclude that the larger $h(X)$ is, the larger is the average number of bits required to describe a uniformly quantized X within a fixed accuracy.

Example 5.5 A continuous random variable X with support $S_X = [0, 1)$ and pdf $f_X(x) = 2x$ for $x \in S_X$ has differential entropy equal to

$$\int_0^1 -2x \cdot \log_2(2x) dx = \left. \frac{x^2(\log_2 e - 2\log_2(2x))}{2} \right|_0^1$$

$$= \frac{1}{2 \ln 2} - \log_2(2) = -0.278652 \text{ bits}.$$

We herein illustrate Lemma 5.2 by uniformly quantizing X to an n-bit accuracy and computing the entropy $H(q_n(X))$ and $H(q_n(X)) - n$ for increasing values of n, where $q_n(X)$ is the quantized version of X.
We have that $q_n(X)$ is given by

$$q_n(X) = \frac{i}{2^n}, \quad \text{if } \frac{i-1}{2^n} \le X < \frac{i}{2^n},$$

for $1 \le i \le 2^n$. Hence,

$$\Pr\left\{q_n(X) = \frac{i}{2^n}\right\} = \frac{(2i-1)}{2^{2n}},$$

which yields

$$H(q_n(X)) = - \sum_{i=1}^{2^n} \frac{2i-1}{2^{2n}} \log_2 \left(\frac{2i-1}{2^{2n}} \right)$$

$$= -\frac{1}{2^{2n}} \sum_{i=1}^{2^n} (2i-1) \log_2(2i-1) + 2 \log_2(2^n).$$

As shown in Table 5.1, we indeed observe that as n increases, $H(q_n(X))$ tends to infinity while $H(q_n(X)) - n$ converges to $h(X) = -0.278652$ bits.

Thus, a continuous random variable X contains an infinite amount of information; but we can measure the information contained in its n-bit quantized version $q_n(X)$ as: $H(q_n(X)) \approx h(X) + n$ (for n large enough).

Example 5.6 Let us determine the minimum average number of bits required to describe the uniform quantization with 3-digit accuracy of the decay time (in years) of a radium atom assuming that the half-life of the radium (i.e., the median of the decay time) is 80 years and that its pdf is given by $f_X(x) = \lambda e^{-\lambda x}$, where $x > 0$.

Since the median of the decay time is 80, we obtain

$$\int_0^{80} \lambda e^{-\lambda x} dx = 0.5,$$

which implies that $\lambda = 0.00866$. Also, 3-digit accuracy is approximately equivalent to $\log_2 999 = 9.96 \approx 10$ bits accuracy. Therefore, by Theorem 5.3, the number of bits required to describe the quantized decay time is approximately

$$h(X) + 10 = \log_2 \frac{e}{\lambda} + 10 = 18.29 \text{ bits.}$$

Table 5.1 Quantized random variable $q_n(X)$ under an n-bit accuracy: $H(q_n(X))$ and $H(q_n(X)) - n$ versus n

n	$H(q_n(X))$	$H(q_n(X)) - n$
1	0.811278 bits	−0.188722 bits
2	1.748999 bits	−0.251000 bits
3	2.729560 bits	−0.270440 bits
4	3.723726 bits	−0.276275 bits
5	4.722023 bits	−0.277977 bits
6	5.721537 bits	−0.278463 bits
7	6.721399 bits	−0.278600 bits
8	7.721361 bits	−0.278638 bits
9	8.721351 bits	−0.278648 bits

We close this section by computing the differential entropy for two common real-valued random variables: the uniformly distributed random variable and the Gaussian distributed random variable.

Example 5.7 (Differential entropy of a uniformly distributed random variable) Let X be a continuous random variable that is uniformly distributed over the interval (a, b), where $b > a$; i.e., its pdf is given by

$$f_X(x) = \begin{cases} \frac{1}{b-a} & \text{if } x \in (a, b); \\ 0 & \text{otherwise.} \end{cases}$$

So its differential entropy is given by

$$h(X) = -\int_a^b \frac{1}{b-a} \log_2 \frac{1}{b-a} = \log_2(b-a) \quad \text{bits.}$$

Note that if $(b - a) < 1$ in the above example, then $h(X)$ is *negative*, unlike entropy. The above example indicates that although differential entropy has a form analogous to entropy (in the sense that summation and pmf for entropy are replaced by integration and pdf, respectively, for differential entropy), differential entropy does *not* retain all the properties of entropy (one such operational difference was already highlighted in the previous lemma and theorem).[3]

Example 5.8 (Differential entropy of a Gaussian random variable)   Let $X \sim \mathcal{N}(\mu, \sigma^2)$; i.e., X is a Gaussian (or normal) random variable with finite mean μ, variance $\text{Var}(X) = \sigma^2 > 0$ and pdf

$$f_X(x) = \frac{1}{\sqrt{2\pi\sigma^2}} e^{-\frac{(x-\mu)^2}{2\sigma^2}}$$

for $x \in \mathbb{R}$. Then, its differential entropy is given by

$$\begin{aligned} h(X) &= \int_{\mathbb{R}} f_X(x) \left[\frac{1}{2} \log_2(2\pi\sigma^2) + \frac{(x-\mu)^2}{2\sigma^2} \log_2 e \right] dx \\ &= \frac{1}{2} \log_2(2\pi\sigma^2) + \frac{\log_2 e}{2\sigma^2} E[(X-\mu)^2] \\ &= \frac{1}{2} \log_2(2\pi\sigma^2) + \frac{1}{2} \log_2 e \\ &= \frac{1}{2} \log_2(2\pi e\sigma^2) \quad \text{bits.} \end{aligned} \tag{5.1.1}$$

Note that for a Gaussian random variable, its differential entropy is only a function of its variance σ^2 (it is independent from its mean μ). This is similar to the differential

---

[3] By contrast, entropy and differential entropy are sometimes called *discrete entropy* and *continuous entropy*, respectively.

entropy of a uniform random variable, which only depends on difference $(b - a)$ but not the mean $(a + b)/2$.

5.2   Joint and Conditional Differential Entropies, Divergence, and Mutual Information

Definition 5.9 (*Joint differential entropy*) If $X^n = (X_1, X_2, \ldots, X_n)$ is a continuous random vector of size n (i.e., a vector of n continuous random variables) with joint pdf f_{X^n} and support $S_{X^n} \subseteq \mathbb{R}^n$, then its joint differential entropy is defined as

$$h(X^n) := -\int_{S_{X^n}} f_{X^n}(x_1, x_2, \ldots, x_n) \log_2 f_{X^n}(x_1, x_2, \ldots, x_n)\, dx_1\, dx_2\, \cdots\, dx_n$$
$$= E[-\log_2 f_{X^n}(X^n)]$$

when the n-dimensional integral exists.

Definition 5.10 (*Conditional differential entropy*) Let X and Y be two jointly distributed continuous random variables with joint pdf[4] $f_{X,Y}$ and support $S_{X,Y} \subseteq \mathbb{R}^2$ such that the conditional pdf of Y given X, given by

$$f_{Y|X}(y|x) = \frac{f_{X,Y}(x, y)}{f_X(x)},$$

is well defined for all $(x, y) \in S_{X,Y}$, where f_X is the marginal pdf of X. Then, the conditional differential entropy of Y given X is defined as

$$h(Y|X) := -\int_{S_{X,Y}} f_{X,Y}(x, y) \log_2 f_{Y|X}(y|x)\, dx\, dy = E[-\log_2 f_{Y|X}(Y|X)],$$

when the integral exists.

Note that as in the case of (discrete) entropy, the chain rule holds for differential entropy:

$$h(X, Y) = h(X) + h(Y|X) = h(Y) + h(X|Y).$$

Definition 5.11 (*Divergence or relative entropy*) Let X and Y be two continuous random variables with marginal pdfs f_X and f_Y, respectively, such that their supports satisfy $S_X \subseteq S_Y \subseteq \mathbb{R}$. Then, the divergence (or relative entropy or Kullback–Leibler distance) between X and Y is written as $D(X \| Y)$ or $D(f_X \| f_Y)$ and defined by

$$D(X \| Y) := \int_{S_X} f_X(x) \log_2 \frac{f_X(x)}{f_Y(x)}\, dx = E\left[\frac{f_X(X)}{f_Y(X)}\right]$$

---

[4]Note that the joint pdf $f_{X,Y}$ is also commonly written as f_{XY}.

when the integral exists. The definition carries over similarly in the multivariate case: for $X^n = (X_1, X_2, \ldots, X_n)$ and $Y^n = (Y_1, Y_2, \ldots, Y_n)$ two random vectors with joint pdfs f_{X^n} and f_{Y^n}, respectively, and supports satisfying $S_{X^n} \subseteq S_{Y^n} \subseteq \mathbb{R}^n$, the divergence between X^n and Y^n is defined as

$$D(X^n \| Y^n) := \int_{S_{X^n}} f_{X^n}(x_1, x_2, \ldots, x_n) \log_2 \frac{f_{X^n}(x_1, x_2, \ldots, x_n)}{f_{Y^n}(x_1, x_2, \ldots, x_n)} \, dx_1 \, dx_2 \, \cdots \, dx_n$$

when the integral exists.

Definition 5.12 (*Mutual information*) Let X and Y be two jointly distributed continuous random variables with joint pdf $f_{X,Y}$ and support $S_{XY} \subseteq \mathbb{R}^2$. Then, the mutual information between X and Y is defined by

$$I(X; Y) := D(f_{X,Y} \| f_X f_Y) = \int_{S_{X,Y}} f_{X,Y}(x, y) \log_2 \frac{f_{X,Y}(x, y)}{f_X(x) f_Y(y)} \, dx \, dy,$$

assuming the integral exists, where f_X and f_Y are the marginal pdfs of X and Y, respectively.

Observation 5.13 For two jointly distributed continuous random variables X and Y with joint pdf $f_{X,Y}$, support $S_{XY} \subseteq \mathbb{R}^2$ and joint differential entropy

$$h(X, Y) = - \int_{S_{XY}} f_{X,Y}(x, y) \log_2 f_{X,Y}(x, y) \, dx \, dy,$$

then as in Lemma 5.2 and the ensuing discussion, one can write

$$H(q_n(X), q_m(Y)) \approx h(X, Y) + n + m$$

for n and m sufficiently large, where $q_k(Z)$ denotes the (uniformly) quantized version of random variable Z with a k-bit accuracy.

On the other hand, for the above continuous X and Y,

$$\begin{aligned}
I(q_n(X); q_m(Y)) &= H(q_n(X)) + H(q_m(Y)) - H(q_n(X), q_m(Y)) \\
&\approx [h(X) + n] + [h(Y) + m] - [h(X, Y) + n + m] \\
&= h(X) + h(Y) - h(X, Y) \\
&= \int_{S_{X,Y}} f_{X,Y}(x, y) \log_2 \frac{f_{X,Y}(x, y)}{f_X(x) f_Y(y)} \, dx \, dy
\end{aligned}$$

for n and m sufficiently large; in other words,

$$\lim_{n,m \to \infty} I(q_n(X); q_m(Y)) = h(X) + h(Y) - h(X, Y).$$

Furthermore, it can be shown that

$$\lim_{n \to \infty} D(q_n(X) \| q_n(Y)) = \int_{S_X} f_X(x) \log_2 \frac{f_X(x)}{f_Y(x)} \, dx.$$

Thus, mutual information and divergence can be considered as the *true* tools of information theory, as they retain the same operational characteristics and properties for both discrete and continuous probability spaces (as well as general spaces where they can be defined in terms of Radon–Nikodym derivatives (e.g., cf. [196]).[5]

The following lemma illustrates that for continuous systems, $I(\cdot; \cdot)$ and $D(\cdot\|\cdot)$ keep the same properties already encountered for discrete systems, while differential entropy (as already seen with its possibility of being negative) satisfies some different properties from entropy. The proof is left as an exercise.

Lemma 5.14 *The following properties hold for the information measures of continuous systems.*

1. **Nonnegativity of divergence***: Let X and Y be two continuous random variables with marginal pdfs f_X and f_Y, respectively, such that their supports satisfy $S_X \subseteq S_Y \subseteq \mathbb{R}$. Then*

$$D(f_X \| f_Y) \geq 0,$$

   with equality iff $f_X(x) = f_Y(x)$ for all $x \in S_X$ except in a set of f_X-measure zero (i.e., $X = Y$ almost surely).
2. **Nonnegativity of mutual information***: For any two continuous jointly distributed random variables X and Y,*

$$I(X; Y) \geq 0,$$

   with equality iff X and Y are independent.
3. **Conditioning never increases differential entropy***: For any two continuous random variables X and Y with joint pdf $f_{X,Y}$ and well-defined conditional pdf $f_{X|Y}$,*

$$h(X|Y) \leq h(X),$$

   with equality iff X and Y are independent.
4. **Chain rule for differential entropy***: For a continuous random vector $X^n = (X_1, X_2, \ldots, X_n)$,*

$$h(X_1, X_2, \ldots, X_n) = \sum_{i=1}^{n} h(X_i | X_1, X_2, \ldots, X_{i-1}),$$

   where $h(X_i | X_1, X_2, \ldots, X_{i-1}) := h(X_1)$ for $i = 1$.

---

[5]This justifies using identical notations for both $I(\cdot; \cdot)$ and $D(\cdot\|\cdot)$ as opposed to the discerning notations of $H(\cdot)$ for entropy and $h(\cdot)$ for differential entropy.

5. **Chain rule for mutual information**: *For continuous random vector $X^n = (X_1, X_2, \ldots, X_n)$ and random variable Y with joint pdf $f_{X^n, Y}$ and well-defined conditional pdfs $f_{X_i, Y | X^{i-1}}$, $f_{X_i | X^{i-1}}$ and $f_{Y | X^{i-1}}$ for $i = 1, \ldots, n$, we have that*

$$I(X_1, X_2, \ldots, X_n; Y) = \sum_{i=1}^{n} I(X_i; Y | X_{i-1}, \ldots, X_1),$$

where $I(X_i; Y | X_{i-1}, \ldots, X_1) := I(X_1; Y)$ for $i = 1$.

6. **Data processing inequality**: *For continuous random variables X, Y, and Z such that $X \to Y \to Z$, i.e., X and Z are conditional independent given Y,*

$$I(X; Y) \geq I(X; Z).$$

7. **Independence bound for differential entropy**: *For a continuous random vector $X^n = (X_1, X_2, \ldots, X_n)$,*

$$h(X^n) \leq \sum_{i=1}^{n} h(X_i),$$

with equality iff all the X_i's are independent from each other.

8. **Invariance of differential entropy under translation**: *For continuous random variables X and Y with joint pdf $f_{X, Y}$ and well-defined conditional pdf $f_{X|Y}$,*

$$h(X + c) = h(X) \quad \text{for any constant } c \in \mathbb{R},$$

and

$$h(X + Y | Y) = h(X | Y).$$

The results also generalize in the multivariate case: for two continuous random vectors $X^n = (X_1, X_2, \ldots, X_n)$ and $Y^n = (Y_1, Y_2, \ldots, Y_n)$ with joint pdf f_{X^n, Y^n} and well-defined conditional pdf $f_{X^n | Y^n}$,

$$h(X^n + c^n) = h(X^n)$$

for any constant n-tuple $c^n = (c_1, c_2, \ldots, c_n) \in \mathbb{R}^n$, and

$$h(X^n + Y^n | Y^n) = h(X^n | Y^n),$$

where the addition of two n-tuples is performed component-wise.

9. **Differential entropy under scaling**: *For any continuous random variable X and any nonzero real constant a,*

$$h(aX) = h(X) + \log_2 |a|.$$

10. **Joint differential entropy under linear mapping**: *Consider the random (col-umn) vector $\underline{X} = (X_1, X_2, \ldots, X_n)^T$ with joint pdf f_{X^n}, where T denotes trans-position, and let $\underline{Y} = (Y_1, Y_2, \ldots, Y_n)^T$ be a random (column) vector obtained from the linear transformation $\underline{Y} = \mathbf{A}\underline{X}$, where $\mathbf{A}$ is an invertible (non-singular) $n \times n$ real-valued matrix. Then*

$$h(\underline{Y}) = h(Y_1, Y_2, \ldots, Y_n) = h(X_1, X_2, \ldots, X_n) + \log_2 |\det(\mathbf{A})|,$$

where $\det(\mathbf{A})$ is the determinant of the square matrix $\mathbf{A}$.

11. **Joint differential entropy under nonlinear mapping**: *Consider the ran-dom (column) vector $\underline{X} = (X_1, X_2, \ldots, X_n)^T$ with joint pdf f_{X^n}, and let $\underline{Y} = (Y_1, Y_2, \ldots, Y_n)^T$ be a random (column) vector obtained from the non-linear transformation*

$$\underline{Y} = \underline{g}(\underline{X}) := (g_1(X_1), g_2(X_2), \ldots, g_n(X_n))^T,$$

where each $g_i : \mathbb{R} \rightarrow \mathbb{R}$ is a differentiable function, $i = 1, 2, \ldots, n$. Then

$$h(\underline{Y}) = h(Y_1, Y_2, \ldots, Y_n)$$
$$= h(X_1, \ldots, X_n) + \int_{\mathbb{R}^n} f_{X^n}(x_1, \ldots, x_n) \log_2 |\det(\mathbf{J})| \, dx_1 \cdots dx_n,$$

where $\mathbf{J}$ is the $n \times n$ Jacobian matrix given by

$$\mathbf{J} := \begin{bmatrix} \frac{\partial g_1}{\partial x_1} & \frac{\partial g_1}{\partial x_2} & \cdots & \frac{\partial g_1}{\partial x_n} \\ \frac{\partial g_2}{\partial x_1} & \frac{\partial g_2}{\partial x_2} & \cdots & \frac{\partial g_2}{\partial x_n} \\ \vdots & \vdots & \cdots & \vdots \\ \frac{\partial g_n}{\partial x_1} & \frac{\partial g_n}{\partial x_2} & \cdots & \frac{\partial g_n}{\partial x_n} \end{bmatrix}.$$

Observation 5.15 Property 9 of the above Lemma indicates that for a continuous random variable X, $h(X) \neq h(aX)$ (except for the trivial case of $a = 1$) and hence differential entropy is not in general invariant under invertible maps. This is in contrast to entropy, which is always invariant under invertible maps: given a discrete random variable X with alphabet $\mathcal{X}$,

$$H(f(X)) = H(X)$$

for all invertible maps $f : \mathcal{X} \rightarrow \mathcal{Y}$, where $\mathcal{Y}$ is a discrete set; in particular $H(aX) = H(X)$ for all nonzero reals a.

On the other hand, for *both* discrete and continuous systems, mutual information and divergence are invariant under invertible maps:

$$I(X; Y) = I(g(X); Y) = I(g(X); h(Y))$$

and

$$D(X\|Y) = D(g(X)\|g(Y))$$

for all invertible maps g and h properly defined on the alphabet/support of the concerned random variables. This reinforces the notion that mutual information and divergence constitute the true tools of information theory.

Definition 5.16 (*Multivariate Gaussian*) A continuous random vector $\underline{X} = (X_1, X_2, \ldots, X_n)^T$ is called a size-n (multivariate) Gaussian random vector with a finite mean vector $\underline{\mu} := (\mu_1, \mu_2, \ldots, \mu_n)^T$, where $\mu_i := E[X_i] < \infty$ for $i = 1, 2, \ldots, n$, and an $n \times n$ invertible (real-valued) covariance matrix

$$
\begin{aligned}
\mathbf{K}_{\underline{X}} &= [K_{i,j}] \\
&:= E[(\underline{X} - \underline{\mu})(\underline{X} - \underline{\mu})^T] \\
&= \begin{bmatrix}
\mathrm{Cov}(X_1, X_1) & \mathrm{Cov}(X_1, X_2) & \cdots & \mathrm{Cov}(X_1, X_n) \\
\mathrm{Cov}(X_2, X_1) & \mathrm{Cov}(X_2, X_2) & \cdots & \mathrm{Cov}(X_2, X_n) \\
\vdots & \vdots & \ddots & \vdots \\
\mathrm{Cov}(X_n, X_1) & \mathrm{Cov}(X_n, X_2) & \cdots & \mathrm{Cov}(X_n, X_n)
\end{bmatrix},
\end{aligned}
$$

where $K_{i,j} = \mathrm{Cov}(X_i, X_j) := E[(X_i - \mu_i)(X_j - \mu_j)]$ is the covariance[6] between X_i and X_j for $i, j = 1, 2, \ldots, n$, if its joint pdf is given by the multivariate Gaussian pdf

$$f_{X^n}(x_1, x_2, \ldots, x_n) = \frac{1}{(\sqrt{2\pi})^n \sqrt{\det(\mathbf{K}_{\underline{X}})}} \, e^{-\frac{1}{2}(\underline{x}-\underline{\mu})^T \mathbf{K}_{\underline{X}}^{-1}(\underline{x}-\underline{\mu})}$$

for any $(x_1, x_2, \ldots, x_n) \in \mathbb{R}^n$, where $\underline{x} = (x_1, x_2, \ldots, x_n)^T$. As in the scalar case (i.e., for $n = 1$), we write $\underline{X} \sim \mathcal{N}_n(\underline{\mu}, \mathbf{K}_{\underline{X}})$ to denote that $\underline{X}$ is a size-n Gaussian random vector with mean vector $\underline{\mu}$ and covariance matrix $\mathbf{K}_{\underline{X}}$.

Observation 5.17 In light of the above definition, we make the following remarks.

1. Note that a covariance matrix $\mathbf{K}$ is always symmetric (i.e., $\mathbf{K}^T = \mathbf{K}$) and positive-semidefinite.[7] But as we require $\mathbf{K}_{\underline{X}}$ to be invertible in the definition of the multivariate Gaussian distribution above, we will hereafter assume that the covariance

---

[6]Note that the diagonal components of $\mathbf{K}_{\underline{X}}$ yield the variance of the different random variables: $K_{i,i} = \mathrm{Cov}(X_i, X_i) = \mathrm{Var}(X_i) = \sigma_{X_i}^2$, $i = 1, \ldots, n$.

[7]An $n \times n$ real-valued symmetric matrix $\mathbf{K}$ is *positive-semidefinite* (e.g., cf. [128]) if for every real-valued vector $\underline{x} = (x_1, x_2, \ldots, x_n)^T$,

$$\underline{x}^T \mathbf{K} \underline{x} = (x_1, \ldots, x_n)\mathbf{K} \begin{pmatrix} x_1 \\ \vdots \\ x_n \end{pmatrix} \geq 0,$$

with equality holding only when $x_i = 0$ for $i = 1, 2, \ldots, n$. Furthermore, the matrix is positive-definite if $\underline{x}^T \mathbf{K} \underline{x} > 0$ for all real-valued vectors $\underline{x} \neq \underline{0}$, where $\underline{0}$ is the all-zero vector of size n.

matrix of Gaussian random vectors is positive-definite (which is equivalent to having all the eigenvalues of $\mathbf{K}_{\underline{X}}$ positive), thus rendering the matrix invertible.

2. If a random vector $\underline{X} = (X_1, X_2, \ldots, X_n)^T$ has a diagonal covariance matrix $\mathbf{K}_{\underline{X}}$ (i.e., all the off-diagonal components of $\mathbf{K}_{\underline{X}}$ are zero: $K_{i,j} = 0$ for all $i \neq j$, $i, j = 1, \ldots, n$), then all its component random variables are uncorrelated but not necessarily independent. However, if $\underline{X}$ is Gaussian and have a diagonal covariance matrix, then all its component random variables are independent from each other.

3. Any linear transformation of a Gaussian random vector yields another Gaussian random vector. Specifically, if $\underline{X} \sim \mathcal{N}_n(\underline{\mu}, \mathbf{K}_{\underline{X}})$ is a size-n Gaussian random vector with mean vector $\underline{\mu}$ and covariance matrix $\mathbf{K}_{\underline{X}}$, and if $\underline{Y} = \mathbf{A}_{mn}\underline{X}$, where $\mathbf{A}_{mn}$ is a given $m \times n$ real-valued matrix, then

$$\underline{Y} \sim \mathcal{N}_m(\mathbf{A}_{mn}\underline{\mu}, \mathbf{A}_{mn}\mathbf{K}_{\underline{X}}\mathbf{A}_{mn}^T)$$

is a size-m Gaussian random vector with mean vector $\mathbf{A}_{mn}\underline{\mu}$ and covariance matrix $\mathbf{A}_{mn}\mathbf{K}_{\underline{X}}\mathbf{A}_{mn}^T$.

More generally, any affine transformation of a Gaussian random vector yields another Gaussian random vector: if $\underline{X} \sim \mathcal{N}_n(\underline{\mu}, \mathbf{K}_{\underline{X}})$ and $\underline{Y} = \mathbf{A}_{mn}\underline{X} + \underline{b}_m$, where $\mathbf{A}_{mn}$ is a $m \times n$ real-valued matrix and $\underline{b}_m$ is a size-m real-valued vector, then

$$\underline{Y} \sim \mathcal{N}_m(\mathbf{A}_{mn}\underline{\mu} + \underline{b}_m, \mathbf{A}_{mn}\mathbf{K}_{\underline{X}}\mathbf{A}_{mn}^T).$$

Theorem 5.18 (Joint differential entropy of the multivariate Gaussian) *If $\underline{X} \sim \mathcal{N}_n(\underline{\mu}, \mathbf{K}_{\underline{X}})$ is a Gaussian random vector with mean vector $\underline{\mu}$ and (positive-definite) covariance matrix $\mathbf{K}_{\underline{X}}$, then its joint differential entropy is given by*

$$h(\underline{X}) = h(X_1, X_2, \ldots, X_n) = \frac{1}{2}\log_2\left[(2\pi e)^n \det(\mathbf{K}_{\underline{X}})\right]. \tag{5.2.1}$$

In particular, in the univariate case of $n = 1$, (5.2.1) reduces to (5.1.1).

Proof Without loss of generality, we assume that $\underline{X}$ has a zero-mean vector since its differential entropy is invariant under translation by Property 8 of Lemma 5.14:

$$h(\underline{X}) = h(\underline{X} - \underline{\mu});$$

so we assume that $\underline{\mu} = \underline{0}$.

Since the covariance matrix $\mathbf{K}_{\underline{X}}$ is a real-valued symmetric matrix, then it is orthogonally diagonalizable; i.e., there exists a square $(n \times n)$ orthogonal matrix $\mathbf{A}$ (i.e., satisfying $\mathbf{A}^T = \mathbf{A}^{-1}$) such that $\mathbf{A}\mathbf{K}_{\underline{X}}\mathbf{A}^T$ is a diagonal matrix whose entries are given by the eigenvalues of $\mathbf{K}_{\underline{X}}$ ($\mathbf{A}$ is constructed using the eigenvectors of $\mathbf{K}_{\underline{X}}$; e.g., see [128]). As a result, the linear transformation $\underline{Y} = \mathbf{A}\underline{X} \sim \mathcal{N}_n\left(\underline{0}, \mathbf{A}\mathbf{K}_{\underline{X}}\mathbf{A}^T\right)$

is a Gaussian vector with the diagonal covariance matrix $\mathbf{K}_{\underline{Y}} = \mathbf{A}\mathbf{K}_{\underline{X}}\mathbf{A}^T$ and has therefore independent components (as noted in Observation 5.17). Thus

$$\begin{aligned} h(\underline{Y}) &= h(Y_1, Y_2, \ldots, Y_n) \\ &= h(Y_1) + h(Y_2) + \cdots + h(Y_n) \end{aligned}$$ (5.2.2)

$$= \sum_{i=1}^{n} \frac{1}{2} \log_2 [2\pi e \operatorname{Var}(Y_i)]$$ (5.2.3)

$$= \frac{n}{2} \log_2(2\pi e) + \frac{1}{2} \log_2 \left[\prod_{i=1}^{n} \operatorname{Var}(Y_i) \right]$$

$$= \frac{n}{2} \log_2(2\pi e) + \frac{1}{2} \log_2 \left[\det\left(\mathbf{K}_{\underline{Y}}\right) \right]$$ (5.2.4)

$$= \frac{1}{2} \log_2 (2\pi e)^n + \frac{1}{2} \log_2 \left[\det\left(\mathbf{K}_{\underline{X}}\right) \right]$$ (5.2.5)

$$= \frac{1}{2} \log_2 \left[(2\pi e)^n \det\left(\mathbf{K}_{\underline{X}}\right) \right],$$ (5.2.6)

where (5.2.2) follows by the independence of the random variables $Y_1, \ldots, Y_n$ (e.g., see Property 7 of Lemma 5.14), (5.2.3) follows from (5.1.1), (5.2.4) holds since the matrix $\mathbf{K}_{\underline{Y}}$ is diagonal and hence its determinant is given by the product of its diagonal entries, and (5.2.5) holds since

$$\begin{aligned} \det\left(\mathbf{K}_{\underline{Y}}\right) &= \det\left(\mathbf{A}\mathbf{K}_{\underline{X}}\mathbf{A}^T\right) \\ &= \det(\mathbf{A})\det\left(\mathbf{K}_{\underline{X}}\right)\det(\mathbf{A}^T) \\ &= \det(\mathbf{A})^2\det\left(\mathbf{K}_{\underline{X}}\right) \\ &= \det\left(\mathbf{K}_{\underline{X}}\right), \end{aligned}$$

where the last equality holds since $(\det(\mathbf{A}))^2 = 1$, as the matrix $\mathbf{A}$ is orthogonal $(\mathbf{A}^T = \mathbf{A}^{-1} \implies \det(\mathbf{A}) = \det(\mathbf{A}^T) = 1/[\det(\mathbf{A})]$; thus, $\det(\mathbf{A})^2 = 1)$.
 Now invoking Property 10 of Lemma 5.14 and noting that $|\det(\mathbf{A})| = 1$ yield that

$$h(Y_1, Y_2, \ldots, Y_n) = h(X_1, X_2, \ldots, X_n) + \underbrace{\log_2 |\det(\mathbf{A})|}_{=0} = h(X_1, X_2, \ldots, X_n).$$

We therefore obtain using (5.2.6) that

$$h(X_1, X_2, \ldots, X_n) = \frac{1}{2} \log_2 \left[(2\pi e)^n \det\left(\mathbf{K}_{\underline{X}}\right) \right],$$

hence completing the proof.
 An alternate (but rather mechanical) proof to the one presented above consists of directly evaluating the joint differential entropy of $\underline{X}$ by integrating $-f_{X^n}(x^n)$ $\log_2 f_{X^n}(x^n)$ over $\mathbb{R}^n$; it is left as an exercise. □

Corollary 5.19 (Hadamard's inequality) *For any real-valued $n \times n$ positive-definite matrix* $\mathbf{K} = [K_{i,j}]_{i,j=1,\ldots,n}$,

$$\det(\mathbf{K}) \leq \prod_{i=1}^{n} K_{i,i},$$

with equality iff $\mathbf{K}$ *is a diagonal matrix, where* $K_{i,i}$ *are the diagonal entries of* $\mathbf{K}$.

Proof Since every positive-definite matrix is a covariance matrix (e.g., see [162]), let $\underline{X} = (X_1, X_2, \ldots, X_n)^T \sim \mathcal{N}_n\left(\underline{0}, \mathbf{K}\right)$ be a jointly Gaussian random vector with zero-mean vector and covariance matrix $\mathbf{K}$. Then

$$\frac{1}{2} \log_2 \left[(2\pi e)^n \det(\mathbf{K})\right] = h(X_1, X_2, \ldots, X_n) \qquad (5.2.7)$$

$$\leq \sum_{i=1}^{n} h(X_i) \qquad (5.2.8)$$

$$= \sum_{i=1}^{n} \frac{1}{2} \log_2 \left[2\pi e \mathrm{Var}(X_i)\right] \qquad (5.2.9)$$

$$= \frac{1}{2} \log_2 \left[(2\pi e)^n \prod_{i=1}^{n} K_{i,i}\right], \qquad (5.2.10)$$

where (5.2.7) follows from Theorem 5.18, (5.2.8) follows from Property 7 of Lemma 5.14 and (5.2.9)–(5.2.10) hold using (5.1.1) along with the fact that each random variable $X_i \sim \mathcal{N}(0, K_{i,i})$ is Gaussian with zero mean and variance $\mathrm{Var}(X_i) = K_{i,i}$ for $i = 1, 2, \ldots, n$ (as the marginals of a multivariate Gaussian are also Gaussian e.g., cf. [162]).

Finally, from (5.2.10), we directly obtain that

$$\det(\mathbf{K}) \leq \prod_{i=1}^{n} K_{i,i},$$

with equality iff the jointly Gaussian random variables $X_1, X_2, \ldots, X_n$ are independent from each other, or equivalently iff the covariance matrix $\mathbf{K}$ is diagonal. □

The next theorem states that among all real-valued size-n random vectors (of support $\mathbb{R}^n$) with identical mean vector and covariance matrix, the *Gaussian* random vector has the largest differential entropy.

Theorem 5.20 (Maximal differential entropy for real-valued random vectors) *Let* $\underline{X} = (X_1, X_2, \ldots, X_n)^T$ *be a real-valued random vector with a joint pdf of support* $S_{X^n} = \mathbb{R}^n$, *mean vector* μ, *covariance matrix* $\mathbf{K}_{\underline{X}}$ *and finite joint differential entropy* $h(X_1, X_2, \ldots, X_n)$. *Then*

$$h(X_1, X_2, \ldots, X_n) \leq \frac{1}{2} \log_2 \left[(2\pi e)^n \det(\mathbf{K}_{\underline{X}}) \right], \tag{5.2.11}$$

with equality iff $\underline{X}$ is Gaussian; i.e., $\underline{X} \sim \mathcal{N}_n \left(\underline{\mu}, \mathbf{K}_{\underline{X}} \right)$.

Proof We will present the proof in two parts: the scalar or univariate case, and the multivariate case.

(i) *Scalar case* $(n = 1)$: For a real-valued random variable with support $S_X = \mathbb{R}$, mean μ and variance σ^2, let us show that

$$h(X) \leq \frac{1}{2} \log_2 \left(2\pi e \sigma^2 \right), \tag{5.2.12}$$

with equality iff $X \sim \mathcal{N}(\mu, \sigma^2)$.

For a Gaussian random variable $Y \sim \mathcal{N}(\mu, \sigma^2)$, using the nonnegativity of divergence, we can write

$$0 \leq D(X \| Y)$$
$$= \int_{\mathbb{R}} f_X(x) \log_2 \frac{f_X(x)}{\frac{1}{\sqrt{2\pi\sigma^2}} e^{-\frac{(x-\mu)^2}{2\sigma^2}}} \, dx$$
$$= -h(X) + \int_{\mathbb{R}} f_X(x) \left[\log_2 \left(\sqrt{2\pi\sigma^2} \right) + \frac{(x-\mu)^2}{2\sigma^2} \log_2 e \right] dx$$
$$= -h(X) + \frac{1}{2} \log_2(2\pi\sigma^2) + \frac{\log_2 e}{2\sigma^2} \underbrace{\int_{\mathbb{R}} (x-\mu)^2 f_X(x) \, dx}_{=\sigma^2}$$
$$= -h(X) + \frac{1}{2} \log_2 \left[2\pi e \sigma^2 \right].$$

Thus

$$h(X) \leq \frac{1}{2} \log_2 \left[2\pi e \sigma^2 \right],$$

with equality iff $X = Y$ (almost surely); i.e., $X \sim \mathcal{N}(\mu, \sigma^2)$.

(ii). *Multivariate case* $(n > 1)$: As in the proof of Theorem 5.18, we can use an orthogonal square matrix $\mathbf{A}$ (i.e., satisfying $\mathbf{A}^T = \mathbf{A}^{-1}$ and hence $|\det(\mathbf{A})| = 1$) such that $\mathbf{A}\mathbf{K}_{\underline{X}}\mathbf{A}^T$ is diagonal. Therefore, the random vector generated by the linear map

$$\underline{Z} = \mathbf{A}\underline{X}$$

will have a covariance matrix given by $\mathbf{K}_{\underline{Z}} = \mathbf{A}\mathbf{K}_{\underline{X}}\mathbf{A}^T$ and hence have uncorrelated (but not necessarily independent) components. Thus

$$h(\underline{X}) = h(\underline{Z}) - \underbrace{\log_2 |\det(\mathbf{A})|}_{=0} \tag{5.2.13}$$

$$= h(Z_1, Z_2, \ldots, Z_n)$$

$$\leq \sum_{i=1}^{n} h(Z_i) \tag{5.2.14}$$

$$\leq \sum_{i=1}^{n} \frac{1}{2} \log_2 [2\pi e \mathrm{Var}(Z_i)] \tag{5.2.15}$$

$$= \frac{n}{2} \log_2(2\pi e) + \frac{1}{2} \log_2 \left[\prod_{i=1}^{n} \mathrm{Var}(Z_i) \right]$$

$$= \frac{1}{2} \log_2 (2\pi e)^n + \frac{1}{2} \log_2 \left[\det \left(\mathbf{K}_{\underline{Z}} \right) \right] \tag{5.2.16}$$

$$= \frac{1}{2} \log_2 (2\pi e)^n + \frac{1}{2} \log_2 \left[\det \left(\mathbf{K}_{\underline{X}} \right) \right] \tag{5.2.17}$$

$$= \frac{1}{2} \log_2 \left[(2\pi e)^n \det \left(\mathbf{K}_{\underline{X}} \right) \right],$$

where (5.2.13) holds by Property 10 of Lemma 5.14 and since $|\det(\mathbf{A})| = 1$, (5.2.14) follows from Property 7 of Lemma 5.14, (5.2.15) follows from (5.2.12) (the scalar case above), (5.2.16) holds since $\mathbf{K}_{\underline{Z}}$ is diagonal, and (5.2.17) follows from the fact that $\det \left(\mathbf{K}_{\underline{Z}} \right) = \det \left(\mathbf{K}_{\underline{X}} \right)$ (as $\mathbf{A}$ is orthogonal). Finally, equality is achieved in both (5.2.14) and (5.2.15) iff the random variables Z_1, Z_2, ..., Z_n are Gaussian and independent from each other, or equivalently iff $\underline{X} \sim \mathcal{N}_n \left(\underline{\mu}, \mathbf{K}_{\underline{X}} \right)$.   □

Observation 5.21 (*Examples of maximal differential entropy under various constraints*) The following three results can also be shown (the proof is left as an exercise):

1. Among all continuous random variables admitting a pdf with support the interval (a, b), where $b > a$ are real numbers, the uniformly distributed random variable maximizes differential entropy.
2. Among all continuous random variables admitting a pdf with support the interval $[0, \infty)$, finite mean μ, and finite differential entropy, the exponentially distributed random variable with parameter (or rate parameter) $\lambda = 1/\mu$ maximizes differential entropy.
3. Among all continuous random variables admitting a pdf with support $\mathbb{R}$, finite mean μ, and finite differential entropy and satisfying $E[|X - \mu|] = \lambda$, where $\lambda > 0$ is a fixed finite parameter, the Laplacian random variable with mean μ, variance $2\lambda^2$ and pdf

$$f_X(x) = \frac{1}{2\lambda} e^{-\frac{|x-\mu|}{\lambda}} \text{ for } x \in \mathbb{R}$$

maximizes differential entropy.

A systematic approach to finding distributions that maximize differential entropy subject to various support and moments constraints can be found in [83, 415].

Observation 5.22 (*Information rates for stationary Gaussian sources*) We close this section by noting that for stationary zero-mean Gaussian processes $\{X_i\}$ and $\{\hat{X}_i\}$, the differential entropy rate, $\lim_{n\to\infty} \frac{1}{n} h(X^n)$, the divergence rate, $\lim_{n\to\infty} \frac{1}{n} D(X^n \| \hat{X}^n)$, as well as their Rényi counterparts all exist and admit analytical expressions in terms of the source power spectral densities [154, 196, 223, 393], [144, Table 4]. In particular, the differential entropy rate of $\{X_i\}$ and the divergence rate between $\{X_i\}$ and $\{\hat{X}_i\}$ are given (in nats) by

$$\lim_{n\to\infty} \frac{1}{n} h(X^n) = \frac{1}{2} \ln(2\pi e) + \frac{1}{4\pi} \int_{-\pi}^{\pi} \ln \phi_X(\lambda) \, d\lambda, \qquad (5.2.18)$$

and

$$\lim_{n\to\infty} \frac{1}{n} D(X^n \| \hat{X}^n) = \frac{1}{4\pi} \int_{-\pi}^{\pi} \left(\frac{\phi_X(\lambda)}{\phi_{\hat{X}}(\lambda)} - 1 - \ln \frac{\phi_X(\lambda)}{\phi_{\hat{X}}(\lambda)} \right) d\lambda, \qquad (5.2.19)$$

respectively. Here, $\phi_X(\cdot)$ and $\phi_{\hat{X}}(\cdot)$ denote the power spectral densities of the zero-mean stationary Gaussian processes $\{X_i\}$ and $\{\hat{X}_i\}$, respectively. Recall that for a stationary zero-mean process $\{Z_i\}$, its power spectral density $\phi_Z(\cdot)$ is the (discrete-time) Fourier transform of its covariance function $K_Z(\tau) := E[Z_{n+\tau} Z_n] - E[Z_{n+\tau}] E[Z_n] = E[Z_{n+\tau} Z_n]$, $n, \tau = 1, 2, \ldots$; more precisely,

$$\phi_Z(\lambda) = \sum_{\tau=-\infty}^{\infty} K_Z(\tau) e^{-j\tau\lambda}, \quad -\pi \le \lambda \le \pi,$$

where $j = \sqrt{-1}$ is the imaginary unit number. Note that (5.2.18) and (5.2.19) hold under mild integrability and boundedness conditions; see [196, Sect. 2.4] for the details.

5.3   AEP for Continuous Memoryless Sources

The AEP theorem and its consequence for discrete memoryless (i.i.d.) sources reveal to us that the number of elements in the typical set is approximately $2^{nH(X)}$, where $H(X)$ is the source entropy, and that the typical set carries almost all the probability mass asymptotically (see Theorems 3.4 and 3.5). An extension of this result from discrete to continuous memoryless sources by just counting the number of elements in a continuous (typical) set defined via a law of large numbers argument is not possible, since the total number of elements in a continuous set is infinite. However, when considering the *volume* of that continuous typical set (which is a natural analog to the size of a discrete set), such an extension, with differential entropy playing a similar role as entropy, becomes straightforward.

Theorem 5.23 (AEP for continuous memoryless sources) *Let* $\{X_i\}_{i=1}^{\infty}$ *be a continuous memoryless source (i.e., an infinite sequence of continuous i.i.d. random variables) with pdf* $f_X(\cdot)$ *and differential entropy* $h(X)$*. Then*

$$-\frac{1}{n}\log f_X(X_1,\ldots,X_n) \to E[-\log_2 f_X(X)] = h(X) \quad \text{in probability.}$$

Proof The proof is an immediate result of the law of large numbers (e.g., see Theorem 3.4). ☐

Definition 5.24 (*Typical set*) For $\delta > 0$ and any n given, define the typical set for the above continuous source as

$$\mathcal{F}_n(\delta) := \left\{ x^n \in \mathbb{R}^n : \left| -\frac{1}{n}\log_2 f_X(X_1,\ldots,X_n) - h(X) \right| < \delta \right\}.$$

Definition 5.25 (*Volume*) The *volume* of a set $\mathcal{A} \subset \mathbb{R}^n$ is defined as

$$\text{Vol}(\mathcal{A}) := \int_{\mathcal{A}} dx_1 \cdots dx_n.$$

Theorem 5.26 (Consequence of the AEP for continuous memoryless sources) *For a continuous memoryless source* $\{X_i\}_{i=1}^{\infty}$ *with differential entropy* $h(X)$*, the following hold.*

1. *For* n *sufficiently large,* $P_{X^n}\{\mathcal{F}_n(\delta)\} > 1 - \delta$*.*
2. $\text{Vol}(\mathcal{F}_n(\delta)) \leq 2^{n(h(X)+\delta)}$ *for all* n*.*
3. $\text{Vol}(\mathcal{F}_n(\delta)) \geq (1-\delta)2^{n(h(X)-\delta)}$ *for* n *sufficiently large.*

Proof The proof is quite analogous to the corresponding theorem for discrete memoryless sources (Theorem 3.5) and is left as an exercise. ☐

5.4 Capacity and Channel Coding Theorem for the Discrete-Time Memoryless Gaussian Channel

We next study the fundamental limits for error-free communication over the discrete-time memoryless Gaussian channel, which is the most important continuous-alphabet channel and is widely used to model real-world wired and wireless channels. We first state the definition of discrete-time continuous-alphabet memoryless channels.

Definition 5.27 (*Discrete-time continuous memoryless channels*) Consider a discrete-time channel with continuous input and output alphabets given by $\mathcal{X} \subseteq \mathbb{R}$ and $\mathcal{Y} \subseteq \mathbb{R}$, respectively, and described by a sequence of n-dimensional transition (conditional) pdfs $\{f_{Y^n|X^n}(y^n|x^n)\}_{n=1}^{\infty}$ that govern the reception of $y^n =$

$(y_1, y_2, \ldots, y_n) \in \mathcal{Y}^n$ at the channel output when $x^n = (x_1, x_2, \ldots, x_n) \in \mathcal{X}^n$ is sent as the channel input.

The channel (without feedback) is said to be memoryless with a given (marginal) transition pdf $f_{Y|X}$ if its sequence of transition pdfs $f_{Y^n|X^n}$ satisfies

$$f_{Y^n|X^n}(y^n|x^n) = \prod_{i=1}^{n} f_{Y|X}(y_i|x_i) \tag{5.4.1}$$

for every $n = 1, 2, \ldots, x^n \in \mathcal{X}^n$ and $y^n \in \mathcal{Y}^n$.

In practice, the real-valued input to a continuous channel satisfies a certain constraint or limitation on its amplitude or power; otherwise, one would have a realistically implausible situation where the input can take on any value from the uncountably infinite set of real numbers. We will thus impose an *average cost constraint* (t, P) on any input n-tuple $x^n = (x_1, x_2, \ldots, x_n)$ transmitted over the channel by requiring that

$$\frac{1}{n} \sum_{i=1}^{n} t(x_i) \leq P, \tag{5.4.2}$$

where $t(\cdot)$ is a given nonnegative real-valued function describing the cost for transmitting an input symbol, and P is a given positive number representing the maximal average amount of available resources per input symbol.

Definition 5.28 The capacity (or capacity-cost function) of a discrete-time continuous memoryless channel with input average cost constraint (t, P) is denoted by $C(P)$ and defined as

$$C(P) := \sup_{F_X:\, E[t(X)] \leq P} I(X; Y) \quad \text{(in bits/channel use)}, \tag{5.4.3}$$

where the supremum is over all input distributions F_X.

Lemma 5.29 (Concavity of capacity) *If $C(P)$ as defined in (5.4.3) is finite for any $P > 0$, then it is concave, continuous, and strictly increasing in P.*

Proof Fix $P_1 > 0$ and $P_2 > 0$. Then since $C(P)$ is finite for any $P > 0$, then by the third property in Property A.4, there exist two input distributions F_{X_1} and F_{X_2} such that for all $\epsilon > 0$,

$$I(X_i; Y_i) \geq C(P_i) - \epsilon \tag{5.4.4}$$

and

$$E[t(X_i)] \leq P_i, \tag{5.4.5}$$

where X_i denotes the input with distribution F_{X_i} and Y_i is the corresponding channel output for $i = 1, 2$. Now, for $0 \leq \lambda \leq 1$, let X_λ be a random variable with distribution $F_{X_\lambda} := \lambda F_{X_1} + (1 - \lambda) F_{X_2}$. Then by (5.4.5)

$$E_{X_\lambda}[t(X)] = \lambda E_{X_1}[t(X)] + (1 - \lambda)E_{X_2}[t(X)] \leq \lambda P_1 + (1 - \lambda)P_2. \qquad (5.4.6)$$

Furthermore,

$$
\begin{aligned}
C(\lambda P_1 + (1 - \lambda)P_2) &= \sup_{F_X:\; E[t(X)] \leq \lambda P_1 + (1-\lambda)P_2} I(F_X, f_{Y|X}) \\
&\geq I(F_{X_\lambda}, f_{Y|X}) \\
&\geq \lambda I(F_{X_1}, f_{Y|X}) + (1 - \lambda)I(F_{X_2}, f_{Y|X}) \\
&= \lambda I(X_1; Y_1) + (1 - \lambda)I(X_2; Y_2) \\
&\geq \lambda C(P_1) - \epsilon + (1 - \lambda)C(P_2) - \epsilon,
\end{aligned}
$$

where the first inequality holds by (5.4.6), the second inequality follows from the concavity of the mutual information with respect to its first argument (cf. Lemma 2.46), and the third inequality follows from (5.4.4). Letting $\epsilon \to 0$ yields that

$$C(\lambda P_1 + (1 - \lambda)P_2) \geq \lambda C(P_1) + (1 - \lambda)C(P_2)$$

and hence $C(P)$ is concave in P.

Finally, it can directly be seen by definition that $C(\cdot)$ is nondecreasing, which, together with its concavity, imply that it is continuous and strictly increasing. □

The most commonly used cost function is the power cost function, $t(x) = x^2$, resulting in the *average power constraint* P for each transmitted input n-tuple:

$$\frac{1}{n}\sum_{i=1}^{n} x_i^2 \leq P. \qquad (5.4.7)$$

Throughout this chapter, we will adopt this average power constraint on the channel input.

We herein focus on the discrete-time memoryless Gaussian channel[8] with average input power constraint P and establish an operational meaning for the channel capacity $C(P)$ as the largest coding rate for achieving reliable communication over the channel. The channel is described by the following additive noise equation:

$$Y_i = X_i + Z_i, \quad \text{for } i = 1, 2, \ldots, \qquad (5.4.8)$$

where Y_i, X_i, and Z_i are the channel output, input and noise at time i. The input and noise processes are assumed to be independent from each other and the noise source $\{Z_i\}_{i=1}^{\infty}$ is i.i.d. Gaussian with each Z_i having mean zero and variance σ^2, $Z_i \sim \mathcal{N}(0, \sigma^2)$. Since the noise process is i.i.d, we directly get that the channel satisfies (5.4.1) and is hence memoryless, where the channel transition pdf is explicitly given in terms of the noise pdf as follows:

---

[8]This channel is also commonly referred to as the discrete-time additive white Gaussian noise (AWGN) channel.

$$f_{Y|X}(y|x) = f_Z(y-x) = \frac{1}{\sqrt{2\pi\sigma^2}} e^{-\frac{(y-x)^2}{2\sigma^2}}.$$

As mentioned above, we impose the average power constraint (5.4.7) on the channel input.

Observation 5.30 The memoryless Gaussian channel is a good approximating model for many practical channels such as radio, satellite, and telephone line channels. The additive noise is usually due to a multitude of causes, whose cumulative effect can be approximated via the Gaussian distribution. This is justified by the central limit theorem which states that for an i.i.d. process $\{U_i\}_{i=1}^{\infty}$ with mean μ and variance σ^2, $\frac{1}{\sqrt{n}}\sum_{i=1}^{n}(U_i - \mu)$ converges in distribution as $n \to \infty$ to a Gaussian distributed random variable with mean zero and variance σ^2 (see Appendix B).[9]

Before proving the channel coding theorem for the above memoryless Gaussian channel with input power constraint P, we first show that its capacity $C(P)$ as defined in (5.4.3) with $t(x) = x^2$ admits a simple expression in terms of P and the channel noise variance σ^2. Indeed, we can write the channel mutual information $I(X;Y)$ between its input and output as follows:

$$
\begin{aligned}
I(X;Y) &= h(Y) - h(Y|X) \\
&= h(Y) - h(X + Z|X) & (5.4.9) \\
&= h(Y) - h(Z|X) & (5.4.10) \\
&= h(Y) - h(Z) & (5.4.11) \\
&= h(Y) - \frac{1}{2}\log_2\left(2\pi e\sigma^2\right), & (5.4.12)
\end{aligned}
$$

where (5.4.9) follows from (5.4.8), (5.4.10) holds since differential entropy is invariant under translation (see Property 8 of Lemma 5.14), (5.4.11) follows from the independence of X and Z, and (5.4.12) holds since $Z \sim \mathcal{N}(0, \sigma^2)$ is Gaussian (see (5.1.1)). Now since $Y = X + Z$, we have that

$$E[Y^2] = E[X^2] + E[Z^2] + 2E[X]E[Z] = E[X^2] + \sigma^2 + 2E[X](0) \le P + \sigma^2$$

since the input in (5.4.3) is constrained to satisfy $E[X^2] \le P$. Thus, the variance of Y satisfies $\mathrm{Var}(Y) \le E[Y^2] \le P + \sigma^2$, and

$$h(Y) \le \frac{1}{2}\log_2\left(2\pi e \mathrm{Var}(Y)\right) \le \frac{1}{2}\log_2\left(2\pi e(P + \sigma^2)\right),$$

where the first inequality follows by Theorem 5.20 since Y is real-valued (with support $\mathbb{R}$). Noting that equality holds in the first inequality above iff Y is Gaussian and in the second inequality iff $\mathrm{Var}(Y) = P + \sigma^2$, we obtain that choosing the input X as $X \sim \mathcal{N}(0, P)$ yields $Y \sim \mathcal{N}(0, P + \sigma^2)$ and hence maximizes $I(X;Y)$ over

---

[9]The reader is referred to [209] for an information theoretic treatment of the central limit theorem.

all inputs satisfying $E[X^2] \leq P$. Thus, the capacity of the discrete-time memoryless Gaussian channel with input average power constraint P and noise variance (or power) σ^2 is given by

$$C(P) = \frac{1}{2} \log_2 \left(2\pi e (P + \sigma^2) \right) - \frac{1}{2} \log_2 \left(2\pi e \sigma^2 \right)$$
$$= \frac{1}{2} \log_2 \left(1 + \frac{P}{\sigma^2} \right). \tag{5.4.13}$$

Note P/σ^2 is called the channel's *signal-to-noise ratio* (SNR) and is usually measured in decibels (dB).[10]

Definition 5.31 Given positive integers n and M, and a discrete-time memoryless Gaussian channel with input average power constraint P, a fixed-length data transmission code (or block code) $\mathcal{C}_n = (n, M)$ for this channel with blocklength n and rate (or code rate) $\frac{1}{n} \log_2 M$ message bits per channel symbol (or channel use) consists of:

1. M information messages intended for transmission.
2. An encoding function

$$f : \{1, 2, \ldots, M\} \rightarrow \mathbb{R}^n$$

yielding real-valued codewords $c_1 = f(1), c_2 = f(2), \ldots, c_M = f(M)$, where each codeword $c_m = (c_{m1}, \ldots, c_{mn})$ is of length n and satisfies the power constraint P

$$\frac{1}{n} \sum_{i=1}^{n} c_i^2 \leq P,$$

for $m = 1, 2, \ldots, M$. The set of these M codewords is called the codebook and we usually write $\mathcal{C}_n = \{c_1, c_2, \ldots, c_M\}$ to list the codewords.
3. A decoding function $g : \mathbb{R}^n \rightarrow \{1, 2, \ldots, M\}$.

As in Chap. 4, we assume that a message W follows a uniform distribution over the set of messages: $\Pr[W = w] = \frac{1}{M}$ for all $w \in \{1, 2, \ldots, M\}$. Similarly, to convey message W over the channel, the encoder sends its corresponding codeword $X^n = f(W) \in \mathcal{C}_n$ at the channel input. Finally, Y^n is received at the channel output and the decoder yields $\hat{W} = g(Y^n)$ as the message estimate. Also, the average probability of error for this block code used over the memoryless Gaussian channel is defined as

$$P_e(\mathcal{C}_n) := \frac{1}{M} \sum_{w=1}^{M} \lambda_w(\mathcal{C}_n),$$

where

---

[10]More specifically, SNR|$_{dB}$:= $10 \log_{10}$ SNR in dB.

$$\lambda_w(\mathcal{C}_n) := \Pr[\hat{W} \neq W \,|\, W = w]$$
$$= \Pr[g(Y^n) \neq w \,|\, X^n = f(w)]$$
$$= \int_{y^n \in \mathbb{R}^n :\, g(y^n) \neq w} f_{Y^n|X^n}(y^n \,|\, f(w))\, dy^n$$

is the code's conditional probability of decoding error given that message w is sent over the channel. Here

$$f_{Y^n|X^n}(y^n|x^n) = \prod_{i=1}^{n} f_{Y|X}(y_i|x_i)$$

as the channel is memoryless, where $f_{Y|X}$ is the channel's transition pdf.

We next prove that for a memoryless Gaussian channel with input average power constraint P, its capacity $C(P)$ is the channel's operational capacity; i.e., it is the supremum of all rates for which there exists a sequence of data transmission block codes satisfying the power constraint and having a probability of error that vanishes with increasing blocklength.

Theorem 5.32 (Shannon's coding theorem for the memoryless Gaussian channel) *Consider a discrete-time memoryless Gaussian channel with input average power constraint P, channel noise variance σ^2 and capacity $C(P)$ as given by (5.4.13).*

- *Forward part (achievability): For any $\varepsilon \in (0, 1)$, there exist $0 < \gamma < 2\varepsilon$ and a sequence of data transmission block code $\{\mathcal{C}_n = (n, M_n)\}_{n=1}^{\infty}$ satisfying*

$$\frac{1}{n} \log_2 M_n > C(P) - \gamma$$

with each codeword $c = (c_1, c_2, \ldots, c_n)$ in $\mathcal{C}_n$ satisfying

$$\frac{1}{n} \sum_{i=1}^{n} c_i^2 \leq P \tag{5.4.14}$$

such that the probability of error $P_e(\mathcal{C}_n) < \varepsilon$ for sufficiently large n.
- *Converse part: If for any sequence of data transmission block codes $\{\mathcal{C}_n = (n, M_n)\}_{n=1}^{\infty}$ whose codewords satisfy (5.4.14), we have that*

$$\liminf_{n \to \infty} \frac{1}{n} \log_2 M_n > C(P),$$

then the codes' probability of error $P_e(\mathcal{C}_n)$ is bounded away from zero for all n sufficiently large.

Proof of the forward part: The theorem holds trivially when $C(P) = 0$ because we can choose $M_n = 1$ for every n and have $P_e(\mathcal{C}_n) = 0$. Hence, we assume without loss of generality $C(P) > 0$.

Step 0:

Take a positive γ satisfying $\gamma < \min\{2\varepsilon, C(P)\}$. Pick $\xi > 0$ small enough such that $2[C(P) - C(P - \xi)] < \gamma$, where the existence of such ξ is assured by the strictly increasing property of $C(P)$. Hence, we have $C(P - \xi) - \gamma/2 > C(P) - \gamma > 0$. Choose M_n to satisfy

$$C(P - \xi) - \frac{\gamma}{2} > \frac{1}{n}\log_2 M_n > C(P) - \gamma,$$

for which the choice should exist for all sufficiently large n. Take $\delta = \gamma/8$. Let F_X be the distribution that achieves $C(P - \xi)$, where $C(P)$ is given by (5.4.13). In this case, F_X is the Gaussian distribution with mean zero and variance $P - \xi$ and admits a pdf f_X. Hence, $E[X^2] \leq P - \xi$ and $I(X; Y) = C(P - \xi)$.

Step 1: Random coding with average power constraint.

Randomly draw M_n codewords according to pdf f_{X^n} with

$$f_{X^n}(x^n) = \prod_{i=1}^{n} f_X(x_i).$$

By the law of large numbers, each randomly selected codeword

$$\boldsymbol{c}_m = (c_{m1}, \ldots, c_{mn})$$

satisfies

$$\lim_{n \to \infty} \frac{1}{n} \sum_{i=1}^{n} c_{mi}^2 = E[X^2] \leq P - \xi$$

for $m = 1, 2, \ldots, M_n$.

Step 2: Code construction.

For M_n selected codewords $\{\boldsymbol{c}_1, \ldots, \boldsymbol{c}_{M_n}\}$, replace the codewords that violate the power constraint (i.e., (5.4.14)) by an all-zero (default) codeword $\boldsymbol{0}$. Define the encoder as

$$f_n(m) = \boldsymbol{c}_m \quad \text{for} \quad 1 \leq m \leq M_n.$$

Given a received output sequence y^n, the decoder $g_n(\cdot)$ is given by

$$g_n(y^n) = \begin{cases} m, & \text{if } (\boldsymbol{c}_m, y^n) \in \mathcal{F}_n(\delta) \\ & \quad \text{and } (\forall\, m' \neq m)\, (\boldsymbol{c}_{m'}, y^n) \notin \mathcal{F}_n(\delta), \\ \text{arbitrary,} & \text{otherwise,} \end{cases}$$

where the set

$$\mathcal{F}_n(\delta) := \left\{ (x^n, y^n) \in \mathcal{X}^n \times \mathcal{Y}^n : \left| -\frac{1}{n} \log_2 f_{X^n Y^n}(x^n, y^n) - h(X, Y) \right| < \delta, \right.$$

$$\left| -\frac{1}{n} \log_2 f_{X^n}(x^n) - h(X) \right| < \delta,$$

$$\left. \text{and } \left| -\frac{1}{n} \log_2 f_{Y^n}(y^n) - h(Y) \right| < \delta \right\}$$

is generated by $f_{X^n Y^n}(x^n, y^n) = \prod_{i=1}^{n} f_{X,Y}(x_i, y_i)$ where $f_{X^n Y^n}(x^n, y^n)$ is the joint input–output pdf realized when the memoryless Gaussian channel (with n-fold transition pdf $f_{Y^n|X^n}(y^n|x^n) = \prod_{i=1}^{n} f_{Y|X}(y_i|x_i)$) is driven by input X^n with pdf $f_{X^n}(x^n) = \prod_{i=1}^{n} f_X(x_i)$ (where f_X achieves $C(P - \xi)$).

Step 3: Conditional probability of error.

Let λ_m denote the conditional error probability given codeword m is transmitted. Define

$$\mathcal{E}_0 := \left\{ x^n \in \mathcal{X}^n : \frac{1}{n} \sum_{i=1}^{n} x_i^2 > P \right\}.$$

Then by following similar argument as (4.3.4),[11] we get:

---

[11] In this proof, specifically, (4.3.3) becomes

$$\lambda_m(\mathcal{C}_n) \leq \int_{y^n \notin \mathcal{F}_n(\delta|c_m)} f_{Y^n|X^n}(y^n|c_m) \, dy^n + \sum_{\substack{m'=1 \\ m' \neq m}}^{M_n} \int_{y^n \in \mathcal{F}_n(\delta|c_{m'})} f_{Y^n|X^n}(y^n|c_m) \, dy^n.$$

By taking expectation with respect to the mth codeword-selecting distribution $f_{X^n}(c_m)$, we obtain

$$E[\lambda_m] = \int_{c_m \in \mathcal{X}^n} f_{X^n}(c_m) \lambda_m(\mathcal{C}_n) \, dc_m$$

$$= \int_{c_m \in \mathcal{X}^n \cap \mathcal{E}_0} f_{X^n}(c_m) \lambda_m(\mathcal{C}_n) \, dc_m + \int_{c_m \in \mathcal{X}^n \cap \mathcal{E}_0^c} f_{X^n}(c_m) \lambda_m(\mathcal{C}_n) \, dc_m$$

$$\leq \int_{c_m \in \mathcal{E}_0} f_{X^n}(c_m) \, dc_m + \int_{c_m \in \mathcal{X}^n} f_{X^n}(c_m) \lambda_m(\mathcal{C}_n) \, dc_m$$

$$\leq P_{X^n}(\mathcal{E}_0) + \int_{c_m \in \mathcal{X}^n} \int_{y^n \notin \mathcal{F}_n(\delta|c_m)} f_{X^n}(c_m) f_{Y^n|X^n}(y^n|c_m) \, dy^n \, dc_m$$

$$+ \int_{c_m \in \mathcal{X}^n} \sum_{\substack{m'=1 \\ m' \neq m}}^{M_n} \int_{y^n \in \mathcal{F}_n(\delta|c_{m'})} f_{X^n}(c_m) f_{Y^n|X^n}(y^n|c_m) \, dy^n \, dc_m.$$

$$E[\lambda_m] \le P_{X^n}(\mathcal{E}_0) + P_{X^n,Y^n}\left(\mathcal{F}_n^c(\delta)\right)$$

$$+ \sum_{\substack{m'=1 \\ m' \ne m}}^{M_n} \int_{c_m \in \mathcal{X}^n} \int_{y^n \in \mathcal{F}_n(\delta|c_{m'})} f_{X^n,Y^n}(c_m, y^n)\, dy^n dc_m, \qquad (5.4.15)$$

where

$$\mathcal{F}_n(\delta|x^n) := \left\{ y^n \in \mathcal{Y}^n : (x^n, y^n) \in \mathcal{F}_n(\delta) \right\}.$$

Note that the additional term $P_{X^n}(\mathcal{E}_0)$ in (5.4.15) is to cope with the errors due to all-zero codeword replacement, which will be less than δ for all sufficiently large n by the law of large numbers. Finally, by carrying out a similar procedure as in the proof of the channel coding theorem for discrete channels (cf. page 123), we obtain

$$\begin{aligned} E[P_e(\mathcal{C}_n)] &\le P_{X^n}(\mathcal{E}_0) + P_{X^n,Y^n}\left(\mathcal{F}_n^c(\delta)\right) \\ &\quad + M_n \cdot 2^{n(h(X,Y)+\delta)} 2^{-n(h(X)-\delta)} 2^{-n(h(Y)-\delta)} \\ &\le P_{X^n}(\mathcal{E}_0) + P_{X^n,Y^n}\left(\mathcal{F}_n^c(\delta)\right) + 2^{n(C(P-\xi)-4\delta)} \cdot 2^{-n(I(X;Y)-3\delta)} \\ &= P_{X^n}(\mathcal{E}_0) + P_{X^n,Y^n}\left(\mathcal{F}_n^c(\delta)\right) + 2^{-n\delta}. \end{aligned}$$

Accordingly, we can make the average probability of error, $E[P_e(\mathcal{C}_n)]$, less than $3\delta = 3\gamma/8 < 3\varepsilon/4 < \varepsilon$ for all sufficiently large n.   □

Proof of the converse part: Consider an (n, M_n) block data transmission code satisfying the power constraint (5.4.14) with encoding function

$$f_n : \{1, 2, \dots, M_n\} \to \mathcal{X}^n$$

and decoding function

$$g_n : \mathcal{Y}^n \to \{1, 2, \dots, M_n\}.$$

Since the message W is uniformly distributed over $\{1, 2, \dots, M_n\}$, we have $H(W) = \log_2 M_n$. Since $W \to X^n = f_n(W) \to Y^n$ forms a Markov chain (as Y^n only depends on X^n), we obtain by the data processing lemma that $I(W; Y^n) \le I(X^n; Y^n)$. We can also bound $I(X^n; Y^n)$ by $C(P)$ as follows:

$$\begin{aligned} I(X^n; Y^n) &\le \sup_{F_{X^n}:\, (1/n)\sum_{i=1}^n E[X_i^2]\le P} I(X^n; Y^n) \\ &\le \sup_{F_{X^n}:\, (1/n)\sum_{i=1}^n E[X_i^2]\le P} \sum_{j=1}^n I(X_j; Y_j) \quad \text{(by Theorem 2.21)} \\ &= \sup_{(P_1,P_2,\dots,P_n):\, (1/n)\sum_{i=1}^n P_i=P} \sup_{F_{X^n}:\, (\forall\, i)\, E[X_i^2]\le P_i} \sum_{j=1}^n I(X_j; Y_j) \end{aligned}$$

$$\leq \sup_{(P_1, P_2, \ldots, P_n):\, (1/n)\sum_{i=1}^{n} P_i = P} \sum_{j=1}^{n} \sup_{F_{X^n}:\, (\forall\, i)\, E[X_i^2] \leq P_i} I(X_j; Y_j)$$

$$= \sup_{(P_1, P_2, \ldots, P_n):\, (1/n)\sum_{i=1}^{n} P_i = P} \sum_{j=1}^{n} \sup_{F_{X_j}:\, E[X_j^2] \leq P_j} I(X_j; Y_j)$$

$$= \sup_{(P_1, P_2, \ldots, P_n):\, (1/n)\sum_{i=1}^{n} P_i = P} \sum_{j=1}^{n} C(P_j)$$

$$= \sup_{(P_1, P_2, \ldots, P_n):\, (1/n)\sum_{i=1}^{n} P_i = P} n \sum_{j=1}^{n} \frac{1}{n} C(P_j)$$

$$\leq \sup_{(P_1, P_2, \ldots, P_n):\, (1/n)\sum_{i=1}^{n} P_i = P} nC\left(\frac{1}{n}\sum_{j=1}^{n} P_j\right) \quad \text{(by concavity of } C(P)\text{)}$$

$$= nC(P).$$

Consequently, recalling that $P_e(\mathscr{C}_n)$ is the average error probability incurred by guessing W from observing Y^n via the decoding function $g_n: \mathcal{Y}^n \to \{1, 2, \ldots, M_n\}$, we get

$$\begin{aligned}
\log_2 M_n &= H(W) \\
&= H(W|Y^n) + I(W; Y^n) \\
&\leq H(W|Y^n) + I(X^n; Y^n) \\
&\leq h_b(P_e(\mathscr{C}_n)) + P_e(\mathscr{C}_n) \cdot \log_2(|\mathcal{W}| - 1) + nC(P) \\
&\quad \text{(by Fano's inequality)} \\
&\leq 1 + P_e(\mathscr{C}_n) \cdot \log_2(M_n - 1) + nC(P), \\
&\quad \text{(by the fact that } (\forall\, t \in [0, 1])\, h_b(t) \leq 1) \\
&< 1 + P_e(\mathscr{C}_n) \cdot \log_2 M_n + nC(P),
\end{aligned}$$

which implies that

$$P_e(\mathscr{C}_n) > 1 - \frac{C(P)}{(1/n)\log_2 M_n} - \frac{1}{\log_2 M_n}.$$

So if $\liminf_{n\to\infty}(1/n)\log_2 M_n > C(P)$, then there exist $\delta > 0$ and an integer N such that for $n \geq N$,
$$\frac{1}{n}\log_2 M_n > C(P) + \delta.$$

Hence, for $n \geq N_0 := \max\{N, 2/\delta\}$,

$$P_e(\mathscr{C}_n) \geq 1 - \frac{C(P)}{C(P) + \delta} - \frac{1}{n(C(P) + \delta)} \geq \frac{\delta}{2(C(P) + \delta)}.$$

$\square$

We next show that among all power-constrained continuous memoryless channels with additive noise admitting a pdf, choosing a Gaussian distributed noise yields the smallest channel capacity. In other words, the memoryless Gaussian model results in the most pessimistic (smallest) capacity within the class of additive noise continuous memoryless channels.

Theorem 5.33 (*Gaussian noise minimizes capacity of additive noise channels*) *Every discrete-time continuous memoryless channel with additive noise (admitting a pdf) of mean zero and variance σ^2 and input average power constraint P has its capacity $C(P)$ lower bounded by the capacity of the memoryless Gaussian channel with identical input constraint and noise variance:*

$$C(P) \geq \frac{1}{2} \log_2 \left(1 + \frac{P}{\sigma^2} \right).$$

Proof Let $f_{Y|X}$ and $f_{Y_g|X_g}$ denote the transition pdfs of the additive noise channel and the Gaussian channel, respectively, where both channels satisfy input average power constraint P. Let Z and Z_g respectively denote their zero-mean noise variables of identical variance σ^2.

Writing the mutual information in terms of the channel's transition pdf and input distribution as in Lemma 2.46, then for any Gaussian input with pdf f_{X_g} with corresponding outputs Y and Y_g when applied to channels $f_{Y|X}$ and $f_{Y_g|X_g}$, respectively, we have that

$$
\begin{aligned}
&I(f_{X_g}, f_{Y|X}) - I(f_{X_g}, f_{Y_g|X_g}) \\
&= \int_x \int_y f_{X_g}(x) f_Z(y - x) \log_2 \frac{f_Z(y - x)}{f_Y(y)} dy dx \\
&\quad - \int_x \int_y f_{X_g}(x) f_{Z_g}(y - x) \log_2 \frac{f_{Z_g}(y - x)}{f_{Y_g}(y)} dy dx \\
&= \int_x \int_y f_{X_g}(x) f_Z(y - x) \log_2 \frac{f_Z(y - x)}{f_Y(y)} dy dx \\
&\quad - \int_x \int_y f_{X_g}(x) f_Z(y - x) \log_2 \frac{f_{Z_g}(y - x)}{f_{Y_g}(y)} dy dx \\
&= \int_x \int_y f_{X_g}(x) f_Z(y - x) \log_2 \frac{f_Z(y - x) f_{Y_g}(y)}{f_{Z_g}(y - x) f_Y(y)} dy dx \\
&\geq \int_x \int_y f_{X_g}(x) f_Z(y - x) (\log_2 e) \left(1 - \frac{f_{Z_g}(y - x) f_Y(y)}{f_Z(y - x) f_{Y_g}(y)} \right) dy dx \\
&= (\log_2 e) \left[1 - \int_y \frac{f_Y(y)}{f_{Y_g}(y)} \left(\int_x f_{X_g}(x) f_{Z_g}(y - x) dx \right) dy \right] \\
&= 0,
\end{aligned}
$$

with equality holding in the inequality iff

$$\frac{f_Y(y)}{f_{Y_g}(y)} = \frac{f_Z(y-x)}{f_{Z_g}(y-x)}$$

for all x and y. Therefore,

$$
\begin{aligned}
\frac{1}{2}\log_2\left(1+\frac{P}{\sigma^2}\right) &= \sup_{F_X:\,E[X^2]\le P} I(F_X, f_{Y_g|X_g}) \\
&= I(f_{X_g}^*, f_{Y_g|X_g}) \\
&\le I(f_{X_g}^*, f_{Y|X}) \\
&\le \sup_{F_X:\,E[X^2]\le P} I(F_X, f_{Y|X}) \\
&= C(P),
\end{aligned}
$$

thus completing the proof.   □

Observation 5.34 (*Shannon's channel coding theorem for continuous memoryless channels*) We point out that Theorem 5.32 can be generalized to a wide class of discrete-time continuous memoryless channels with input cost constraint (5.4.2) where the cost function $t(\cdot)$ is arbitrary, by showing that

$$C(P) := \sup_{F_X:\,E[t(X)]\le P} I(X; Y)$$

is the largest rate for which there exist block codes for the channel satisfying (5.4.2) which are reliably good (i.e., with asymptotically vanishing error probability).

The proof is quite similar to that of Theorem 5.32, except that some modifications are needed in the forward part as for a general (non-Gaussian) channel, the input distribution F_X used to construct the random code may not admit a pdf (e.g., cf. [135, Chap. 7], [415, Theorem 11.14]).

Observation 5.35 (*Capacity of memoryless fading channels*) We briefly examine the capacity of the memoryless fading channel, which is widely used to model wireless communications channels [151, 307, 387]. The channel is described by the following multiplicative and additive noise equation:

$$Y_i = A_i X_i + Z_i, \qquad \text{for } i = 1, 2, \ldots, \tag{5.4.16}$$

where Y_i, X_i, Z_i, and A_i are the channel output, input, additive noise, and amplitude fading coefficient (or gain) at time i. It is assumed that the fading process $\{A_i\}$ and the noise process $\{Z_i\}$ are each i.i.d. and that they are independent of each other and of the input process. As in the case of the memoryless Gaussian (AWGN) channel, the input power constraint is given by P and the noise $\{Z_i\}$ is Gaussian with $Z_i \sim \mathcal{N}(0, \sigma^2)$. The fading coefficients A_i are typically Rayleigh or Rician distributed [151]. In both cases, we assume that $E[A_i^2] = 1$ so that the channel SNR is unchanged as P/σ^2.

Note setting $A_i = 1$ for all i in (5.4.16) reduces the channel to the AWGN channel in (5.4.8). We next examine the effect of the random fading coefficient on the channel's capacity. We consider two scenarios regularly considered in the literature: (1) the fading coefficients are known at the receiver, and (2) the fading coefficients are known at both the receiver and the transmitter.[12]

1. *Capacity of the fading channel with decoder side information*: A common assumption used in many wireless communication systems is that the decoder knows the values of the fading coefficients at each time instant; in this case, we say that the channel has decoder side information (DSI). This assumption is realistic for wireless systems where the fading amplitudes change slowly with respect to the transmitted codeword so that the decoder can acquire knowledge of the fading coefficients via the the use of prearranged pilots signals. In this case, as both A and Y are known at the receiver, we can consider (Y, A) as the channel's output and thus aim to maximize

$$I(X; A, Y) = I(X; A) + I(X; Y|A) = I(X; Y|A),$$

where $I(X; A) = 0$ since X and A are independent from each other. Thus the channel capacity in this case, $C_{DSI}(P)$, can be solved as in the case of the AWGN channel (with the minor change of having the input scaled by the fading) to obtain:

$$\begin{aligned}
C_{DSI}(P) &= \sup_{F_X : E[X^2] \leq P} I(X; Y|A) \\
&= \sup_{F_X : E[X^2] \leq P} [h(Y|A) - h(Y|X, A)] \\
&= E_A \left[\frac{1}{2} \log_2 \left(1 + \frac{A^2 P}{\sigma^2}, \right) \right]
\end{aligned} \tag{5.4.17}$$

where the expectation is taken with respect to the fading distribution. Note that the capacity-achieving distribution here is also Gaussian with mean zero and variance P and is independent of the fading coefficient.

At this point, it is natural to compare the capacity in (5.4.17) with that of the AWGN channel in (5.4.13). In light of the concavity of the logarithm and using Jensen's inequality (in Theorem B.18), we readily obtain that

$$\begin{aligned}
C_{DSI}(P) &= E_A \left[\frac{1}{2} \log_2 \left(1 + \frac{A^2 P}{\sigma^2} \right) \right] \\
&\leq \frac{1}{2} \log_2 \left(1 + \frac{E[A^2] P}{\sigma^2} \right) \\
&= \frac{1}{2} \log_2 \left(1 + \frac{P}{\sigma^2} \right) := C_G(P)
\end{aligned} \tag{5.4.18}$$

---

[12]For other scenarios, see [151, 387].

which is the capacity of the AWGN channel with identical SNR, and where the last step follows since $E[A^2] = 1$. Thus, we conclude that fading degrades capacity as $C_{DSI}(P) \leq C_G(P)$.

2. *Capacity of the fading channel with full side information*: We next assume that both the receiver and the transmitter have knowledge of the fading coefficients; this is the case of the fading channel with full side information (FSI). This assumption applies to situations where there exists a reliable and fast feedback channel in the reverse direction where the decoder can communicate its knowledge of the fading process to the encoder. In this case, the transmitter can adaptively adjust its input power according to the value of the fading coefficient. It can be shown (e.g., see [387]) using Lagrange multipliers that the capacity in this case is given by

$$C_{FSI}(P) = E_A \left[\sup_{p(\cdot):\, E_A[p(A)]=P} \frac{1}{2} \log_2 \left(1 + \frac{A^2 p(A)}{\sigma^2} \right) \right]$$

$$= E_A \left[\frac{1}{2} \log_2 \left(1 + \frac{A^2 p^*(A)}{\sigma^2} \right) \right] \qquad (5.4.19)$$

where

$$p^*(a) = \max \left(0, \frac{1}{\lambda} - \frac{\sigma^2}{a^2} \right)$$

and λ satisfies $E_A[p(A)] = P$. The optimal power allotment $p^*(A)$ above is a so-called *water-filling* allotment, which we examine in more detail in the next section in the case of parallel AWGN channels.

Finally, we note that real-world wireless channels are often not memoryless; they exhibit statistical temporal *memory* in their fading process [80] and as a result signals traversing the channels are distorted in a bursty fashion. We refer the reader to [12, 108, 125, 135, 145, 211, 277, 298–301, 325, 334, 389, 420, 421] and the references therein for models of channels with memory and for finite-state Markov channel models which characterize the behavior of time-correlated fading channels in various settings.

5.5 Capacity of Uncorrelated Parallel Gaussian Channels: The Water-Filling Principle

Consider a network of k mutually independent discrete-time memoryless Gaussian channels with respective positive noise powers (variances) $\sigma_1^2, \sigma_2^2, \ldots$ and σ_k^2. If one wants to transmit information using these channels simultaneously (in parallel), what will be the system's channel capacity, and how should the signal powers for each

channel be apportioned given a fixed overall power budget ? The answer to the above
question lies in the so-called *water-filling* or *water-pouring* principle.

Theorem 5.36 (Capacity of uncorrelated parallel Gaussian channels) *The capacity
of k uncorrelated parallel Gaussian channels under an overall input power constraint
P is given by*

$$C(P) = \sum_{i=1}^{k} \frac{1}{2} \log_2 \left(1 + \frac{P_i}{\sigma_i^2}\right),$$

where σ_i^2 is the noise variance of channel i,

$$P_i = \max\{0, \theta - \sigma_i^2\},$$

*and θ is chosen to satisfy $\sum_{i=1}^{k} P_i = P$. This capacity is achieved by a tuple of
independent Gaussian inputs $(X_1, X_2, \ldots, X_k)$, where $X_i \sim \mathcal{N}(0, P_i)$ is the input
to channel i, for $i = 1, 2, \ldots, k$.*

Proof By definition,

$$C(P) = \sup_{F_{X^k}: \ \sum_{i=1}^{k} E[X_k^2] \leq P} I(X^k; Y^k).$$

Since the noise random variables $Z_1, \ldots, Z_k$ are independent from each other,

$$
\begin{aligned}
I(X^k; Y^k) &= h(Y^k) - h(Y^k|X^k) \\
&= h(Y^k) - h(Z^k + X^k|X^k) \\
&= h(Y^k) - h(Z^k|X^k) \\
&= h(Y^k) - h(Z^k) \\
&= h(Y^k) - \sum_{i=1}^{k} h(Z_i) \\
&\leq \sum_{i=1}^{k} h(Y_i) - \sum_{i=1}^{k} h(Z_i) \\
&\leq \sum_{i=1}^{k} \frac{1}{2} \log_2 \left(1 + \frac{P_i}{\sigma_i^2}\right),
\end{aligned}
$$

where the first inequality follows from the chain rule for differential entropy and the
fact that conditioning cannot increase differential entropy, and the second inequality
holds since output Y_i of channel i due to input X_i with $E[X_i^2] = P_i$ has its differential
entropy maximized if it is Gaussian distributed with zero-mean and variance $P_i + \sigma_i^2$.
Equalities hold above if all the X_i inputs are independent of each other with each
input $X_i \sim \mathcal{N}(0, P_i)$ such that $\sum_{i=1}^{k} P_i = P$.

Thus, the problem is reduced to finding the power allotment that maximizes the overall capacity subject to the constraint $\sum_{i=1}^{k} P_i = P$ with $P_i \geq 0$. By using the Lagrange multipliers technique and verifying the KKT conditions (see Example B.21 in Appendix B.8), the maximizer $(P_1, \ldots, P_k)$ of

$$\max \left\{ \sum_{i=1}^{k} \frac{1}{2} \log_2 \left(1 + \frac{P_i}{\sigma_i^2} \right) + \sum_{i=1}^{k} \lambda_i P_i - \nu \left(\sum_{i=1}^{k} P_i - P \right) \right\}$$

can be found by taking the derivative of the above equation (with respect to P_i) and setting it to zero, which yields

$$\lambda_i = \begin{cases} -\dfrac{1}{2\ln(2)} \dfrac{1}{P_i + \sigma_i^2} + \nu = 0, & \text{if } P_i > 0; \\ -\dfrac{1}{2\ln(2)} \dfrac{1}{P_i + \sigma_i^2} + \nu \geq 0, & \text{if } P_i = 0. \end{cases}$$

Hence,

$$\begin{cases} P_i = \theta - \sigma_i^2, & \text{if } P_i > 0; \\ P_i \geq \theta - \sigma_i^2, & \text{if } P_i = 0, \end{cases} \quad \text{(equivalently, } P_i = \max\{0, \theta - \sigma_i^2\}),$$

where $\theta := \log_2 e/(2\nu)$ is chosen to satisfy $\sum_{i=1}^{k} P_i = P$. ☐

We illustrate the above result in Fig. 5.1 and elucidate why the P_i power allotments form a *water-filling* (or *water-pouring*) scheme. In the figure, we have a vessel where the height of each of the solid bins represents the noise power of each channel (while the width is set to unity so that the area of each bin yields the noise power of the corresponding Gaussian channel). We can thus visualize the system as a vessel with an uneven bottom where the optimal input signal allocation P_i to each channel is realized by pouring an amount P units of water into the vessel (with the resulting overall area of filled water equal to P). Since the vessel has an uneven bottom, water is unevenly distributed among the bins: noisier channels are allotted less signal power (note that in this example, channel 3, whose noise power is largest, is given no input power at all and is hence not used).

Observation 5.37 (*Practical considerations*) According to the water-filling principle, one needs to use capacity-achieving Gaussian inputs and allocate more power to less noisy channels for the optimization of channel capacity. However, Gaussian inputs do not fit digital communication systems in practice. One may then wonder what is the optimal power allocation scheme when the channel inputs are practically dictated to be discrete in value, such as inputs used in conjunction with binary phase-shift keying (BPSK), quadrature phase-shift keying (QPSK), or 16 quadrature-amplitude modulation (16-QAM) signaling. Surprisingly under certain conditions,

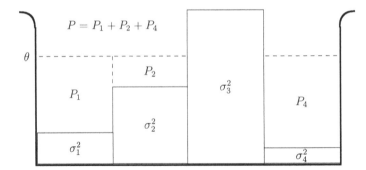

Fig. 5.1 The water-pouring scheme for uncorrelated parallel Gaussian channels. The horizontal dashed line, which indicates the level where the water rises to, indicates the value of θ for which $\sum_{i=1}^{k} P_i = P$

the answer is different from the water-filling principle. By characterizing the relationship between mutual information and minimum mean square error (MMSE) [165], the optimal power allocation for parallel AWGN channels with inputs constrained to be discrete is established in [250], resulting in a new graphical power allocation interpretation called the *mercury/water-filling principle*: mercury of proper amounts [250, Eq. (43)] must be individually poured into each channel bin before water of amount $P = \sum_{i=1}^{k} P_i$ is added to the vessel. It is thus named because mercury is heavier than water and does not dissolve in it; so it can play the role of pre-adjuster of bin heights. This line of inquiry concludes with the observation that when the total transmission power P is small, the strategy that maximizes capacity follows approximately the equal SNR principle; i.e., a larger power should be allotted to a noisier channel to optimize capacity.

Furthermore, it was found in [400] that when the channel's additive noise is no longer Gaussian, the mercury adjustment fails to interpret the optimal power allocation scheme. For additive Gaussian noise with arbitrary discrete inputs, the pre-adjustment before the water pouring step is always upward; hence, the mercury-filling scheme is used to increase bin heights. However, since the pre-adjustment of bin heights can generally be in both upward and downward directions for channels with non-Gaussian noise, the use of the name mercury/water filling becomes inappropriate (see [400, Example 1] for quaternary-input additive Laplacian noise channels). In this case, the graphical interpretation of the optimal power allocation scheme is simply named *two-phase water-filling principle* [400].

We end this observation by emphasizing that a vital measure for practical digital communication systems is the effective transmission rate subject to an acceptably small decoding error rate (e.g., an overall bit error probability $\leq 10^{-5}$). Instead, researchers typically adopt channel capacity as a design criterion in order to make the analysis tractable and obtain a simple reference scheme for practical systems.

5.6 Capacity of Correlated Parallel Gaussian Channels

In the previous section, we considered a network of k parallel discrete-time memoryless Gaussian channels in which the noise samples from different channels are independent from each other. We found out that the power allocation strategy that maximizes the system's capacity is given by the water-filling scheme. We next study a network of k parallel memoryless Gaussian channels where the noise variables from different channels are correlated. Surprisingly, we obtain that water-filling provides also the optimal power allotment policy.

Let $\mathbf{K}_Z$ denote the covariance matrix of the noise tuple $(Z_1, Z_2, \ldots, Z_k)$, and let $\mathbf{K}_X$ denote the covariance matrix of the system input $(X_1, \ldots, X_k)$, where we assume (without loss of the generality) that each X_i has zero mean. We assume that $\mathbf{K}_Z$ is positive-definite. The input power constraint becomes

$$\sum_{i=1}^{k} E[X_i^2] = \mathrm{tr}(\mathbf{K}_X) \leq P,$$

where $\mathrm{tr}(\cdot)$ denotes the trace of the $k \times k$ matrix $\mathbf{K}_X$. Since in each channel, the input and noise variables are independent from each other, we have

$$
\begin{aligned}
I(X^k; Y^k) &= h(Y^k) - h(Y^k|X^k) \\
&= h(Y^k) - h(Z^k + X^k|X^k) \\
&= h(Y^k) - h(Z^k|X^k) \\
&= h(Y^k) - h(Z^k).
\end{aligned}
$$

Since $h(Z^k)$ is not determined by the input, determining the system's capacity reduces to maximizing $h(Y^k)$ over all possible inputs $(X_1, \ldots, X_k)$ satisfying the power constraint.

Now observe that the covariance matrix of Y^k is equal to $\mathbf{K}_Y = \mathbf{K}_X + \mathbf{K}_Z$, which implies by Theorem 5.20 that the differential entropy of Y^k is upper bounded by

$$h(Y^k) \leq \frac{1}{2} \log_2 \left[(2\pi e)^k \det(\mathbf{K}_X + \mathbf{K}_Z) \right],$$

with equality iff Y^k Gaussian. It remains to find out whether we can find inputs $(X_1, \ldots, X_k)$ satisfying the power constraint which achieve the above upper bound and maximize it.

As in the proof of Theorem 5.18, we can orthogonally diagonalize $\mathbf{K}_Z$ as

$$\mathbf{K}_Z = \mathbf{A}\Lambda\mathbf{A}^T,$$

where $\mathbf{A}\mathbf{A}^T = \mathbf{I}_k$ (and thus $\det(\mathbf{A})^2 = 1$), $\mathbf{I}_k$ is the $k \times k$ identity matrix, and Λ is a diagonal matrix with positive diagonal components consisting of the eigenvalues of $\mathbf{K}_Z$ (as $\mathbf{K}_Z$ is positive-definite). Then

$$
\begin{aligned}
\det(\mathbf{K}_X + \mathbf{K}_Z) &= \det(\mathbf{K}_X + \mathbf{A}\Lambda\mathbf{A}^T) \\
&= \det(\mathbf{A}\mathbf{A}^T\mathbf{K}_X\mathbf{A}\mathbf{A}^T + \mathbf{A}\Lambda\mathbf{A}^T) \\
&= \det(\mathbf{A}) \cdot \det(\mathbf{A}^T\mathbf{K}_X\mathbf{A} + \Lambda) \cdot \det(\mathbf{A}^T) \\
&= \det(\mathbf{A}^T\mathbf{K}_X\mathbf{A} + \Lambda) \\
&= \det(\mathbf{B} + \Lambda),
\end{aligned}
$$

where $\mathbf{B} := \mathbf{A}^T\mathbf{K}_X\mathbf{A}$. Since for any two matrices $\mathbf{C}$ and $\mathbf{D}$, $\mathrm{tr}(\mathbf{CD}) = \mathrm{tr}(\mathbf{DC})$, we have that

$$
\mathrm{tr}(\mathbf{B}) = \mathrm{tr}(\mathbf{A}^T\mathbf{K}_X\mathbf{A}) = \mathrm{tr}(\mathbf{A}\mathbf{A}^T\mathbf{K}_X) = \mathrm{tr}(\mathbf{I}_k\mathbf{K}_X) = \mathrm{tr}(\mathbf{K}_X).
$$

Thus, the capacity problem is further transformed to maximizing $\det(\mathbf{B} + \Lambda)$ subject to $\mathrm{tr}(\mathbf{B}) \le P$.

By observing that $\mathbf{B} + \Lambda$ is positive-definite (because Λ is positive-definite) and using Hadamard's inequality given in Corollary 5.19, we have

$$
\det(\mathbf{B} + \Lambda) \le \prod_{i=1}^{k}(B_{ii} + \lambda_i),
$$

where λ_i is the component of matrix Λ locating at ith row and ith column, which is exactly the ith eigenvalue of $\mathbf{K}_Z$. Thus, the maximum value of $\det(\mathbf{B} + \Lambda)$ under $\mathrm{tr}(\mathbf{B}) \le P$ is realized by a diagonal matrix $\mathbf{B}$ (to achieve equality in Hadamard's inequality) with

$$
\sum_{i=1}^{k} B_{ii} = P.
$$

Finally, as in the proof of Theorem 5.36, we obtain a water-filling allotment for the optimal diagonal elements of $\mathbf{B}$:

$$
B_{ii} = \max\{0, \theta - \lambda_i\},
$$

where θ is chosen to satisfy $\sum_{i=1}^{k} B_{ii} = P$. We summarize this result in the next theorem.

Theorem 5.38 (Capacity of correlated parallel Gaussian channels) *The capacity of k correlated parallel Gaussian channels with positive-definite noise covariance matrix $\mathbf{K}_Z$ under overall input power constraint P is given by*

$$
C(P) = \sum_{i=1}^{k} \frac{1}{2} \log_2\left(1 + \frac{P_i}{\lambda_i}\right),
$$

where λ_i is the ith eigenvalue of $\mathbf{K}_Z$,

$$P_i = \max\{0, \theta - \lambda_i\},$$

and θ is chosen to satisfy $\sum_{i=1}^{k} P_i = P$. This capacity is achieved by a tuple of zero-mean Gaussian inputs $(X_1, X_2, \ldots, X_k)$ with covariance matrix $\mathbf{K}_X$ having the same eigenvectors as $\mathbf{K}_Z$, where the ith eigenvalue of $\mathbf{K}_X$ is P_i, for $i = 1, 2, \ldots, k$.

We close this section by briefly examining the capacity of two important systems used in wireless communications.

Observation 5.39 (*Capacity of memoryless MIMO channels*) As today's wireless communication systems persistently demand higher data rates, the exploitation of multiple-input multiple-output (MIMO) systems has become a vital technological option. By employing a transmitter with M transmit antennas and a receiver with N receive antennas, a single-user memoryless MIMO channel, whose radom fading gains (or coefficients) are represented by a sequence of i.i.d. $N \times M$ matrices $\{\mathbf{H}_i\}$, is described by

$$\underline{Y}_i = \mathbf{H}_i \underline{X}_i + \underline{Z}_i, \qquad \text{for } i = 1, 2, \ldots, \tag{5.6.1}$$

where $\underline{X}_i$ is the $M \times 1$ transmitted vector, $\underline{Y}_i$ is the $N \times 1$ received vector, and $\underline{Z}_i$ is the $N \times 1$ AWGN vector. In general, $\underline{Y}_i$, $\mathbf{H}_i$, $\underline{X}_i$, and $\underline{Z}_i$ are complex-valued.[13] The MIMO channel model in (5.6.1) can be regarded as a vector extension of the scalar system in (5.4.16). It is also a generalization of the Gaussian systems of Theorems 5.36 and 5.38 with $M = N$ and $\mathbf{H}_i$ being deterministically equal to the $N \times N$ identity matrix.

The noise covariance matrix $\mathbf{K}_Z$ is often assumed to be given by the identity matrix $\mathbf{I}_N$.[14] Assume that both the transmitter and the receiver are not only aware of the distribution of $\mathbf{H}_i = \mathbf{H}$ but know perfectly its value at each time instance i

---

[13]For a multivariate Gaussian vector $\underline{Z}$, its pdf has a slightly different form when it is complex-valued as opposed to when it is real-valued. For example, when $\underline{Z} = (Z_1, Z_2, \ldots, Z_N)^T$ is Gaussian with zero-mean and covariance matrix $\mathbf{K}_{\underline{Z}} = \sigma^2 \mathbf{I}_N$, we have

$$f_{\underline{Z}}(\underline{z}) = \begin{cases} \left(\frac{1}{\sqrt{2\pi\sigma^2}}\right)^N \exp\left(-\frac{1}{2\sigma^2} \sum_{j=1}^{N} Z_i^2\right), & \text{if } \underline{Z} \text{ real-valued} \\ \left(\frac{1}{\pi\sigma^2}\right)^N \exp\left(-\frac{1}{\sigma^2} \sum_{j=1}^{N} |Z_j|^2\right), & \text{if } \underline{Z} \text{ complex-valued.} \end{cases}$$

Thus, in parallel to Theorem 5.18, the joint differential entropy for a complex-valued Gaussian $\underline{Z}$ is equal to

$$h(\underline{Z}) = h(Z_1, Z_2, \ldots, Z_N) = \log_2\left[(\pi e)^N \det(\mathbf{K}_{\underline{Z}})\right],$$

where the multiplicative factors $1/2$ and 2 in the differential entropy formula in Theorem 5.18 are removed. Accordingly, the multiplicative factor $1/2$ in the capacity formula for real-valued AWGN channels is no longer necessary when a complex-valued AWGN channel is considered (e.g., see (5.6.2) and (5.6.3)).

[14]This assumption can be made valid as long as a whitening (i.e., decorrelation) matrix $\mathbf{W}$ of $\underline{Z}_i$ exists. One can thus multiply the received vector $\underline{Y}_i$ with $\mathbf{W}$ to yield the desired equivalent channel model with $\mathbf{I}_N$ as the noise covariance matrix (see [153, Eq. (1)] and the ensuing description).

(as in the FSI case in Observation (5.35)). Then, we can follow a similar approach to the one carried earlier in this section and obtain

$$\det(\mathbf{K}_{\underline{Y}}) = \det(\mathbf{H}\,\mathbf{K}_{\underline{X}}\,\mathbf{H}^{\dagger} + \mathbf{I}_N),$$

where "$\dagger$" is the Hermitian (conjugate) transposition operation. Thus, the capacity problem is transformed to maximizing $\det(\mathbf{H}\,\mathbf{K}_{\underline{X}}\,\mathbf{H}^{\dagger} + \mathbf{I}_N)$ subject to the power constraint $\mathrm{tr}(\mathbf{K}_{\underline{X}}) \leq P$. As a result, the fading MIMO channel capacity assuming perfect channel knowledge at both the transmitter and the receiver is

$$C_{FSI}(P) = E_{\mathbf{H}}\left[\max_{\mathbf{K}_{\underline{X}}:\mathrm{tr}(\mathbf{K}_{\underline{X}})\leq P} \log_2\left(\det(\mathbf{H}\,\mathbf{K}_{\underline{X}}\,\mathbf{H}^{\dagger} + \mathbf{I}_N)\right)\right]. \qquad (5.6.2)$$

If, however, only the decoder has perfect knowledge of $\mathbf{H}_i$ while the transmitter only knows its distribution (as in the DSI scenario in Observation (5.35)), then

$$C_{DSI}(P) = \max_{\mathbf{K}_{\underline{X}}:\mathrm{tr}(\mathbf{K}_{\underline{X}})\leq P} E_{\mathbf{H}}\left[\log_2\left(\det(\mathbf{H}\,\mathbf{K}_{\underline{X}}\,\mathbf{H}^{\dagger} + \mathbf{I}_N)\right)\right]. \qquad (5.6.3)$$

It directly follows from (5.6.2) and (5.6.3) above (and the property of the maximum) that in general,
$$D_{DSI}(P) \leq C_{FSI}(P).$$

A key find emanating from the analysis of MIMO channels is that in virtue of their spatial (multi-antenna) diversity, such channels can provide significant capacity gains vis-a-vis the traditional single-antenna (with $M = N = 1$) channel. For example, it can be shown that when the receiver knows the channel fading coefficients perfectly with the latter governed by a Rayleigh distribution, then MIMO channel capacity scales *linearly* in the minimum of the number of receive and transmit antennas at high channel SNR values, and thus it can be significantly larger than in the single-antenna case [380, 387]. Detailed studies about MIMO systems, including their capacity benefits under various conditions and configurations, can be found in [151, 153, 387] and the references therein. MIMO technology has become an essential component of mobile communication standards, such as IEEE 802.11 Wi-Fi, 4th generation (4G) Worldwide Interoperability for Microwave Access (WiMax), 4G Long Term Evolution (LTE) and others; see for example [110].

Observation 5.40 (*Capacity of memoryless OFDM systems*) We have so far considered discrete-time "narrowband flat" fading channels (e.g., in Observations 5.35 and 5.39) in the sense that the underlying continuous-time channel has a constant fading gain over a bandwidth which is larger than that of the transmitted signal (see Sect. 5.8 for more details on band-limited continuous-time channels). There are however many situations such as in wideband systems where the reverse property holds: the channel has a constant fading gain over a bandwidth which is smaller than the bandwidth of the transmitted signal. In this case, the sent signal undergoes frequency selective fading with different frequency components of the signal affected by different fading due to multipath propagation effects (which occur when the signal arrives at the receiver via several paths). It has been shown that such fading channels are well handled by multi-carrier modulation schemes such as orthogonal frequency division multiplexing (OFDM) which deftly exploits the channels' frequency diversity to provide resilience against the deleterious consequences of fading and interference.

OFDM transforms a single-user frequency selective channel into k parallel narrowband fading channels, where k is the number of OFDM subcarriers. It can be modeled as a memoryless multivariate channel:

$$\underline{Y}_i = \mathbf{H}_i \, \underline{X}_i + \underline{Z}_i, \qquad \text{for } i = 1, 2, \ldots,$$

where $\underline{X}_i$, $\underline{Y}_i$, and $\underline{Z}_i$ are respectively the $k \times 1$ transmitted vector, the $k \times 1$ received vector and the $k \times 1$ AWGN vector at time instance i. Furthermore, $\mathbf{H}_i$ is a $k \times k$ diagonal channel gain matrix at time instance i. As in the case of MIMO systems (see Observation 5.39), the vectors $\underline{Y}_i$, $\mathbf{H}_i$, $\underline{X}_i$, and $\underline{Z}_i$ are in general complex-valued. Under the assumption that $\mathbf{H}_i$ can be perfectly estimated and remains constant over the entire code transmission block, the (sum rate) capacity for a given power allocation vector $\mathbf{P} = (P_1, P_2, \cdots, P_k)$ is given by

$$C(\mathbf{P}) = \sum_{\ell=1}^{k} \log_2 \left(1 + \frac{|h_\ell|^2 P_\ell}{\sigma_\ell^2} \right) = \sum_{\ell=1}^{k} \log_2 \left(1 + \frac{P_\ell}{\sigma_\ell^2 / |h_\ell|^2} \right),$$

where σ_ℓ^2 is the variance of the ℓth component of $\underline{Z}_i$ and h_ℓ is the ℓth diagonal entry of $\mathbf{H}_i$. Thus the overall system capacity, $C(P)$, optimized subject to the power constraint $\sum_{\ell=1}^{k} P_\ell \leq P$, can be obtained via the water-filling principle of Sect. 5.5:

$$C(P) = \max_{\mathbf{P} = (P_1, P_2, \cdots, P_k) : \sum_{\ell=1}^{k} P_\ell \leq P} C(\mathbf{P})$$

$$= \sum_{\ell=1}^{k} \log_2 \left(1 + \frac{P_\ell^*}{\sigma_\ell^2 / |h_\ell|^2} \right),$$

where

$$P_\ell^* = \max\{0, \, \theta - \sigma_\ell^2 / |h_\ell|^2\},$$

and the parameter θ is chosen to satisfy $\sum_{\ell=1}^{k} P_\ell = P$.

Like MIMO, OFDM has been adopted by many communication standards, including Digital Video Broadcasting (DVB-S/T), Digital Subscriber Line (DSL), Wi-Fi, WiMax, and LTE. The reader is referred to wireless communication textbooks such as [151, 255, 387] for a thorough examination of OFDM systems.

5.7 Non-Gaussian Discrete-Time Memoryless Channels

If a discrete-time channel has an additive but non-Gaussian memoryless noise and an input power constraint, then it is often hard to calculate its capacity. Hence, in this section, we introduce an upper bound and a lower bound on the capacity of such a channel (we assume that the noise admits a pdf).

Definition 5.41 (*Entropy power*) For a continuous random variable Z with (well-defined) differential entropy $h(Z)$ (measured in bits), its *entropy power* is denoted by Z_e and defined as

$$Z_e := \frac{1}{2\pi e} 2^{2 \cdot h(Z)}.$$

Lemma 5.42 *For a discrete-time continuous-alphabet memoryless additive noise channel with input power constraint P and noise variance σ^2, its capacity satisfies*

$$\frac{1}{2} \log_2 \frac{P + \sigma^2}{Z_e} \geq C(P) \geq \frac{1}{2} \log_2 \frac{P + \sigma^2}{\sigma^2}. \tag{5.7.1}$$

Proof The lower bound in (5.7.1) is already proved in Theorem 5.33. The upper bound follows from

$$I(X; Y) = h(Y) - h(Z)$$
$$\leq \frac{1}{2} \log_2[2\pi e(P + \sigma^2)] - \frac{1}{2} \log_2[2\pi e Z_e].$$

$\square$

The entropy power of Z can be viewed as the variance of a corresponding Gaussian random variable with the same differential entropy as Z. Indeed, if Z is Gaussian, then its entropy power is equal to

$$Z_e = \frac{1}{2\pi e} 2^{2h(Z)} = \mathrm{Var}(Z),$$

as expected.

Whenever two independent Gaussian random variables, Z_1 and Z_2, are added, the power (variance) of the sum is equal to the sum of the powers (variances) of Z_1 and Z_2. This relationship can then be written as

$$2^{2h(Z_1 + Z_2)} = 2^{2h(Z_1)} + 2^{2h(Z_2)},$$

or equivalently

$$\mathrm{Var}(Z_1 + Z_2) = \mathrm{Var}(Z_1) + \mathrm{Var}(Z_2).$$

However, when two independent random variables are non-Gaussian, the relationship becomes

$$2^{2h(Z_1+Z_2)} \geq 2^{2h(Z_1)} + 2^{2h(Z_2)}, \tag{5.7.2}$$

or equivalently

$$\mathcal{Z}_e(Z_1 + Z_2) \geq \mathcal{Z}_e(Z_1) + \mathcal{Z}_e(Z_2). \tag{5.7.3}$$

Inequality (5.7.2) (or equivalently 5.7.3), whose proof can be found in [83, Sect. 17.8] or [51, Theorem 7.10.4], is called the *entropy power inequality*. It reveals that the sum of two independent random variables may introduce more entropy power than the sum of each individual entropy power, except in the Gaussian case.

Observation 5.43 (*Capacity bounds in terms of Gaussian capacity and non-Gaussianness*) It can be readily verified that

$$\frac{1}{2} \log_2 \frac{P + \sigma^2}{\mathcal{Z}_e} = \frac{1}{2} \log_2 \frac{P + \sigma^2}{\sigma^2} + D(Z \| Z_G),$$

where $D(Z \| Z_G)$ is the divergence between Z and a Gaussian random variable Z_G of mean zero and variance σ^2. Note that $D(Z \| Z_G)$ is called the *non-Gaussianness* of Z (e.g., see [388]) and is a measure of the "non-Gaussianity" of the noise Z. Thus recalling from (5.4.18)

$$C_G(P) := \frac{1}{2} \log_2 \frac{P + \sigma^2}{\sigma^2}$$

as the capacity of the channel when the additive noise is Gaussian, we obtain the following equivalent form for (5.7.1):

$$C_G(P) + D(Z \| Z_G) \geq C(P) \geq C_G(P). \tag{5.7.4}$$

5.8 Capacity of the Band-Limited White Gaussian Channel

We have so far considered discrete-time channels (with discrete or continuous alphabets). We close this chapter by briefly presenting the capacity expression of the continuous-time (waveform) band-limited channel with additive white Gaussian noise. The reader is referred to [411], [135, Chap. 8], [30, Sects. 8.2 and 8.3] and [196, Chap. 6] for rigorous and detailed treatments (including coding theorems) of waveform channels.

The continuous-time band-limited AWGN channel is a common model for a radio network or a telephone line. For such a channel, illustrated in Fig. 5.2, the output waveform is given by

$$Y(t) = (X(t) + Z(t)) * h(t), \qquad t \geq 0,$$

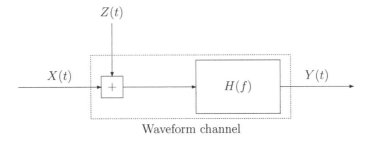

Fig. 5.2 Band-limited waveform channel with additive white Gaussian noise

where "$*$" represents the convolution operation (recall that the convolution between two signals $a(t)$ and $b(t)$ is defined as $a(t) * b(t) = \int_{-\infty}^{\infty} a(\tau)b(t - \tau)d\tau$). Here, $X(t)$ is the channel input waveform with average power constraint

$$\lim_{T \to \infty} \frac{1}{T} \int_{-T/2}^{T/2} E[X^2(t)]dt \leq P \qquad (5.8.1)$$

and bandwidth W cycles per second or Hertz (Hz); i.e., its spectrum or Fourier transform $X(f) := \mathcal{F}[X(t)] = \int_{-\infty}^{+\infty} X(t)e^{-j2\pi ft}dt = 0$ for all frequencies $|f| > W$, where $j = \sqrt{-1}$ is the imaginary unit number. $Z(t)$ is the noise waveform of a zero-mean stationary white Gaussian process with power spectral density $N_0/2$; i.e., its power spectral density $\text{PSD}_Z(f)$, which is the Fourier transform of the process covariance (equivalently, correlation) function $K_Z(\tau) := E[Z(s)Z(s + \tau)]$, $s, \tau \in \mathbb{R}$, is given by

$$\text{PSD}_Z(f) = \mathcal{F}[K_Z(t)] = \int_{-\infty}^{+\infty} K_Z(t)e^{-j2\pi ft}dt = \frac{N_0}{2} \qquad \forall f.$$

Finally, $h(t)$ is the impulse response of an ideal bandpass filter with cutoff frequencies at $\pm W$ Hz:

$$H(f) = \mathcal{F}[(h(t)] = \begin{cases} 1 & \text{if } -W \leq f \leq W, \\ 0 & \text{otherwise.} \end{cases}$$

Recall that one can recover $h(t)$ by taking the inverse Fourier transform of $H(f)$; this yields

$$h(t) = \mathcal{F}^{-1}[H(f)] = \int_{-\infty}^{+\infty} H(f)e^{j2\pi ft}df = 2W\text{sinc}(2Wt),$$

where

$$\text{sinc}(t) := \frac{\sin(\pi t)}{\pi t}$$

is the sinc function and is defined to equal 1 at $t = 0$ by continuity.

Note that we can write the channel output as

$$Y(t) = X(t) + \tilde{Z}(t),$$

where $\tilde{Z}(t) := Z(t) * h(t)$ is the filtered noise waveform. The input $X(t)$ is not affected by the ideal unit-gain bandpass filter since it has an identical bandwidth as $h(t)$. Note also that the power spectral density of the filtered noise is given by

$$\text{PSD}_{\tilde{Z}}(f) = \text{PSD}_Z(f)|H(f)|^2 = \begin{cases} \frac{N_0}{2} & \text{if } -W \le f \le W, \\ 0 & \text{otherwise.} \end{cases}$$

Taking the inverse Fourier transform of $\text{PSD}_{\tilde{Z}}(f)$ yields the covariance function of the filtered noise process:

$$K_{\tilde{Z}}(\tau) = \mathcal{F}^{-1}[\text{PSD}_{\tilde{Z}}(f)] = N_0 W \text{sinc}(2W\tau) \quad \tau \in \mathbb{R}. \tag{5.8.2}$$

To determine the capacity (in bits per second) of this continuous-time band-limited white Gaussian channel with parameters, P, W, and N_0, we convert it to an "equivalent" discrete-time channel with power constraint P by using the well-known sampling theorem (due to Nyquist, Kotelnikov and Shannon), which states that sampling a band-limited signal with bandwidth W at a rate of $1/(2W)$ is sufficient to reconstruct the signal from its samples. Since $X(t)$, $\tilde{Z}(t)$, and $Y(t)$ are all band-limited to $[-W, W]$, we can thus represent these signals by their samples taken $\frac{1}{2W}$ seconds apart and model the channel by a discrete-time channel described by:

$$Y_n = X_n + \tilde{Z}_n, \quad n = 0, \pm 1, \pm 2, \ldots,$$

where $X_n := X(\frac{n}{2W})$ are the input samples and $\tilde{Z}_n = Z(\frac{n}{2W})$ and $Y_n = Y(\frac{n}{2W})$ are the random samples of the noise $\tilde{Z}(t)$ and output $Y(t)$ signals, respectively.

Since $\tilde{Z}(t)$ is a filtered version of $Z(t)$, which is a zero-mean stationary Gaussian process, we obtain that $\tilde{Z}(t)$ is also zero-mean, stationary and Gaussian. This directly implies that the samples $\tilde{Z}_n$, $n = 1, 2, \ldots$, are zero-mean Gaussian identically distributed random variables. Now an examination of the expression of $K_{\tilde{Z}}(\tau)$ in (5.8.2) reveals that

$$K_{\tilde{Z}}(\tau) = 0$$

for $\tau = \frac{n}{2W}$, $n = 1, 2, \ldots$, since $\text{sinc}(t) = 0$ for all nonzero integer values of t. Hence, the random variables $\tilde{Z}_n$, $n = 1, 2, \ldots$, are uncorrelated and hence independent (since they are Gaussian) and their variance is given by $E[\tilde{Z}_n^2] = K_{\tilde{Z}}(0) =$

$N_0 W$. We conclude that the discrete-time process $\{\tilde{Z}_n\}_{n=1}^{\infty}$ is i.i.d. Gaussian with each $\tilde{Z}_n \sim \mathcal{N}(0, N_0 W)$. As a result, the above discrete-time channel is a discrete-time memoryless Gaussian channel with power constraint P and noise variance $N_0 W$; thus the capacity of the band-limited white Gaussian channel in bits per channel use is given using (5.4.13) by

$$\frac{1}{2} \log_2 \left(1 + \frac{P}{N_0 W}\right) \qquad \text{bits/channel use.}$$

Given that we are using the channel (with inputs X_n) every $\frac{1}{2W}$ seconds, we obtain that the capacity in bits/second of the band-limited white Gaussian channel is given by

$$C(P) = W \log_2 \left(1 + \frac{P}{N_0 W}\right) \qquad \text{bits/second,} \qquad (5.8.3)$$

where $\frac{P}{N_0 W}$ is the channel SNR.[15]

We emphasize that the above derivation of (5.8.3) is heuristic as we have not rigorously shown the equivalence between the original band-limited Gaussian channel and its discrete-time version and we have not established a coding theorem for the original channel. We point the reader to the references mentioned at the beginning of the section for a full development of this subject.

Example 5.44 (Telephone line channel) Suppose telephone signals are band-limited to 4 kHz. Given an SNR of 40 decibels (dB), i.e.,

$$10 \log_{10} \frac{P}{N_0 W} = 40 \text{ dB,}$$

then from (5.8.3), we calculate that the capacity of the telephone line channel (when modeled via the band-limited white Gaussian channel) is given by

$$4000 \log_2 (1 + 10000) = 53151.4 \qquad \text{bits/second.}$$

Example 5.45 (Infinite bandwidth white Gaussian channel) As the channel bandwidth W grows without bound, we obtain from (5.8.3) that

---

[15]Note that (5.8.3) is achieved by zero-mean i.i.d. Gaussian $\{X_n\}_{n=-\infty}^{\infty}$ with $E[X_n^2] = P$, which can be obtained by sampling a zero-mean, stationary and Gaussian $X(t)$ with

$$\text{PSD}_X(f) = \begin{cases} \frac{P}{2W} & \text{if } -W \leq f \leq W, \\ 0 & \text{otherwise.} \end{cases}$$

Examining this $X(t)$ confirms that it satisfies (5.8.1):

$$\frac{1}{T} \int_{-T/2}^{T/2} E[X^2(t)] dt = E[X^2(t)] = K_X(0) = P \cdot \text{sinc}(2W \cdot 0) = P.$$

$$\lim_{W \to \infty} C(P) = \frac{P}{N_0} \log_2 e \quad \text{bits/second,}$$

which indicates that in the infinite-bandwidth regime, capacity grows linearly with power.

Observation 5.46 (*Band-limited colored Gaussian channel*) If the above band-limited channel has a stationary *colored* (nonwhite) additive Gaussian noise, then it can be shown (e.g., see [135]) that the capacity of this channel becomes

$$C(P) = \frac{1}{2} \int_{-W}^{W} \max \left[0, \log_2 \frac{\theta}{\mathrm{PSD}_Z(f)} \right] df,$$

where θ is the solution of

$$P = \int_{-W}^{W} \max \left[0, \theta - \mathrm{PSD}_Z(f) \right] df.$$

The above capacity formula is indeed reminiscent of the water-pouring scheme we saw in Sects. 5.5 and 5.6, albeit it is herein applied in the spectral domain. In other words, we can view the curve of $\mathrm{PSD}_Z(f)$ as a bowl, and water is imagined being poured into the bowl up to level θ under which the area of the water is equal to P (see Fig. 5.3a). Furthermore, the distributed water indicates the shape of the optimum transmission power spectrum (see Fig. 5.3b).

Problems

1. *Differential entropy under translation and scaling*: Let X be a continuous random variable with a pdf defined on its support S_X.

    (a) *Translation*: Show that differential entropy is invariant under translations:

    $$h(X) = h(X + c)$$

    for any real constant c.

    (b) *Scaling*: Show that
    $$h(aX) = h(X) + \log_2 |a|$$

    for any nonzero real constant a.

2. Determine the differential entropy (in nats) of random variable X for each of the following cases.

    (a) X is exponential with parameter $\lambda > 0$ and pdf $f_X(x) = \lambda e^{-\lambda x}$, $x \geq 0$.
    (b) X is Laplacian with parameter $\lambda > 0$, mean zero and pdf $f_X(x) = \frac{1}{2\lambda} e^{-\frac{|x|}{\lambda}}$, $x \in \mathbb{R}$.
    (c) X is log-normal with parameters $\mu \in \mathbb{R}$ and $\sigma > 0$; i.e., $X = e^Y$, where $Y \sim \mathcal{N}(\mu, \sigma^2)$ is a Gaussian random variable with mean μ and variance σ^2.

The pdf of X is given by

$$f_X(x) = \frac{1}{\sigma x \sqrt{2\pi}} e^{-\frac{(\ln x - \mu)^2}{2\sigma^2}}, \quad x > 0.$$

(d) The source $X = aX_1 + bX_2$, where a and b are nonzero constants and X_1 and X_2 are independent Gaussian random variables such that $X_1 \sim \mathcal{N}(\mu_1, \sigma_1^2)$ and $X_2 \sim \mathcal{N}(\mu_2, \sigma_2^2)$.

3. *Generalized Gaussian*: Let X be a *generalized Gaussian* random variable with mean zero, variance σ^2 and pdf given by

$$f_X(x) = \frac{\alpha \eta}{2\Gamma(\frac{1}{\alpha})} e^{-\eta^\alpha |x|^\alpha}, \quad x \in \mathbb{R},$$

where $\alpha > 0$ is a parameter describing the distribution's exponential rate of decay,

$$\eta = \sigma^{-1} \left[\frac{\Gamma(\frac{3}{\alpha})}{\Gamma(\frac{1}{\alpha})} \right]^{1/2}$$

and

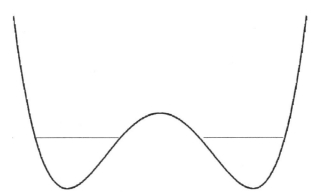

(a) The spectrum of $\mathrm{PSD}_Z(f)$ where the horizontal line represents θ, the level at which water rises to.

(b) The input spectrum that achieves capacity.

Fig. 5.3 Water-pouring for the band-limited colored Gaussian channel

$$\Gamma(x) = \int_0^\infty t^{x-1}e^{-t}dt, \quad x > 0,$$

is the gamma function (recall that $\Gamma(1/2) = \sqrt{\pi}$, $\Gamma(1) = 1$ and that $\Gamma(x+1) = x\Gamma(x)$ for any positive x). The generalized Gaussian distribution (also called the *exponential power distribution*) is well known to provide a good model for symmetrically distributed random processes, including image wavelet transform coefficients [258] and broad-tailed processes such as atmospheric impulsive noise [18]. Note that when $\alpha = 2$, f_X reduces to the Gaussian pdf with mean zero and variance σ^2, and when $\alpha = 1$, it reduces to the Laplacian pdf with parameter $\sigma/\sqrt{2}$ (i.e., variance σ^2).

(a) Show that the differential entropy of X is given by

$$h(X) = \frac{1}{\alpha} + \ln\left(\frac{2\Gamma(\frac{1}{\alpha})}{\alpha\eta}\right) \quad \text{(in nats)}.$$

(b) Show that when $\alpha = 2$ and $\alpha = 1$, $h(X)$ reduces to the differential entropy of the Gaussian and Laplacian distributions, respectively.

4. Prove that, of all pdfs with support $[0, 1]$, the uniform density function has the largest differential entropy.
5. Of all pdfs with continuous support $[0, K]$, where $K > 1$ is finite, which pdf has the largest differential entropy?
   Hint: If f_X is the pdf that maximizes differential entropy among all pdfs with support $[0, K]$, then $E[\log f_X(X)] = E[\log f_X(Y)]$ for any random variable Y of support $[0, K]$.
6. Show that the exponential distribution has the largest differential entropy among all pdfs with mean μ and support $[0, \infty)$. (Recall that the pdf of the exponential distribution with mean μ is given by $f_X(x) = \frac{1}{\mu}\exp(-\frac{x}{\mu})$ for $x \geq 0$.)
7. Show that among all continuous random variables X admitting a pdf with support $\mathbb{R}$ and finite differential entropy and satisfying $E[X] = 0$ and $E[|X|] = \lambda$, where $\lambda > 0$ is a fixed parameter, the Laplacian random variable with pdf

$$f_X(x) = \frac{1}{2\lambda}e^{-\frac{|x|}{\lambda}} \text{ for } x \in \mathbb{R}$$

maximizes differential entropy.
8. Find the mutual information between the dependent Gaussian zero-mean random variables X and Y with covariance matrix

$$\begin{pmatrix} \sigma^2 & \rho\sigma^2 \\ \rho\sigma^2 & \sigma^2 \end{pmatrix},$$

where $\rho \in [-1, 1]$ is the correlation coefficient. Evaluate the value of $I(X; Y)$ when $\rho = 1$, $\rho = 0$ and $\rho = -1$, and explain the results.

9. *A variant of the fundamental inequality for the logarithm*: For any $x > 0$ and $y > 0$, show that

$$y \ln \left(\frac{y}{x} \right) \geq y - x,$$

with equality iff $x = y$.

10. *Nonnegativity of divergence*: Let X and Y be two continuous random variables with pdfs f_X and f_Y, respectively, such that their supports satisfy $S_X \subseteq S_Y \subseteq \mathbb{R}$. Use Problem 4.9 to show that

$$D(f_X \| f_Y) \geq 0,$$

with equality iff $f_X = f_Y$ almost everywhere.

11. *Divergence between Laplacians*: Let S be a zero-mean random variable admitting a pdf with support $\mathbb{R}$ and satisfying $E[|S|] = \lambda$, where $\lambda > 0$ is a fixed parameter.

   (a) Let f_S and g_S be two zero-mean Laplacian pdfs for S with parameters λ and $\tilde{\lambda}$, respectively (see Problem 7 above for the Laplacian pdf expression). Determine $D(f_S \| g_S)$ in nats in terms of λ and $\tilde{\lambda}$.
   (b) Show that for any valid pdf f for S, we have

$$D(f \| g_S) \geq D(f_S \| g_S),$$

   with equality iff $f = f_S$ (almost everywhere).

12. Let X, Y, and Z be jointly Gaussian random variables, each with mean 0 and variance 1; let the correlation coefficient of X and Y as well as that of Y and Z be ρ, while X and Z are uncorrelated. Determine $h(X, Y, Z)$.

13. Let random variables Z_1 and Z_2 have Gaussian joint distribution with $E[Z_1] = E[Z_2] = 0$, $E[Z_1^2] = E[Z_2^2] = 1$ and $E[Z_1 Z_2] = \rho$, where $0 < \rho < 1$. Also, let U be a uniformly distributed random variable over the interval $(0, 2\pi e)$. Determine whether or not the following inequality holds:

$$h(U) > h(Z_1, Z_2 - 3Z_1).$$

14. Let Z_1, Z_2, and Z_3 be independent continuous random variables with identical pdfs. Show that

$$I(Z_1 + Z_2; Z_1 + Z_2 + Z_3) \geq \frac{1}{2} \log_2(3) \quad \text{(in bits)}.$$

Hint: Use the entropy power inequality in (5.7.2).

15. *An alternative form of the entropy power inequality*: Show that the entropy power inequality in (5.7.2) can be written as

$$h(Z_1 + Z_2) \geq h(Y_1 + Y_2),$$

where Z_1 and Z_2 are two independent continuous random variables, and Y_1 and Y_2 are two independent Gaussian random variables such that

$$h(Y_1) = h(Z_1)$$

and

$$h(Y_2) = h(Z_2).$$

16. *A relation between differential entropy and estimation error*: Consider a continuous random variable Z with a finite variance and admitting a pdf. It is desired to estimate Z by observing a correlated random variable Y (assume that the joint pdf of Z and Y and their conditional pdfs are well-defined). Let $\hat{Z} = \hat{Z}(Y)$ be such estimate of Z.

   (a) Show that the mean square estimation error satisfies

$$E[(Z - \hat{Z}(Y))^2] \geq \frac{2^{2h(Z|Y)}}{2\pi e}.$$

   Hint: Note that $\hat{Z}^*(Y) = E[Z|Y]$ is the optimal (MMSE) estimate of Z.

   (b) Assume now that Z and Y are zero-mean unit-variance jointly Gaussian random variables with correlation parameter ρ. Also assume that a simple linear estimator is used: $\hat{Z}(Y) = aY + b$, where a and b are chosen so that the estimation error is minimal. Evaluate the tightness of the bound in (a) and comment on the result.

17. Consider continuous real-valued random variables X and $Y_1, Y_2, \ldots, Y_n$, admitting a joint pdf and conditional pdfs among them such that $Y_1, Y_2, \ldots, Y_n$ are conditionally independent and conditionally identical distributed given X.

   (a) Show $I(X; Y_1, Y_2, \ldots, Y_n) \leq n \cdot I(X; Y_1)$.
   (b) Show that the capacity of the channel $X \rightarrow (Y_1, Y_2, \ldots, Y_n)$ with input power constraint P is less than n times the capacity of the channel $X \rightarrow Y_1$, where the notation $U \rightarrow V$ refers to a channel with input U and output V.

18. Consider the channel $X \rightarrow (Y_1, Y_2, \ldots, Y_n)$ with $Y_i = X + Z_i$ for $1 \leq i \leq n$, where X is independent of $\{Z_i\}_{i=1}^n$. Assume $\{Z_i\}_{i=1}^n$ are zero-mean equally correlated Gaussian random variables with

$$E[Z_i Z_j] = \begin{cases} \sigma^2, & \text{for } i = j; \\ \sigma^2 \rho, & \text{for } i \neq j, \end{cases}$$

where $\sigma^2 > 0$ and $-\frac{1}{n-1} \leq \rho \leq 1$.

   (a) By applying a power constraint on the input, i.e., $E[X^2] \leq P$, where $P > 0$, show the channel capacity $C_n(P)$ of this channel is given by

$$C_n(P) = \frac{1}{2}(n-1)\log(1-\rho) + \frac{1}{2}\log\left((1-\rho) + n\left(\frac{P}{\sigma^2} + \rho\right)\right).$$

(b) Prove that if $P > \frac{\sigma^2}{n-1}$, then $C_n(P) < nC_1(P)$.

Hint: It suffices to prove that $\exp\{2C_n(P)\} < \exp\{2nC_1(P)\}$. Use also the following identity regarding the determinant of an $n \times n$ matrix with identical diagonal entries and identical non-diagonal entries:

$$\det \begin{bmatrix} a & b & b & \cdots & b \\ b & a & b & \cdots & b \\ \vdots & \vdots & & \ddots & \vdots \\ b & b & \cdots & b & a \end{bmatrix} = (a-b)^{n-1}(a + (n-1)b).$$

19. Consider a continuous-alphabet channel with a vectored output for a scalar input as follows.

$$X \rightarrow \boxed{\quad\text{Channel}\quad} \rightarrow Y_1, Y_2$$

Suppose that the channel's transition pdf satisfies

$$f_{Y_1,Y_2|X}(y_1, y_2|x) = f_{Y_1|X}(y_1|x)f_{Y_2|X}(y_2|x)$$

for every y_1, y_2 and x.

(a) Show that $I(X; Y_1, Y_2) = \left(\sum_{i=1}^{2} I(X; Y_i)\right) - I(Y_1; Y_2)$.

Hint: $I(X; Y_1, Y_2) = h(Y_1, Y_2) - h(Y_1|X) - h(Y_2|X)$.

(b) Prove that the channel capacity $C_{\text{two}}(S)$ of using two outputs (Y_1, Y_2) is less than $C_1(S) + C_2(S)$ under an input power constraint S, where $C_j(S)$ is the channel capacity of using one output Y_j and ignoring the other output.

(c) Further assume that $f_{Y_i|X}(\cdot|x)$ is Gaussian with mean x and variance σ_j^2. In fact, these channels can be expressed as $Y_1 = X + N_1$ and $Y_2 = X + N_2$, where (N_1, N_2) are independent Gaussian distributed with mean zero and covariance matrix $\begin{bmatrix} \sigma_1^2 & 0 \\ 0 & \sigma_2^2 \end{bmatrix}$. Using the fact that $h(Y_1, Y_2) \leq \frac{1}{2}\log(2\pi e)^2 \left|K_{Y_1,Y_2}\right|$ with equality holding when (Y_1, Y_2) are joint Gaussian, where K_{Y_1,Y_2} is the covariance matrix of (Y_1, Y_2), derive $C_{\text{two}}(S)$ for the two-output channel under the power constraint $E[X^2] \leq S$.

Hint: $I(X; Y_1, Y_2) = h(Y_1, Y_2) - h(N_1, N_2) = h(Y_1, Y_2) - h(N_1) - h(N_2)$.

20. Consider the three-input three-output memoryless additive Gaussian channel

$$Y = X + Z,$$

where $X = [X_1, X_2, X_3]$, $Y = [Y_1, Y_2, Y_3]$, and $Z = [Z_1, Z_2, Z_3]$ are all three-dimensional real vectors. Assume that X is independent of Z, and the input power constraint is S (i.e., $E(X_1^2 + X_2^2 + X_3^2) \leq S$). Also, assume that Z is Gaussian distributed with zero-mean and covariance matrix $\mathbb{K}$, where

$$\mathbb{K} = \begin{bmatrix} 1 & 0 & 0 \\ 0 & 1 & \rho \\ 0 & \rho & 1 \end{bmatrix}.$$

(a) Determine the capacity-cost function of the channel, if $\rho = 0$.
    Hint: Directly apply Theorem 5.36.
(b) Determine the capacity-cost function of the channel, if $0 < \rho < 1$.
    Hint: Directly apply Theorem 5.38.

Chapter 6
Lossy Data Compression and Transmission

6.1 Preliminaries

6.1.1 Motivation

In a number of situations, one may need to compress a source to a rate less than the source entropy, which as we saw in Chap. 3 is the minimum lossless data compression rate. In this case, some sort of data loss is inevitable and the resultant code is referred to as a *lossy data compression code*. The following are examples for requiring the use of lossy data compression.

Example 6.1 (Digitization or quantization of continuous signals) The information content of continuous-alphabet signals , such as voice or analog images, is typically infinite, requiring an unbounded number of bits to digitize them without incurring any loss, which is not feasible. Therefore, a lossy data compression code must be used to reduce the output of a continuous source to a finite number of bits.

Example 6.2 (Extracting useful information) In some scenarios, the source information may not be operationally useful in its entirety. A quick example is the hypothesis testing problem where the system designer is only concerned with knowing the likelihood ratio of the null hypothesis distribution against the alternative hypothesis distribution (see Chap. 2). Therefore, any two distinct source letters which produce the same likelihood ratio are not encoded into distinct codewords and the resultant code is lossy.

Example 6.3 (Channel capacity bottleneck) Transmitting at a rate of r source symbol/channel symbol a discrete (memoryless) source with entropy H over a channel with capacity C such that $rH > C$ is problematic. Indeed as stated by the lossless joint source–channel coding theorem (see Theorem 4.32), if $rH > C$, then the system's error probability is bounded away from zero (in fact, in many cases, it grows exponentially fast to one with increasing blocklength). Hence, unmanageable error

© Springer Nature Singapore Pte Ltd. 2018
F. Alajaji and P.-N. Chen, *An Introduction to Single-User Information Theory*,
Springer Undergraduate Texts in Mathematics and Technology,
https://doi.org/10.1007/978-981-10-8001-2_6

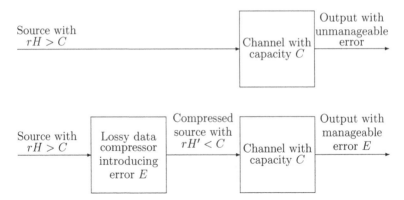

Fig. 6.1 Example for the application of lossy data compression

or distortion will be introduced at the destination (beyond the control of the system designer). A more viable approach would be to reduce the source's information content via a lossy compression step so that the entropy H' of the resulting source satisfies $r H' < C$ (this can, for example, be achieved by grouping the symbols of the original source and thus reducing its alphabet size). By Theorem 4.32, the compressed source can then be reliably sent at rate r over the channel. With this approach, error is only incurred (under the control of the system designer) in the lossy compression stage (cf. Fig. 6.1).

Note that another solution that avoids the use of lossy compression would be to reduce the source–channel transmission rate in the system from r to r' source symbol/channel symbol such that $r'H < C$ holds; in this case, again by Theorem 4.32, lossless reproduction of the source is guaranteed at the destination, albeit at the price of slowing the system.

6.1.2 Distortion Measures

A discrete-time source is modeled as a random process $\{Z_n\}_{n=1}^{\infty}$. To simplify the analysis, we assume that the source discussed in this section is memoryless and with finite alphabet $\mathcal{Z}$. Our objective is to compress the source with rate less than entropy under a prespecified criterion given by a distortion measure.

Definition 6.4 (*Distortion measure*) A distortion measure is a mapping

$$\rho: \mathcal{Z} \times \widehat{\mathcal{Z}} \to \mathbb{R}^+,$$

where $\mathcal{Z}$ is the source alphabet, $\widehat{\mathcal{Z}}$ is a reproduction alphabet, and $\mathbb{R}^+$ is the set of nonnegative real number.

From the above definition, the distortion measure $\rho(z, \hat{z})$ can be viewed as the cost of representing the source symbol $z \in \mathcal{Z}$ by a reproduction symbol $\hat{z} \in \widehat{\mathcal{Z}}$. It is then expected to choose a certain number of (typical) reproduction letters in $\widehat{\mathcal{Z}}$ that represent the source letters with the least cost.

When $\widehat{\mathcal{Z}} = \mathcal{Z}$, the selection of typical reproduction letters is similar to partitioning the source alphabet into several groups, and then choosing one element in each group to represent all group members. For example, suppose that $\widehat{\mathcal{Z}} = \mathcal{Z} = \{1, 2, 3, 4\}$ and that, due to some constraints, we need to reduce the number of outcomes to 2, and we require that the resulting expected cost cannot be larger than 0.5. Assume that the source is uniformly distributed and that the distortion measure is given by the following matrix:

$$[\rho(i, j)] := \begin{bmatrix} 0 & 1 & 2 & 2 \\ 1 & 0 & 2 & 2 \\ 2 & 2 & 0 & 1 \\ 2 & 2 & 1 & 0 \end{bmatrix}.$$

We see that the two groups in $\mathcal{Z}$ which cost least in terms of expected distortion $E[\rho(Z, \widehat{Z})]$ should be $\{1, 2\}$ and $\{3, 4\}$. We may choose, respectively, 1 and 3 as the typical elements for these two groups (cf. Fig. 6.2). The expected cost of such selection is

$$\frac{1}{4}\rho(1, 1) + \frac{1}{4}\rho(2, 1) + \frac{1}{4}\rho(3, 3) + \frac{1}{4}\rho(4, 3) = \frac{1}{2}.$$

Note that the entropy of the source is reduced from $H(Z) = 2$ bits to $H(\widehat{Z}) = 1$ bit. Sometimes, it is convenient to have $|\widehat{\mathcal{Z}}| = |\mathcal{Z}| + 1$. For example,

$$|\mathcal{Z} = \{1, 2, 3\}| = 3, \quad |\widehat{\mathcal{Z}} = \{1, 2, 3, E\}| = 4,$$

where E can be regarded as an *erasure symbol*, and the distortion measure is defined by

Fig. 6.2 "Grouping" as a form of lossy data compression

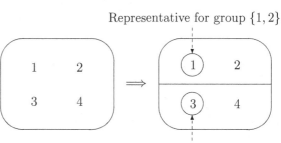

Representative for group $\{1, 2\}$

Representative for group $\{3, 4\}$

$$[\rho(i, j)] := \begin{bmatrix} 0\ 2\ 2\ 0.5 \\ 2\ 0\ 2\ 0.5 \\ 2\ 2\ 0\ 0.5 \end{bmatrix}.$$

In this case, assume again that the source is uniformly distributed and that to represent source letters by distinct letters in $\{1, 2, 3\}$ will yield four times the cost incurred when representing them by E. Therefore, if only 2 outcomes are allowed, and the expected distortion cannot be greater than $1/3$, then employing typical elements 1 and E to represent groups $\{1\}$ and $\{2, 3\}$, respectively, is an optimal choice. The resultant entropy is reduced from $\log_2(3)$ bits to $[\log_2(3) - 2/3]$ bits. It needs to be pointed out that having $|\widehat{\mathcal{Z}}| > |\mathcal{Z}| + 1$ is usually not advantageous.

6.1.3 Frequently Used Distortion Measures

Example 6.5 (Hamming distortion measure) Let the source and reproduction alphabets be identical, i.e., $\mathcal{Z} = \widehat{\mathcal{Z}}$. Then, the Hamming distortion measure is given by

$$\rho(z, \hat{z}) := \begin{cases} 0, \text{ if } z = \hat{z}; \\ 1, \text{ if } z \neq \hat{z}. \end{cases}$$

It is also named the *probability-of-error distortion measure* because

$$E[\rho(Z, \widehat{Z})] = \Pr(Z \neq \widehat{Z}).$$

Example 6.6 (Absolute error distortion measure) Assuming that $\mathcal{Z} = \widehat{\mathcal{Z}} = \mathbb{R}$, the absolute error distortion measure is given by

$$\rho(z, \hat{z}) := |z - \hat{z}|.$$

Example 6.7 (Squared error distortion measure) Again assuming that $\mathcal{Z} = \widehat{\mathcal{Z}} = \mathbb{R}$, the squared error distortion measure is given by

$$\rho(z, \hat{z}) := (z - \hat{z})^2.$$

The squared error distortion measure is perhaps the most popular distortion measure used for continuous alphabets.

Note that all above distortion measures belong to the class of so-called *difference distortion measures*, which have the form $\rho(z, \hat{z}) = d(x - \hat{x})$ for some nonnegative function $d(\cdot, \cdot)$. The squared error distortion measure has the advantages of simplicity and having a closed-form solution for most cases of interest, such as when using least squares prediction. Yet, this measure is not ideal for practical situations involving data operated by human observers (such as image and speech data) as it is inadequate in measuring perceptual quality. For example, two speech waveforms in which one is a marginally time-shifted version of the other may have large square error distortion; however, they sound quite similar to the human ear.

The above definition for distortion measures can be viewed as a *single-letter* distortion measure since they consider only one random variable Z which draws a single letter. For sources modeled as a sequence of random variables $\{Z_n\}$, some extension needs to be made. A straightforward extension is the additive distortion measure.

Definition 6.8 (*Additive distortion measure between vectors*) The additive distortion measure ρ_n between vectors z^n and $\hat{z}^n$ of size n (or n-sequences or n-tuples) is defined by

$$\rho_n(z^n, \hat{z}^n) = \sum_{i=1}^{n} \rho(z_i, \hat{z}_i).$$

Another example that is also based on a per-symbol distortion is the maximum distortion measure:

Definition 6.9 (*Maximum distortion measure*)

$$\rho_n(z^n, \hat{z}^n) = \max_{1 \leq i \leq n} \rho(z_i, \hat{z}_i).$$

After defining the distortion measures for source sequences, a natural question to ask is whether to reproduce source sequence z^n by sequence $\hat{z}^n$ of the same length is a must or not. To be more precise, can we use $\tilde{z}^k$ to represent z^n for $k \neq n$? The answer is certainly *yes* if a distortion measure for z^n and $\tilde{z}^k$ is defined. A quick example will be that the source is a ternary sequence of length n, while the (fixed-length) data compression result is a set of binary indices of length k, which is taken as small as possible subject to some given constraints. Hence, k is not necessarily equal to n. One of the problems for taking $k \neq n$ is that the distortion measure for sequences can no longer be defined based on per-letter distortions, and hence a per-letter formula for the best lossy data compression rate cannot be rendered.

In order to alleviate the aforementioned ($k \neq n$) problem, we claim that for most cases of interest, it is reasonable to assume $k = n$. This is because one can actually implement lossy data compression from $\mathcal{Z}^n$ to $\widetilde{\mathcal{Z}}^k$ in two steps. The first step corresponds to a lossy compression mapping $h_n : \mathcal{Z}^n \to \widehat{\mathcal{Z}}^n$, and the second step performs indexing $h_n(\mathcal{Z}^n)$ into $\widetilde{\mathcal{Z}}^k$:

Step 1: Find the data compression mapping

$$h_n : \mathcal{Z}^n \to \widehat{\mathcal{Z}}^n$$

for which the prespecified distortion and rate constraints are satisfied.

Step 2: Derive the (asymptotically) lossless data compression block code for source $h_n(Z^n)$. When n is sufficiently large, the existence of such code with blocklength

$$k > H(h_n(Z^n)) \quad \left(\text{equivalently, } R = \frac{k}{n} > \frac{1}{n}H(h_n(Z^n))\right)$$

is guaranteed by Shannon's source coding theorem (Theorem 3.6).

Through the above two steps, a lossy data compression code from

$$\underbrace{Z^n \to \widehat{Z}^n}_{\text{Step 1}} \to \{0, 1\}^k$$

is established. Since the second step is already discussed in the (asymptotically) lossless data compression context, we can say that the theorem regarding lossy data compression is basically a theorem on the first step.

6.2   Fixed-Length Lossy Data Compression Codes

Similar to the lossless source coding theorem, the objective is to find the theoretical limit of the compression rate for lossy source codes. Before introducing the main theorem, we first formally define lossy data compression codes, the rate–distortion region, and the (operational) rate–distortion function.

Definition 6.10 (*Fixed-length lossy data compression code subject to an average distortion constraint*) An (n, M, D) fixed-length lossy data compression code for a source $\{Z_n\}$ with alphabet Z and reproduction alphabet $\widehat{Z}$ consists of a compression function

$$h: Z^n \to \widehat{Z}^n$$

with a codebook size (i.e., the image $h(Z^n)$) equal to $|h(Z^n)| = M = M_n$ and an average (expected) distortion no larger than distortion threshold $D \geq 0$:

$$E\left[\frac{1}{n}\rho_n(Z^n, h(Z^n))\right] \leq D.$$

The compression rate of the code is defined as $(1/n)\log_2 M$ bits/source symbol, as $\log_2 M$ bits can be used to represent a sourceword of length n. Indeed, an equivalent description of the above compression code can be made via an encoder–decoder pair (f, g), where

$$f: Z^n \to \{1, 2, \ldots, M\}$$

is an encoding function mapping each sourceword in Z^n to an index in $\{1, \ldots, M\}$ (which can be represented using $\log_2 M$ bits), and

$$g: \{1, 2, \ldots, M\} \to \widehat{Z}^n$$

is a decoding function mapping each index to a reconstruction (or reproduction) vector in $\widehat{\mathcal{Z}}^n$ such that the composition of the encoding and decoding functions yields the above compression function h: $g(f(z^n)) = h(z^n)$ for $z^n \in \mathcal{Z}^n$.

Definition 6.11 (*Achievable rate–distortion pair*) For a given source $\{Z_n\}$ and a sequence of distortion measures $\{\rho_n\}_{n \geq 1}$, a rate–distortion pair (R, D) is achievable if there exists a sequence of fixed-length lossy data compression codes (n, M_n, D) for the source with asymptotic code rate satisfying

$$\limsup_{n \to \infty} \frac{1}{n} \log M_n < R.$$

Definition 6.12 (*Rate–distortion region*) The rate–distortion region $\mathcal{R}$ of a source $\{Z_n\}$ is the closure of the set of all achievable rate–distortion pair (R, D).

Lemma 6.13 (Time-sharing principle) *Under an additive distortion measure ρ_n, the rate–distortion region $\mathcal{R}$ is a convex set; i.e., if $(R_1, D_1) \in \mathcal{R}$ and $(R_2, D_2) \in \mathcal{R}$, then $(\lambda R_1 + (1 - \lambda)R_2, \lambda D_1 + (1 - \lambda)D_2) \in \mathcal{R}$ for all $0 \leq \lambda \leq 1$.*

Proof It is enough to show that the set of all achievable rate–distortion pairs is convex (since the closure of a convex set is convex). Also, assume without loss of generality that $0 < \lambda < 1$.

We will prove convexity of the set of all achievable rate–distortion pairs using a *time-sharing* argument, which basically states that if we can use an (n, M_1, D_1) code $\mathcal{C}_1$ to achieve (R_1, D_1) and an (n, M_2, D_2) code $\mathcal{C}_2$ to achieve (R_2, D_2), then for any rational number $0 < \lambda < 1$, we can use $\mathcal{C}_1$ for a fraction λ of the time and use $\mathcal{C}_2$ for a fraction $1 - \lambda$ of the time to achieve (R_λ, D_λ), where $R_\lambda = \lambda R_1 + (1 - \lambda)R_2$ and $D_\lambda = \lambda D_1 + (1 - \lambda)D_2$; hence, the result holds for any real number $0 < \lambda < 1$ by the density of the rational numbers in $\mathbb{R}$ and the continuity of R_λ and D_λ in λ.

Let r and s be positive integers and let $\lambda = \frac{r}{r+s}$; then $0 < \lambda < 1$. Now assume that the pairs (R_1, D_1) and (R_2, D_2) are achievable. Then, there exist a sequence of (n, M_1, D_1) codes $\mathcal{C}_1$ and a sequence of (n, M_2, D_2) codes $\mathcal{C}_2$ such that for n sufficiently large,

$$\frac{1}{n} \log_2 M_1 \leq R_1$$

and

$$\frac{1}{n} \log_2 M_2 \leq R_2.$$

Now construct a sequence of new codes $\mathcal{C}$ of blocklength $n_\lambda = (r + s)n$, codebook size $M = M_1^r \times M_2^s$ and compression function h: $\mathcal{Z}^{(r+s)n} \to \widehat{\mathcal{Z}}^{(r+s)n}$ such that

$$h(z^{(r+s)n}) = \left(h_1(z_1^n), \ldots, h_1(z_r^n), h_2(z_{r+1}^n), \ldots, h_2(z_{r+s}^n)\right),$$

where

$$z^{(r+s)n} = (z_1^n, \ldots, z_r^n, z_{r+1}^n, \ldots, z_{r+s}^n)$$

and h_1 and h_2 are the compression functions of $\mathscr{C}_1$ and $\mathscr{C}_2$, respectively. In other words, each reconstruction vector $h(z^{(r+s)n})$ of code $\mathscr{C}$ is a concatenation of r reconstruction vectors of code $\mathscr{C}_1$ and s reconstruction vectors of code $\mathscr{C}_2$.

The average (or expected) distortion under the additive distortion measure ρ_n and the rate of code $\mathscr{C}$ are given by

$$
E\left[\frac{\rho_{(r+s)n}(z^{(r+s)n}, h(z^{(r+s)n}))}{(r+s)n} \right]
$$

$$
= \frac{1}{r+s}\left(E\left[\frac{\rho_n(z_1^n, h_1(z_1^n))}{n} \right] + \cdots + E\left[\frac{\rho_n(z_r^n, h_1(z_r^n))}{n} \right] \right.
$$

$$
\left. + E\left[\frac{\rho_n(z_{r+1}^n, h_2(z_{r+1}^n))}{n} \right] + \cdots + E\left[\frac{\rho_n(z_{r+s}^n, h_2(z_{r+s}^n))}{n} \right] \right)
$$

$$
\leq \frac{1}{r+s}(rD_1 + sD_2)
$$

$$
= \lambda D_1 + (1-\lambda)D_2 = D_\lambda
$$

and

$$
\frac{1}{(r+s)n}\log_2 M = \frac{1}{(r+s)n}\log_2(M_1^r \times M_2^s)
$$

$$
= \frac{r}{(r+s)}\frac{1}{n}\log_2 M_1 + \frac{s}{(r+s)}\frac{1}{n}\log_2 M_2
$$

$$
\leq \lambda R_1 + (1-\lambda)R_2 = R_\lambda,
$$

respectively, for n sufficiently large. Thus, (R_λ, D_λ) is achievable by $\mathscr{C}$. $\qquad\square$

Definition 6.14 (*Rate–distortion function*) The rate–distortion function, denoted by $R(D)$, of source $\{Z_n\}$ is the smallest $\hat{R}$ for a given distortion threshold D such that $(\hat{R}, D)$ is an achievable rate–distortion pair; i.e.,

$$
R(D) := \inf\{\hat{R} \geq 0 : (\hat{R}, D) \in \mathcal{R}\}.
$$

Observation 6.15 (*Monotonicity and convexity of $R(D)$*) Note that, under an additive distortion measure ρ_n, the rate–distortion function $R(D)$ is nonincreasing and convex in D (the proof is left as an exercise).

6.3 Rate–Distortion Theorem

We herein derive the rate–distortion theorem for an arbitrary discrete memoryless source (DMS) using a *bounded* additive distortion measure $\rho_n(\cdot, \cdot)$; i.e., given a single-letter distortion measure $\rho(\cdot, \cdot)$, $\rho_n(\cdot, \cdot)$ satisfies the additive property of Definition 6.8 and

$$\max_{(z,\hat{z})\in\mathcal{Z}\times\widehat{\mathcal{Z}}} \rho(z,\hat{z}) < \infty.$$

The basic idea for identifying good data compression reproduction words from the set of sourcewords emanating from a DMS is to draw them from the so-called *distortion typical* set. This set is defined analogously to the jointly typical set studied in channel coding (cf. Definition 4.7).

Definition 6.16 (*Distortion typical set*) The *distortion δ-typical set* with respect to the memoryless (product) distribution $P_{Z,\hat{Z}}$ on $\mathcal{Z}^n \times \widehat{\mathcal{Z}}^n$ (i.e., when pairs of n-tuples $(z^n, \hat{z}^n)$ are drawn i.i.d. from $\mathcal{Z} \times \widehat{\mathcal{Z}}$ according to $P_{Z,\hat{Z}}$) and a bounded additive distortion measure $\rho_n(\cdot,\cdot)$ is defined by

$$\mathcal{D}_n(\delta) := \Big\{ (z^n, \hat{z}^n) \in \mathcal{Z}^n \times \widehat{\mathcal{Z}}^n :$$

$$\Big| -\frac{1}{n}\log_2 P_{Z^n}(z^n) - H(Z) \Big| < \delta,$$

$$\Big| -\frac{1}{n}\log_2 P_{\widehat{Z}^n}(\hat{z}^n) - H(\widehat{Z}) \Big| < \delta,$$

$$\Big| -\frac{1}{n}\log_2 P_{Z^n,\widehat{Z}^n}(z^n, \hat{z}^n) - H(Z,\widehat{Z}) \Big| < \delta,$$

$$\text{and } \Big| \frac{1}{n}\rho_n(z^n, \hat{z}^n) - E[\rho(Z,\widehat{Z})] \Big| < \delta \Big\}.$$

Note that this is the definition of the jointly typical set with an additional constraint on the normalized distortion on sequences of length n being close to the expected value. Since the additive distortion measure between two joint i.i.d. random sequences is actually the sum of the i.i.d. random variables $\rho(Z_i, \widehat{Z}_i)$, i.e.,

$$\rho_n(Z^n, \widehat{Z}^n) = \sum_{i=1}^{n} \rho(Z_i, \widehat{Z}_i),$$

then the (weak) law of large numbers holds for the distortion typical set. Therefore, an AEP-like theorem can be derived for distortion typical set.

Theorem 6.17 *If $(Z_1, \widehat{Z}_1), (Z_2, \widehat{Z}_2), \ldots, (Z_n, \widehat{Z}_n), \ldots$ are i.i.d., and $\rho_n(\cdot,\cdot)$ is a bounded additive distortion measure, then as $n \to \infty$,*

$$-\frac{1}{n}\log_2 P_{Z^n}(Z_1, Z_2, \ldots, Z_n) \to H(Z) \quad \text{in probability,}$$

$$-\frac{1}{n}\log_2 P_{\widehat{Z}^n}(\widehat{Z}_1, \widehat{Z}_2, \ldots, \widehat{Z}_n) \to H(\widehat{Z}) \quad \text{in probability,}$$

$$-\frac{1}{n}\log_2 P_{Z^n,\widehat{Z}^n}((Z_1, \widehat{Z}_1), \ldots, (Z_n, \widehat{Z}_n)) \to H(Z, \widehat{Z}) \quad \text{in probability,}$$

and

$$\frac{1}{n}\rho_n(Z^n, \widehat{Z}^n) \rightarrow E[\rho(Z, \widehat{Z})] \quad \text{in probability.}$$

Proof Functions of i.i.d. random variables are also i.i.d. random variables. Thus by the weak law of large numbers, we have the desired result. □

It needs to be pointed out that without the bounded property assumption on ρ, the normalized sum of an i.i.d. sequence does not necessarily converge in probability to a finite mean, hence the need for requiring that ρ be bounded.

Theorem 6.18 (AEP for distortion measure) *Given a DMS $\{(Z_n, \widehat{Z}_n)\}$ with generic joint distribution $P_{Z,\widehat{Z}}$ and any $\delta > 0$, the distortion δ-typical set satisfies*

1. $P_{Z^n,\widehat{Z}^n}(\mathcal{D}_n^c(\delta)) < \delta$ *for n sufficiently large.*
2. *For all $(z^n, \hat{z}^n)$ in $\mathcal{D}_n(\delta)$,*

$$P_{\widehat{Z}^n}(\hat{z}^n) \geq P_{\widehat{Z}^n|Z^n}(\hat{z}^n|z^n)2^{-n[I(Z;\widehat{Z})+3\delta]}. \tag{6.3.1}$$

Proof The first result follows directly from Theorem 6.17 and the definition of the distortion typical set $\mathcal{D}_n(\delta)$. The second result can be proved as follows:

$$
\begin{aligned}
P_{\widehat{Z}^n|Z^n}(\hat{z}^n|z^n) &= \frac{P_{Z^n,\widehat{Z}^n}(z^n, \hat{z}^n)}{P_{Z^n}(z^n)} \\
&= P_{\widehat{Z}^n}(\hat{z}^n)\frac{P_{Z^n,\widehat{Z}^n}(z^n, \hat{z}^n)}{P_{Z^n}(z^n)P_{\widehat{Z}^n}(\hat{z}^n)} \\
&\leq P_{\widehat{Z}^n}(\hat{z}^n)\frac{2^{-n[H(Z,\widehat{Z})-\delta]}}{2^{-n[H(Z)+\delta]}2^{-n[H(\widehat{Z})+\delta]}} \\
&= P_{\widehat{Z}^n}(\hat{z}^n)2^{n[I(Z;\widehat{Z})+3\delta]},
\end{aligned}
$$

where the inequality follows from the definition of $\mathcal{D}_n(\delta)$. □

Before presenting the lossy data compression theorem, we need the following inequality.

Lemma 6.19 *For $0 \leq x \leq 1, 0 \leq y \leq 1$, and $n > 0$,*

$$(1 - xy)^n \leq 1 - x + e^{-yn}, \tag{6.3.2}$$

with equality holding iff $(x, y) = (1, 0)$.

Proof Let $g_y(t) := (1 - yt)^n$. It can be shown by taking the second derivative of $g_y(t)$ with respect to t that this function is strictly convex for $t \in [0, 1]$. Hence, using $\vee$ to denote disjunction, we have for any $x \in [0, 1]$ that

$$(1 - xy)^n = g_y\big((1 - x) \cdot 0 + x \cdot 1\big)$$
$$\leq (1 - x) \cdot g_y(0) + x \cdot g_y(1)$$

with equality holding iff $(x = 0) \vee (x = 1) \vee (y = 0)$

$$= (1 - x) + x \cdot (1 - y)^n$$
$$\leq (1 - x) + x \cdot \left(e^{-y}\right)^n$$

with equality holding iff $(x = 0) \vee (y = 0)$

$$\leq (1 - x) + e^{-ny}$$

with equality holding iff $(x = 1)$.

From the above derivation, we know that equality holds in (6.3.2) iff

$$[(x = 0) \vee (x = 1) \vee (y = 0)] \wedge [(x = 0) \vee (y = 0)] \wedge [(x = 1)] = (x = 1, y = 0),$$

where $\wedge$ denotes conjunction. (Note that $(x = 0)$ represents $\{(x, y) \in \mathbb{R}^2 : x = 0$ and $y \in [0, 1]\}$; a similar definition applies to the other sets.) $\qquad\square$

Theorem 6.20 (Shannon's rate–distortion theorem for memoryless sources) *Consider a DMS $\{Z_n\}_{n=1}^{\infty}$ with alphabet $\mathcal{Z}$, reproduction alphabet $\widehat{\mathcal{Z}}$ and a bounded additive distortion measure $\rho_n(\cdot, \cdot)$; i.e.,*

$$\rho_n(z^n, \hat{z}^n) = \sum_{i=1}^{n} \rho(z_i, \hat{z}_i) \quad \text{and} \quad \rho_{\max} := \max_{(z,\hat{z}) \in \mathcal{Z} \times \widehat{\mathcal{Z}}} \rho(z, \hat{z}) < \infty,$$

where $\rho(\cdot, \cdot)$ is a given single-letter distortion measure. Then, the source's rate–distortion function satisfies the following expression:

$$R(D) = \min_{P_{\widehat{Z}|Z} : E[\rho(Z,\widehat{Z})] \leq D} I(Z; \widehat{Z}).$$

Proof Define

$$R^{(I)}(D) := \min_{P_{\widehat{Z}|Z} : E[\rho(Z,\widehat{Z})] \leq D} I(Z; \widehat{Z}); \qquad (6.3.3)$$

this quantity is typically called Shannon's *information rate–distortion function*. We will then show that the (operational) rate–distortion function $R(D)$ given in Definition 6.14 equals $R^{(I)}(D)$.

1. *Achievability Part* (i.e., $R(D + \varepsilon) \leq R^{(I)}(D) + 4\varepsilon$ for arbitrarily small $\varepsilon > 0$):
We need to show that for any $\varepsilon > 0$, there exist $0 < \gamma < 4\varepsilon$ and a sequence of lossy data compression codes $\{(n, M_n, D + \varepsilon)\}_{n=1}^{\infty}$ with

$$\limsup_{n \to \infty} \frac{1}{n} \log_2 M_n \leq R^{(I)}(D) + \gamma < R^{(I)}(D) + 4\varepsilon.$$

The proof is as follows.

Step 1: Optimizing conditional distribution. Let $P_{\widetilde{Z}|Z}$ be the conditional distribution that achieves $R^{(I)}(D)$, i.e.,

$$R^{(I)}(D) = \min_{P_{\widehat{Z}|Z}:E[\rho(Z,\widetilde{Z})]\leq D} I(Z; \widehat{Z}) = I(Z; \widetilde{Z}).$$

Then

$$E[\rho(Z, \widetilde{Z})] \leq D.$$

Choose M_n to satisfy

$$R^{(I)}(D) + \frac{1}{2}\gamma \leq \frac{1}{n}\log_2 M_n \leq R^{(I)}(D) + \gamma$$

for some γ in $(0, 4\varepsilon)$, for which the choice should exist for all sufficiently large $n > N_0$ for some N_0. Define

$$\delta := \min\left\{\frac{\gamma}{8}, \frac{\varepsilon}{1 + 2\rho_{\max}}\right\}.$$

Step 2: Random coding. Independently select M_n words from $\widehat{\mathcal{Z}}^n$ according to

$$P_{\widetilde{Z}^n}(\tilde{z}^n) = \prod_{i=1}^{n} P_{\widetilde{Z}}(\tilde{z}_i),$$

and denote this random codebook by $\mathcal{C}_n$, where

$$P_{\widetilde{Z}}(\tilde{z}) = \sum_{z\in\mathcal{Z}} P_Z(z)P_{\widetilde{Z}|Z}(\tilde{z}|z).$$

Step 3: Encoding rule. Define a subset of $\mathcal{Z}^n$ as

$$\mathcal{J}(\mathcal{C}_n) := \left\{z^n \in \mathcal{Z}^n : \exists\, \tilde{z}^n \in \mathcal{C}_n \text{ such that } (z^n, \tilde{z}^n) \in \mathcal{D}_n(\delta)\right\},$$

where $\mathcal{D}_n(\delta)$ is defined under $P_{\widetilde{Z}|Z}$. Based on the codebook

$$\mathcal{C}_n = \{c_1, c_2, \ldots, c_{M_n}\},$$

define the encoding rule as

$$h_n(z^n) = \begin{cases} c_m, & \text{if } (z^n, c_m) \in \mathcal{D}_n(\delta); \\ & \text{(when more than one satisfying the requirement,} \\ & \text{just pick any.)} \\ \text{any word in } \mathcal{C}_n, & \text{otherwise.} \end{cases}$$

Note that when $z^n \in \mathcal{J}(\mathcal{C}_n)$, we have $(z^n, h_n(z^n)) \in \mathcal{D}_n(\delta)$ and

$$\frac{1}{n} \rho_n(z^n, h_n(z^n)) \leq E[\rho(Z, \tilde{Z})] + \delta \leq D + \delta.$$

Step 4: Calculation of the probability of the complement of $\mathcal{J}(\mathcal{C}_n)$. Let N_1 be chosen such that for $n > N_1$,

$$P_{Z^n, \tilde{Z}^n}(\mathcal{D}_n^c(\delta)) < \delta.$$

Let

$$\Omega := P_{Z^n}(\mathcal{J}^c(\mathcal{C}_n)).$$

Then, the expected probability of source n-tuples not belonging to $\mathcal{J}(\mathcal{C}_n)$, averaged over all randomly generated codebooks, is given by

$$E[\Omega] = \sum_{\mathcal{C}_n} P_{\tilde{Z}^n}(\mathcal{C}_n) \left(\sum_{z^n \notin \mathcal{J}(\mathcal{C}_n)} P_{Z^n}(z^n) \right)$$

$$= \sum_{z^n \in \mathcal{Z}^n} P_{Z^n}(z^n) \left(\sum_{\mathcal{C}_n : z^n \notin \mathcal{J}(\mathcal{C}_n)} P_{\tilde{Z}^n}(\mathcal{C}_n) \right).$$

For any z^n given, to select a codebook $\mathcal{C}_n$ satisfying $z^n \notin \mathcal{J}(\mathcal{C}_n)$ is equivalent to *independently* draw M_n n-tuples from $\widehat{\mathcal{Z}}^n$ which are not jointly distortion typical with z^n. Hence,

$$\sum_{\mathcal{C}_n : z^n \notin \mathcal{J}(\mathcal{C}_n)} P_{\tilde{Z}^n}(\mathcal{C}_n) = \left(\Pr\left[(z^n, \tilde{Z}^n) \notin \mathcal{D}_n(\delta) \right] \right)^{M_n}.$$

For convenience, we let $K(z^n, \tilde{z}^n)$ denote the indicator function of $\mathcal{D}_n(\delta)$, i.e.,

$$K(z^n, \tilde{z}^n) = \begin{cases} 1, & \text{if } (z^n, \tilde{z}^n) \in \mathcal{D}_n(\delta); \\ 0, & \text{otherwise.} \end{cases}$$

Then

$$\sum_{\mathcal{C}_n : z^n \notin \mathcal{J}(\mathcal{C}_n)} P_{\tilde{Z}^n}(\mathcal{C}_n) = \left(1 - \sum_{\tilde{z}^n \in \widehat{\mathcal{Z}}^n} P_{\tilde{Z}^n}(\tilde{z}^n) K(z^n, \tilde{z}^n) \right)^{M_n}.$$

Continuing the computation of $E[\Omega]$, we get

$$E[\Omega] = \sum_{z^n \in \mathcal{Z}^n} P_{Z^n}(z^n) \left(1 - \sum_{\tilde{z}^n \in \widehat{\mathcal{Z}}^n} P_{\tilde{Z}^n}(\tilde{z}^n) K(z^n, \tilde{z}^n)\right)^{M_n}$$

$$\leq \sum_{z^n \in \mathcal{Z}^n} P_{Z^n}(z^n) \left(1 - \sum_{\tilde{z}^n \in \widehat{\mathcal{Z}}^n} P_{\tilde{Z}^n|Z^n}(\tilde{z}^n|z^n) 2^{-n(I(Z;\tilde{Z})+3\delta)} K(z^n, \tilde{z}^n)\right)^{M_n}$$

$$\text{(by 6.3.1)}$$

$$= \sum_{z^n \in \mathcal{Z}^n} P_{Z^n}(z^n) \left(1 - 2^{-n(I(Z;\tilde{Z})+3\delta)} \sum_{\tilde{z}^n \in \widehat{\mathcal{Z}}^n} P_{\tilde{Z}^n|Z^n}(\tilde{z}^n|z^n) K(z^n, \tilde{z}^n)\right)^{M_n}$$

$$\leq \sum_{z^n \in \mathcal{Z}^n} P_{Z^n}(z^n) \Bigg(1 - \sum_{\tilde{z}^n \in \widehat{\mathcal{Z}}^n} P_{\tilde{Z}^n|Z^n}(\tilde{z}^n|z^n) K(z^n, \tilde{z}^n)$$

$$+ \exp\left\{-M_n \cdot 2^{-n(I(Z;\tilde{Z})+3\delta)}\right\}\Bigg) \quad \text{(from 6.3.2)}$$

$$\leq \sum_{z^n \in \mathcal{Z}^n} P_{Z^n}(z^n) \Bigg(1 - \sum_{\tilde{z}^n \in \widehat{\mathcal{Z}}^n} P_{\tilde{Z}^n|Z^n}(\tilde{z}^n|z^n) K(z^n, \tilde{z}^n)$$

$$+ \exp\left\{-2^{n(R^{(I)}(D)+\gamma/2)} \cdot 2^{-n(I(Z;\tilde{Z})+3\delta)}\right\}\Bigg),$$

$$\text{(for } R^{(I)}(D) + \gamma/2 < (1/n)\log_2 M_n)$$

$$\leq 1 - P_{Z^n, \tilde{Z}^n}(\mathcal{D}_n(\delta)) + \exp\left\{-2^{n\delta}\right\},$$

$$\text{(for } R^{(I)}(D) = I(Z; \tilde{Z}) \text{ and } \delta \leq \gamma/8)$$

$$= P_{Z^n, \tilde{Z}^n}(\mathcal{D}_n^c(\delta)) + \exp\left\{-2^{n\delta}\right\}$$

$$\leq \delta + \delta = 2\delta$$

for all $n > N := \max\left\{N_0, N_1, \frac{1}{\delta}\log_2 \log\left(\frac{1}{\min\{\delta,1\}}\right)\right\}$.

Since $E[\Omega] = E\left[P_{Z^n}\left(\mathcal{J}^c(\mathcal{C}_n)\right)\right] \leq 2\delta$, there must exist a codebook $\mathcal{C}_n^*$ such that $P_{Z^n}\left(\mathcal{J}^c(\mathcal{C}_n^*)\right)$ is no greater than 2δ for n sufficiently large.

Step 5: Calculation of distortion. The distortion of the optimal codebook $\mathcal{C}_n^*$ (from the previous step) satisfies for $n > N$:

$$\frac{1}{n}E[\rho_n(Z^n, h_n(Z^n))] = \sum_{z^n \in \mathcal{J}(\mathcal{C}_n^*)} P_{Z^n}(z^n)\frac{1}{n}\rho_n(z^n, h_n(z^n))$$

$$+ \sum_{z^n \notin \mathcal{J}(\mathcal{C}_n^*)} P_{Z^n}(z^n)\frac{1}{n}\rho_n(z^n, h_n(z^n))$$

$$\leq \sum_{z^n \in \mathcal{J}(\mathcal{C}_n^*)} P_{Z^n}(z^n)(D+\delta) + \sum_{z^n \notin \mathcal{J}(\mathcal{C}_n^*)} P_{Z^n}(z^n)\rho_{max}$$

$$\leq (D + \delta) + 2\delta \cdot \rho_{max}$$
$$\leq D + \delta(1 + 2\rho_{max})$$
$$\leq D + \varepsilon.$$

This concludes the proof of the achievability part.

2. *Converse Part* (i.e., $R(D + \varepsilon) \geq R^{(I)}(D)$ for arbitrarily small $\varepsilon > 0$ and any $D \in \{D \geq 0 : R^{(I)}(D) > 0\}$): We need to show that for any sequence of $\{(n, M_n, D_n)\}_{n=1}^{\infty}$ code with

$$\limsup_{n \to \infty} \frac{1}{n} \log_2 M_n < R^{(I)}(D),$$

there exists $\varepsilon > 0$ such that

$$D_n = \frac{1}{n} E[\rho_n(Z^n, h_n(Z^n))] > D + \varepsilon$$

for n sufficiently large. The proof is as follows.

Step 1: Convexity of mutual information. By the convexity of mutual information $I(Z; \widehat{Z})$ with respect to $P_{\widehat{Z}|Z}$ for a fixed P_Z, we have

$$I(Z; \widehat{Z}_\lambda) \leq \lambda \cdot I(Z; \widehat{Z}_1) + (1 - \lambda) \cdot I(Z; \widehat{Z}_2),$$

where $\lambda \in [0, 1]$, and

$$P_{\widehat{Z}_\lambda|Z}(\hat{z}|z) := \lambda P_{\widehat{Z}_1|Z}(\hat{z}|z) + (1 - \lambda) P_{\widehat{Z}_2|Z}(\hat{z}|z).$$

Step 2: Convexity of $R^{(I)}(D)$. Let $P_{\widehat{Z}_1|Z}$ and $P_{\widehat{Z}_2|Z}$ be two distributions achieving $R^{(I)}(D_1)$ and $R^{(I)}(D_2)$, respectively. Since

$$E[\rho(Z, \widehat{Z}_\lambda)] = \sum_{z \in \mathcal{Z}} P_Z(z) \sum_{\hat{z} \in \widehat{\mathcal{Z}}} P_{\widehat{Z}_\lambda|Z}(\hat{z}|z) \rho(z, \hat{z})$$

$$= \sum_{z \in \mathcal{Z}, \hat{z} \in \widehat{\mathcal{Z}}} P_Z(z) \left[\lambda P_{\widehat{Z}_1|Z}(\hat{z}|z) + (1 - \lambda) P_{\widehat{Z}_2|Z}(\hat{z}|z) \right] \rho(z, \hat{z})$$

$$= \lambda D_1 + (1 - \lambda) D_2,$$

we have

$$R^{(I)}(\lambda D_1 + (1 - \lambda) D_2) \leq I(Z; \widehat{Z}_\lambda)$$
$$\leq \lambda I(Z; \widehat{Z}_1) + (1 - \lambda) I(Z; \widehat{Z}_2)$$
$$= \lambda R^{(I)}(D_1) + (1 - \lambda) R^{(I)}(D_2).$$

Therefore, $R^{(I)}(D)$ is a convex function.

Step 3: Strictly decreasing and continuity properties of $R^{(I)}(D)$.

By definition, $R^{(I)}(D)$ is nonincreasing in D. Also,

$$R^{(I)}(D) = 0 \quad \text{iff} \quad D \geq D_{\max} := \min_{\hat{z}} \sum_{z \in \mathcal{Z}} P_Z(z)\rho(z, \hat{z}) \tag{6.3.4}$$

which is finite by the boundedness of the distortion measure. Thus, since $R^{(I)}(D)$ is nonincreasing and convex, it directly follows that it is strictly decreasing and continuous over $\{D \geq 0 : R^{(I)}(D) > 0\}$.

Step 4: Main proof.

$$
\begin{aligned}
\log_2 M_n &\geq H(h_n(Z^n)) \\
&= H(h_n(Z^n)) - H(h_n(Z^n)|Z^n), \quad \text{since } H(h_n(Z^n)|Z^n) = 0; \\
&= I(Z^n; h_n(Z^n)) \\
&= H(Z^n) - H(Z^n|h_n(Z^n)) \\
&= \sum_{i=1}^{n} H(Z_i) - \sum_{i=1}^{n} H(Z_i|h_n(Z^n), Z_1, \dots, Z_{i-1}) \\
&\qquad \text{by the independence of } Z^n, \\
&\qquad \text{and the chain rule for conditional entropy;} \\
&\geq \sum_{i=1}^{n} H(Z_i) - \sum_{i=1}^{n} H(Z_i|\widehat{Z}_i) \\
&\qquad \text{where } \widehat{Z}_i \text{ is the } i\text{th component of } h_n(Z^n); \\
&= \sum_{i=1}^{n} I(Z_i; \widehat{Z}_i) \\
&\geq \sum_{i=1}^{n} R^{(I)}(D_i), \quad \text{where } D_i := E[\rho(Z_i, \widehat{Z}_i)]; \\
&= n \sum_{i=1}^{n} \frac{1}{n} R^{(I)}(D_i) \\
&\geq n R^{(I)} \left(\sum_{i=1}^{n} \frac{1}{n} D_i \right), \quad \text{by convexity of } R^{(I)}(D); \\
&= n R^{(I)} \left(\frac{1}{n} E[\rho_n(Z^n, h_n(Z^n))] \right),
\end{aligned}
$$

where the last step follows since the distortion measure is additive. Finally, $\limsup_{n \to \infty} (1/n) \log_2 M_n < R^{(I)}(D)$ implies the existence of N and $\gamma > 0$ such that $(1/n) \log M_n < R^{(I)}(D) - \gamma$ for all $n > N$. Therefore, for $n > N$,

$$R^{(I)}\left(\frac{1}{n}E[\rho_n(Z^n, h_n(Z^n))]\right) < R^{(I)}(D) - \gamma,$$

which, together with the fact that $R^{(I)}(D)$ is strictly decreasing, implies that

$$\frac{1}{n}E[\rho_n(Z^n, h_n(Z^n))] > D + \varepsilon$$

for some $\varepsilon = \varepsilon(\gamma) > 0$ and for all $n > N$.

3. *Summary*: For $D \in \{D \geq 0 : R^{(I)}(D) > 0\}$, the achievability and converse parts jointly imply that $R^{(I)}(D) + 4\varepsilon \geq R(D + \varepsilon) \geq R^{(I)}(D)$ for arbitrarily small $\varepsilon > 0$. These inequalities together with the continuity of $R^{(I)}(D)$ yield that $R(D) = R^{(I)}(D)$ for $D \in \{D \geq 0 : R^{(I)}(D) > 0\}$.

For $D \in \{D \geq 0 : R^{(I)}(D) = 0\}$, the achievability part gives us $R^{(I)}(D) + 4\varepsilon = 4\varepsilon \geq R(D + \varepsilon) \geq 0$ for arbitrarily small $\varepsilon > 0$. This immediately implies that $R(D) = 0 (= R^{(I)}(D))$ as desired. □

As in the case of block source coding in Chap. 3 (compare Theorem 3.6 with Theorem 3.15), the above rate–distortion theorem can be extended for the case of stationary ergodic sources (e.g., see [42, 135]).

Theorem 6.21 (Shannon's rate–distortion theorem for stationary ergodic sources) *Consider a stationary ergodic source $\{Z_n\}_{n=1}^{\infty}$ with alphabet $\mathcal{Z}$, reproduction alphabet $\widehat{\mathcal{Z}}$ and a bounded additive distortion measure $\rho_n(\cdot, \cdot)$; i.e.,*

$$\rho_n(z^n, \hat{z}^n) = \sum_{i=1}^{n} \rho(z_i, \hat{z}_i) \quad \text{and} \quad \rho_{\max} := \max_{(z,\hat{z}) \in \mathcal{Z} \times \widehat{\mathcal{Z}}} \rho(z, \hat{z}) < \infty,$$

where $\rho(\cdot, \cdot)$ is a given single-letter distortion measure. Then, the source's rate–distortion function is given by

$$R(D) = \overline{R}^{(I)}(D),$$

where

$$\overline{R}^{(I)}(D) := \lim_{n \to \infty} R_n^{(I)}(D) \tag{6.3.5}$$

is called the asymptotic information rate–distortion function. *And*

$$R_n^{(I)}(D) := \min_{P_{\widehat{Z}^n | Z^n} : \frac{1}{n}E[\rho_n(Z^n, \widehat{Z}^n)] \leq D} \frac{1}{n} I(Z^n; \widehat{Z}^n) \tag{6.3.6}$$

is the nth-order information rate–distortion function.

Observation 6.22 (*Notes on the asymptotic information rate–distortion function*)

- Note that the quantity $\overline{R}^{(I)}(D)$ in (6.3.5) is well defined as long as the source is stationary; furthermore, $\overline{R}^{(I)}(D)$ satisfies (see [135, p. 492])

$$\overline{R}^{(I)}(D) = \lim_{n\to\infty} R_n^{(I)}(D) = \inf_n R_n^{(I)}(D).$$

Hence,

$$\overline{R}^{(I)}(D) \le R_n^{(I)}(D) \qquad\qquad\qquad (6.3.7)$$

holds for any $n = 1, 2, \ldots$.
- *Wyner–Ziv lower bound on* $\overline{R}^{(I)}(D)$: The following lower bound on $\overline{R}^{(I)}(D)$, due to Wyner and Ziv [412] (see also [42]), holds for stationary sources:

$$\overline{R}^{(I)}(D) \ge R_n^{(I)}(D) - \mu_n, \qquad\qquad\qquad (6.3.8)$$

where

$$\mu_n := \frac{1}{n}H(Z^n) - H(\mathcal{Z})$$

represents the amount of memory in the source and $H(\mathcal{Z}) = \lim_{n\to\infty} \frac{1}{n}H(Z^n)$ is the source entropy rate. Note that as $n \to \infty$, $\mu_n \to 0$; thus, the above Wyner–Ziv lower bound is asymptotically tight in n.[1]
- When the source $\{Z_n\}$ is a DMS, it can readily be verified from (6.3.5) and (6.3.6) that

$$\overline{R}^{(I)}(D) = R_1^{(I)}(D) = R^{(I)}(D),$$

where $R^{(I)}(D)$ is given in (6.3.3).
- $\overline{R}^{(I)}(D)$ *for binary symmetric Markov sources*: When $\{Z_n\}$ is a stationary binary symmetric Markov source, $\overline{R}^{(I)}(D)$ is partially explicitly known. More specifically, it is shown by Gray [156] that

$$\overline{R}^{(I)}(D) = h_b(q) - h_b(D) \qquad \text{for } 0 \le D \le D_c, \qquad\qquad (6.3.9)$$

where

$$D_c = \frac{1}{2}\left(1 - \sqrt{1 - \frac{(1-q)^2}{q}}\right),$$

---

[1] A twin result to the above Wyner–Ziv lower bound, which consists of an upper bound on the capacity-cost function of channels with stationary additive noise, is shown in [16, Corollary 1]. This result, which is expressed in terms of the nth-order capacity-cost function and the amount of memory in the channel noise, illustrates the "natural duality" between the information rate–distortion and capacity-cost functions originally pointed out by Shannon [345].

$q = P\{Z_n = 1 | Z_{n-1} = 0\} = P\{Z_n = 0 | Z_{n-1} = 1\} > 1/2$ is the source's transition probability and $h_b(p) = -p \log_2(p) - (1-p) \log_2(1-p)$ is the binary entropy function. Determining $\overline{R}^{(I)}(D)$ for $D > D_c$ is still an open problem, although $\overline{R}^{(I)}(D)$ can be estimated in this region via lower and upper bounds. Indeed, the right-hand side of (6.3.9) still serves as a lower bound on $\overline{R}^{(I)}(D)$ for $D > D_c$ [156]. Another lower bound on $\overline{R}^{(I)}(D)$ is the above Wyner–Ziv bound (6.3.8), while (6.3.7) gives an upper bound. Various bounds on $\overline{R}^{(I)}(D)$ are studied in [43] and calculated (in particular, see [43, Fig. 1]).

The formula of the rate–distortion function obtained in the previous theorems is also valid for the squared error distortion over the real numbers, even if it is *unbounded*. Here, we put the boundedness assumption just to facilitate the exposition of the current proof.[2] The discussion on lossy data compression, especially for continuous-alphabet sources, will continue in Sect. 6.4. Examples of the calculation of the rate–distortion function for memoryless sources will also be given in the same section.

After introducing *Shannon's source coding theorem* for block codes, *Shannon's channel coding theorem* for block codes and the *rate–distortion theorem* in the memoryless system setting (or even stationary ergodic setting in the case of the source coding and rate–distortion theorems), we briefly elucidate the "key concepts or techniques" behind these lengthy proofs, in particular the notions of a *typical set* and of *random coding*. The *typical set* construct—specifically, δ-typical set for source coding, joint δ-typical set for channel coding, and distortion typical set for rate–distortion—uses a law of large numbers or AEP argument to claim the existence of a set with very high probability; hence, the respective information manipulation can just focus on the set with negligible performance loss. The random coding technique shows that the *expectation* of the desired performance over all possible information manipulation schemes (randomly drawn according to some *properly chosen* statistics) is already acceptably good, and hence the existence of at least one good scheme that fulfills the desired performance index is validated. As a result, in situations where the above two techniques apply, a similar theorem can often be established. A natural question is whether we can extend the theorems to cases where these two techniques fail. It is obvious that only when new methods (other than the above two) are developed can the question be answered in the affirmative; see [73, 74, 157, 172, 364] for examples of more general rate–distortion theorems involving sources with memory.

---

[2] For example, the boundedness assumption in the theorems can be replaced with assuming that there exists a reproduction symbol $\hat{z}_0 \in \hat{\mathcal{Z}}$ such that $E[\rho(Z, \hat{z}_0)] < \infty$ [42, Theorems 7.2.4 and 7.2.5]. This assumption can accommodate the squared error distortion measure and a source with finite second moment (including continuous-alphabet sources such as Gaussian sources); see also [135, Theorem 9.6.2 and p. 479].

6.4   Calculation of the Rate–Distortion Function

In light of Theorem 6.20 and the discussion at the end of the previous section, we know that for a wide class of memoryless sources

$$R(D) = R^{(I)}(D)$$

as given in (6.3.3).

We first note that, like channel capacity, $R(D)$ cannot in general be explicitly determined in a closed-form expression, and thus optimization-based algorithms can be used for its efficient numerical computation [27, 49, 51, 88]. In the following, we consider simple examples involving the Hamming and squared error distortion measures where bounds or exact expressions for $R(D)$ can be obtained.

6.4.1   Rate–Distortion Function for Discrete Memoryless Sources Under the Hamming Distortion Measure

A specific application of the rate–distortion function that is useful in practice is when the Hamming additive distortion measure is used with a (finite alphabet) DMS with $\widehat{\mathcal{Z}} = \mathcal{Z}$.

We first assume that the DMS is binary-valued (i.e., it is a Bernoulli source) with $\widehat{\mathcal{Z}} = \mathcal{Z} = \{0, 1\}$. In this case, the Hamming additive distortion measure satisfies

$$\rho_n(z^n, \hat{z}^n) = \sum_{i=1}^{n} z_i \oplus \hat{z}_i,$$

where $\oplus$ denotes modulo two addition. In such case, $\rho(z^n, \hat{z}^n)$ is exactly the number of bit errors or changes after compression. Therefore, the distortion bound D becomes a bound on the average probability of bit error. Specifically, among n compressed bits, it is expected to have $E[\rho(Z^n, \widehat{Z}^n)]$ bit errors; hence, the expected value of bit error rate is $(1/n)E[\rho(Z^n, \widehat{Z}^n)]$. The rate–distortion function for binary sources and Hamming additive distortion measure is given by the next theorem.

Theorem 6.23 *Fix a binary DMS* $\{Z_n\}_{n=1}^{\infty}$ *with marginal distribution* $P_Z(0) = 1 - P_Z(1) = p$, *where* $0 < p < 1$. *Then the source's rate–distortion function under the Hamming additive distortion measure is given by*

$$R(D) = \begin{cases} h_b(p) - h_b(D) & \text{if } 0 \leq D < \min\{p, 1-p\}; \\ 0 & \text{if } D \geq \min\{p, 1-p\}, \end{cases}$$

where $h_b(\cdot)$ *is the binary entropy function.*

Proof Assume without loss of generality that $p \le 1/2$.

We first prove the theorem for $0 \le D < \min\{p, 1 - p\} = p$. Observe that

$$H(Z|\widehat{Z}) = H(Z \oplus \widehat{Z}|\widehat{Z}).$$

Also, observe that

$$E[\rho(Z, \widehat{Z})] \le D \text{ implies that } \Pr\{Z \oplus \widehat{Z} = 1\} \le D.$$

Then, examining (6.3.3), we have

$$
\begin{aligned}
I(Z; \widehat{Z}) &= H(Z) - H(Z|\widehat{Z}) \\
&= h_b(p) - H(Z \oplus \widehat{Z}|\widehat{Z}) \\
&\ge h_b(p) - H(Z \oplus \widehat{Z}) \quad \text{(conditioning never increases entropy)} \\
&\ge h_b(p) - h_b(D),
\end{aligned}
$$

where the last inequality follows since the binary entropy function $h_b(x)$ is increasing for $x \le 1/2$, and $\Pr\{Z \oplus \widehat{Z} = 1\} \le D$. Since the above derivation is true for any $P_{\widehat{Z}|Z}$, we have

$$R(D) \ge h_b(p) - h_b(D).$$

It remains to show that the lower bound is achievable by some $P_{\widehat{Z}|Z}$, or equivalently, $H(Z|\widehat{Z}) = h_b(D)$ for some $P_{\widehat{Z}|Z}$. By defining $P_{Z|\widehat{Z}}(0|0) = P_{Z|\widehat{Z}}(1|1) = 1 - D$, we immediately obtain $H(Z|\widehat{Z}) = h_b(D)$. The desired $P_{\widehat{Z}|Z}$ can be obtained by simultaneously solving the equations

$$
\begin{aligned}
1 = P_{\widehat{Z}}(0) + P_{\widehat{Z}}(1) &= \frac{P_Z(0)}{P_{Z|\widehat{Z}}(0|0)} P_{\widehat{Z}|Z}(0|0) + \frac{P_Z(0)}{P_{Z|\widehat{Z}}(0|1)} P_{\widehat{Z}|Z}(1|0) \\
&= \frac{p}{1 - D} P_{\widehat{Z}|Z}(0|0) + \frac{p}{D}(1 - P_{\widehat{Z}|Z}(0|0))
\end{aligned}
$$

and

$$
\begin{aligned}
1 = P_{\widehat{Z}}(0) + P_{\widehat{Z}}(1) &= \frac{P_Z(1)}{P_{Z|\widehat{Z}}(1|0)} P_{\widehat{Z}|Z}(0|1) + \frac{P_Z(1)}{P_{Z|\widehat{Z}}(1|1)} P_{\widehat{Z}|Z}(1|1) \\
&= \frac{1 - p}{D}(1 - P_{\widehat{Z}|Z}(1|1)) + \frac{1 - p}{1 - D} P_{\widehat{Z}|Z}(1|1),
\end{aligned}
$$

which yield

$$P_{\widehat{Z}|Z}(0|0) = \frac{1 - D}{1 - 2D}\left(1 - \frac{D}{p}\right)$$

and

$$P_{\widehat{Z}|Z}(1|1) = \frac{1-D}{1-2D}\left(1 - \frac{D}{1-p}\right).$$

If $0 \leq D < \min\{p, 1-p\} = p$, then $P_{\widehat{Z}|Z}(0|0) > 0$ and $P_{\widehat{Z}|Z}(1|1) > 0$, completing the proof.

Finally, the fact that $R(D) = 0$ for $D \geq \min\{p, 1-p\} = p$ follows directly from (6.3.4) by noting that $D_{max} = \min\{p, 1-p\} = p$. $\qquad\qquad\square$

The above theorem can be extended for nonbinary (finite alphabet) memoryless sources resulting in a more complicated (but exact) expression for $R(D)$, see [207]. We instead present a simple lower bound on $R(D)$ for a nonbinary DMS under the Hamming distortion measure; this bound is a special case of the so-called *Shannon lower bound* on the rate–distortion function of a DMS [345] (see also [42, 158], [83, Problem 10.6]).

Theorem 6.24 *Fix a DMS* $\{Z_n\}_{n=1}^{\infty}$ *with distribution* P_Z. *Then, the source's rate–distortion function under the Hamming additive distortion measure and* $\widehat{Z} = Z$ *satisfies*

$$R(D) \geq H(Z) - D\log_2(|\mathcal{Z}|-1) - h_b(D) \quad for \; 0 \leq D \leq D_{max} \quad (6.4.1)$$

where D_{max} *is given in (6.3.4) as*

$$D_{max} := \min_{\widehat{z}} \sum_{z \in \mathcal{Z}} P_Z(z)\rho(z, \widehat{z}) = 1 - \max_{\widehat{z}} P_Z(\widehat{z}),$$

$H(Z)$ *is the source entropy and* $h_b(\cdot)$ *is the binary entropy function. Furthermore, equality holds in the above bound for* $D \leq (|\mathcal{Z}|-1)\min_{z \in \mathcal{Z}} P_Z(z)$.

Proof The proof is left as an exercise. (*Hint:* use Fano's inequality and examine the equality condition.)

Observation 6.25 (*Special cases of Theorem* 6.24)

- If the source is binary (i.e., $|\mathcal{Z}| = 2$) with $P_Z(0) = p$, then

$$(|\mathcal{Z}|-1)\min_{z \in \mathcal{Z}} P_Z(z) = \min\{p, 1-p\} = D_{max}.$$

Thus, the condition for equality in (6.4.1) always holds and Theorem 6.24 reduces to Theorem 6.23.

- If the source is *uniformly distributed* (i.e., $P_Z(z) = 1/|\mathcal{Z}|$ for all $z \in \mathcal{Z}$), then

$$(|\mathcal{Z}|-1)\min_{z \in \mathcal{Z}} P_Z(z) = \frac{|\mathcal{Z}|-1}{|\mathcal{Z}|} = D_{max}.$$

Thus, we directly obtain from Theorem 6.24 that

$$R(D) = \begin{cases} \log_2(|\mathcal{Z}|) - D\log_2(|\mathcal{Z}|-1) - h_b(D), & \text{if } 0 \le D \le \frac{|\mathcal{Z}|-1}{|\mathcal{Z}|}; \\ 0, & \text{if } D > \frac{|\mathcal{Z}|-1}{|\mathcal{Z}|}. \end{cases} \quad (6.4.2)$$

6.4.2 Rate–Distortion Function for Continuous Memoryless Sources Under the Squared Error Distortion Measure

We next examine the calculation or bounding of the rate–distortion function for continuous memoryless sources under the additive squared error distortion measure. We first show that the Gaussian source maximizes the rate–distortion function among all continuous sources with identical variance. This result, whose proof uses the fact that the Gaussian distribution maximizes differential entropy among all real-valued sources with the same variance (Theorem 5.20), can be seen as a dual result to Theorem 5.33, which states that Gaussian noise minimizes the capacity of additive noise channels. We also obtain the *Shannon lower bound* on the rate–distortion function of continuous sources under the squared error distortion measure.

Theorem 6.26 (Gaussian sources maximize the rate–distortion function) *Under the additive squared error distortion measure, namely $\rho_n(z^n, \hat{z}^n) = \sum_{i=1}^{n}(z_i - \hat{z}_i)^2$, the rate–distortion function for any continuous memoryless source $\{Z_i\}$ with a pdf of support $\mathbb{R}$, zero mean, variance σ^2, and finite differential entropy satisfies*

$$R(D) \le \begin{cases} \dfrac{1}{2}\log_2\dfrac{\sigma^2}{D}, & \text{for } 0 < D \le \sigma^2 \\ 0, & \text{for } D > \sigma^2 \end{cases}$$

with equality holding when the source is Gaussian.

Proof By Theorem 6.20 (extended to the "unbounded" squared error distortion measure),

$$R(D) = R^{(I)}(D) = \min_{f_{\hat{Z}|Z} : E[(Z-\hat{Z})^2] \le D} I(Z; \hat{Z}).$$

So for any $f_{\hat{Z}|Z}$ satisfying the distortion constraint,

$$R(D) \le I(f_Z, f_{\hat{Z}|Z}).$$

For $0 < D \le \sigma^2$, choose a dummy Gaussian random variable W with zero mean and variance aD, where $a = 1 - D/\sigma^2$, and is independent of Z. Let $\hat{Z} = aZ + W$. Then

$$E[(Z - \hat{Z})^2] = E[(1-a)^2 Z^2] + E[W^2] = (1-a)^2\sigma^2 + aD = D$$

which satisfies the distortion constraint. Note that the variance of $\hat{Z}$ is equal to $E[a^2 Z^2] + E[W^2] = \sigma^2 - D$. Consequently,

$$
\begin{aligned}
R(D) &\leq I(Z; \widehat{Z}) \\
&= h(\widehat{Z}) - h(\widehat{Z}|Z) \\
&= h(\widehat{Z}) - h(W + aZ|Z) \\
&= h(\widehat{Z}) - h(W|Z) \quad \text{(by Lemma 5.14)} \\
&= h(\widehat{Z}) - h(W) \qquad \text{(by the independence of } W \text{ and } Z) \\
&= h(\widehat{Z}) - \frac{1}{2} \log_2(2\pi e(aD)) \\
&\leq \frac{1}{2} \log_2(2\pi e(\sigma^2 - D)) - \frac{1}{2} \log_2(2\pi e(aD)) \quad \text{(by Theorem 5.20)} \\
&= \frac{1}{2} \log_2 \frac{\sigma^2}{D}.
\end{aligned}
$$

For $D > \sigma^2$, let $\widehat{Z}$ satisfy $\Pr\{\widehat{Z} = 0\} = 1$ and be independent of Z. Then $E[(Z - \widehat{Z})^2] = E[Z^2] + E[\widehat{Z}^2] - 2E[Z]E[\widehat{Z}] = \sigma^2 < D$, and $I(Z; \widehat{Z}) = 0$. Hence, $R(D) = 0$ for $D > \sigma^2$.

The achievability of this upper bound by a Gaussian source (with zero mean and variance σ^2) can be proved by showing that under the Gaussian source, $(1/2) \log_2(\sigma^2/D)$ is a lower bound to $R(D)$ for $0 < D \leq \sigma^2$. Indeed, when the source is Gaussian and for any $f_{\widehat{Z}|Z}$ such that $E[(Z - \widehat{Z})^2] \leq D$, we have

$$
\begin{aligned}
I(Z; \widehat{Z}) &= h(Z) - h(Z|\widehat{Z}) \\
&= \frac{1}{2} \log_2(2\pi e \sigma^2) - h(Z - \widehat{Z}|\widehat{Z}) \\
&\geq \frac{1}{2} \log_2(2\pi e \sigma^2) - h(Z - \widehat{Z}) \quad \text{(by Lemma 5.14)} \\
&\geq \frac{1}{2} \log_2(2\pi e \sigma^2) - \frac{1}{2} \log_2 \left(2\pi e \, \mathrm{Var}[(Z - \widehat{Z})]\right) \quad \text{(by Theorem 5.20)} \\
&\geq \frac{1}{2} \log_2(2\pi e \sigma^2) - \frac{1}{2} \log_2 \left(2\pi e \, E[(Z - \widehat{Z})^2]\right) \\
&\geq \frac{1}{2} \log_2(2\pi e \sigma^2) - \frac{1}{2} \log_2 (2\pi e D) \\
&= \frac{1}{2} \log_2 \frac{\sigma^2}{D}.
\end{aligned}
$$

□

Theorem 6.27 (Shannon lower bound on the rate–distortion function: squared error distortion) *Consider a continuous memoryless source $\{Z_i\}$ with a pdf of support $\mathbb{R}$ and finite differential entropy under the additive squared error distortion measure. Then, its rate–distortion function satisfies*

$$
R(D) \geq h(Z) - \frac{1}{2} \log_2(2\pi e D).
$$

Proof The proof, which follows similar steps as in the achievability of the upper bound in the proof of the previous theorem, is left as an exercise. □

The above two theorems yield that for any continuous memoryless source $\{Z_i\}$ with zero mean and variance σ^2, its rate–distortion function under the mean square error distortion measure satisfies

$$R_{SH}(D) \leq R(D) \leq R_G(D) \quad \text{for } 0 < D \leq \sigma^2, \tag{6.4.3}$$

where

$$R_{SH}(D) := h(Z) - \frac{1}{2}\log_2(2\pi e D)$$

and

$$R_G(D) := \frac{1}{2}\log_2 \frac{\sigma^2}{D},$$

and with equality holding when the source is Gaussian. Thus, the difference between the upper and lower bounds on $R(D)$ in (6.4.3) is

$$R_G(D) - R_{SH}(D) = -h(Z) + \frac{1}{2}\log_2(2\pi e \sigma^2)$$
$$= D(Z\|Z_G) \tag{6.4.4}$$

where $D(Z\|Z_G)$ is the non-Gaussianness of Z, i.e., the divergence between Z and a Gaussian random variable Z_G of mean zero and variance σ^2.

Thus, the gap between the upper and lower bounds in (6.4.4) is zero if the source is Gaussian and strictly positive if the source is non-Gaussian. For example, if Z is uniformly distributed on the interval $[-\sqrt{3\sigma^2}, \sqrt{3\sigma^2}]$ and hence with variance σ^2, then $D(Z\|Z_G) = 0.255$ bits for $\sigma^2 = 1$. On the other hand, if Z is Laplacian distributed with variance σ^2 or parameter $\sigma/\sqrt{2}$ (its pdf is given by $f_Z(z) = \frac{1}{\sqrt{2\sigma^2}}\exp\{-\frac{\sqrt{2}|z|}{\sigma}\}$ for $z \in \mathbb{R}$), then $D(Z\|Z_G) = 0.104$ bits for $\sigma^2 = 1$. We can thus deduce that the Laplacian source is *more similar* to the Gaussian source than the uniformly distributed source and hence its rate–distortion function is closer to Gaussian's rate–distortion function $R_G(D)$ than that of a uniform source. Finally in light of (6.4.4), the bounds on $R(D)$ in (6.4.3) can be expressed in terms of the Gaussian rate–distortion function $R_G(D)$ and the non-Gaussianness $D(Z\|Z_G)$, as follows:

$$R_G(D) - D(Z\|Z_G) \leq R(D) \leq R_G(D) \quad \text{for } 0 < D \leq \sigma^2. \tag{6.4.5}$$

Note that (6.4.5) is nothing but the dual of (5.7.4) and is an illustration of the duality between the rate–distortion and capacity-cost functions.

Observation 6.28 (*Rate–distortion function of Gaussian sources with memory*) Recall that Theorem 6.21 also holds for stationary ergodic Gaussian sources with finite second moment under the squared error distortion (e.g., see Footnote 2 in the previous section or [42, Theorems 7.2.4 and 7.2.5]). Note that a zero-mean stationary Gaussian source $\{X_i\}$ is ergodic if its covariance function $K_X(\tau) \to 0$ as $\tau \to \infty$.

For such sources, the rate–distortion function $R(D)$ can be determined paramet- rically; see [42, Theorem 4.5.3]. Furthermore, if the sources are also Markov, then $R(D)$ admits an explicit analytical expression for small values of D. More specifi- cally, consider a zero-mean unit-variance stationary Gauss–Markov source $\{X_i\}$ with covariance function $K_X(\tau) = a^\tau$, where $0 < a < 1$ is the correlation coefficient. Then,

$$R(D) = \frac{1}{2} \log_2 \frac{1 - a^2}{D} \quad \text{for } D \leq \frac{1 - a}{1 + a}.$$

For $D > \frac{1-a}{1+a}$, $R(D)$ can be obtained parametrically [42, p. 114].

6.4.3 Rate–Distortion Function for Continuous Memoryless Sources Under the Absolute Error Distortion Measure

We herein focus on the rate–distortion function of continuous memoryless sources under the absolute error distortion measure. In particular, we show that among all zero-mean real-valued sources with absolute mean λ (i.e., $E[|Z|] = \lambda$), the Laplacian source with parameter λ (i.e., with variance $2\lambda^2$) maximizes the rate–distortion func- tion. This result, which also provides the expression of the rate–distortion function of Laplacian sources, is similar to Theorem 6.26 regarding the maximal rate–distortion function under the squared error distortion measure (which is achieved by Gaussian sources). It is worth pointing out that in image coding applications, the Laplacian distribution is a good model to approximate the statistics of transform coefficients such as discrete cosine and wavelet transform coefficients [315, 375]. Finally, anal- ogously to Theorem 6.27, we obtain a Shannon lower bound on the rate–distortion function under the absolute error distortion.

Theorem 6.29 (Laplacian sources maximize the rate–distortion function) *Under the additive absolute error distortion measure, namely $\rho_n(z^n, \hat{z}^n) = \sum_{i=1}^{n} |z_i - \hat{z}_i|$, the rate–distortion function for any continuous memoryless source $\{Z_i\}$ with a pdf of support $\mathbb{R}$, zero mean and $E|Z| = \lambda$, where $\lambda > 0$ is a fixed parameter, satisfies*

$$R(D) \leq \begin{cases} \log_2 \dfrac{\lambda}{D}, & \text{for } 0 < D \leq \lambda \\ 0, & \text{for } D > \lambda \end{cases}$$

with equality holding when the source is Laplacian with mean zero and variance $2\lambda^2$ (parameter λ); i.e., its pdf is given by $f_Z(z) = \frac{1}{2\lambda} e^{-\frac{|z|}{\lambda}}$, $z \in \mathbb{R}$.

Proof Since

$$R(D) = \min_{f_{\widehat{Z}|Z}:E[|Z-\widehat{Z}|]\leq D} I(Z;\widehat{Z}),$$

we have that for any $f_{\widehat{Z}|Z}$ satisfying the distortion constraint,

$$R(D) \leq I(Z;\widehat{Z}) = I(f_Z, f_{\widehat{Z}|Z}).$$

For $0 < D \leq \lambda$, choose

$$\widehat{Z} = \left(1 - \frac{D}{\lambda}\right)^2 Z + \text{sgn}(Z)|W|,$$

where $\text{sgn}(Z)$ is equal to 1 if $Z \geq 0$ and to -1 if $Z < 0$, and W is a dummy random variable that is independent of Z and has a Laplacian distribution with mean zero and $E[|W|] = (1 - D/\lambda)D$; i.e., with parameter $(1 - D/\lambda)D$. Thus

$$
\begin{aligned}
E[|\widehat{Z}|] &= E\left|\left(1 - \frac{D}{\lambda}\right)^2 Z + \text{sgn}(Z)|W|\right| \\
&= \left(1 - \frac{D}{\lambda}\right)^2 E[|Z|] + E[|W|] \\
&= \left(1 - \frac{D}{\lambda}\right)^2 \lambda + \left(1 - \frac{D}{\lambda}\right) D \\
&= \lambda - D
\end{aligned}
\tag{6.4.6}
$$

and

$$
\begin{aligned}
E[|Z - \widehat{Z}|] &= E\left|\left(1 - \frac{D}{\lambda}\right)^2 Z + \text{sgn}(Z)|W| - Z\right| \\
&= E\left|\frac{D}{\lambda}\left(2 - \frac{D}{\lambda}\right) Z - \text{sgn}(Z)|W|\right| \\
&= \frac{D}{\lambda}\left(2 - \frac{D}{\lambda}\right) E[|Z|] - E[|W|] \\
&= \frac{D}{\lambda}\left(2 - \frac{D}{\lambda}\right) \lambda - \left(1 - \frac{D}{\lambda}\right) D \\
&= D,
\end{aligned}
$$

and hence this choice of $\widehat{Z}$ satisfies the distortion constraint. We can therefore write for this $\widehat{Z}$ that

$$
\begin{aligned}
R(D) &\le I(Z; \widehat{Z}) \\
&= h(\widehat{Z}) - h(\widehat{Z}|Z) \\
&\overset{(a)}{=} h(\widehat{Z}) - h(\text{sgn}(Z)|W||Z) \\
&= h(\widehat{Z}) - h(|W||Z) - \log_2 |\text{sgn}(Z)| \\
&\overset{(b)}{=} h(\widehat{Z}) - h(|W|) \\
&\overset{(c)}{=} h(\widehat{Z}) - h(W) \\
&\overset{(d)}{=} h(\widehat{Z}) - \log_2[2e(1 - D/\lambda)D] \\
&\overset{(e)}{\le} \log_2[2e(\lambda - D)] - \log_2[2e(1 - D/\lambda)D] \\
&= \log_2 \frac{\lambda}{D},
\end{aligned}
$$

where (a) follows from the expression of $\widehat{Z}$ and the fact that differential entropy is invariant under translations (Lemma 5.14), (b) holds since Z and W are independent of each other, (c) follows from the fact that W is Laplacian and that the Laplacian is symmetric, and (d) holds since the differential entropy of a zero-mean Laplacian random variable Z' with $E[|Z'|] = \lambda'$ is given by

$$
h(Z') = \log_2(2e\lambda') \quad \text{(in bits)}.
$$

Finally, (e) follows by noting that $E[|\widehat{Z}|] = \lambda - D$ from (6.4.6) and from the fact among all zero-mean real-valued random variables Z' with $E[|Z'|] = \lambda'$, the Laplacian random variable with zero-mean and parameter λ' maximizes differential entropy (see Observation 5.21, item 3).

For $D > \lambda$, let $\widehat{Z}$ satisfy $\Pr(\widehat{Z} = 0) = 1$ and be independent of Z. Then $E[|Z - \widehat{Z}| \le E[|Z|] + E[|\widehat{Z}|] = \lambda < D$. For this choice of $\widehat{Z}$, $R(D) \le I(Z; \widehat{Z}) = 0$ and hence $R(D) = 0$. This completes the proof of the upper bound.

We next show that the upper bound can be achieved by a Laplacian source with mean zero and parameter λ. This is proved by showing that for a Laplacian source with mean zero and parameter λ, $(1/2) \log_2(\lambda/D)$ is a lower bound to $R(D)$ for $0 < D \le \lambda$: for such a Laplacian source and for any $f_{\widehat{Z}|Z}$ satisfying $E[|Z - \widehat{Z}|] \le D$, we have

$$
\begin{aligned}
I(Z; \widehat{Z}) &= h(Z) - h(Z|\widehat{Z}) \\
&= \log_2(2e\lambda) - h(Z - \widehat{Z}|\widehat{Z}) \\
&\ge \log_2(2e\lambda) - h(Z - \widehat{Z}) \quad \text{(by Lemma 5.14)} \\
&\ge \log_2(2e\lambda) - \log_2(2eD) \\
&= \log_2 \frac{\lambda}{D},
\end{aligned}
$$

where the last inequality follows since

$$h(Z - \widehat{Z}) \leq \log_2 \left(2e E[|Z - \widehat{Z}|]\right) \leq \log_2(2eD)$$

by Observation 5.21 and the fact that $E[|Z - \widehat{Z}|] \leq D$.   □

Theorem 6.30 (Shannon lower bound on the rate–distortion function: absolute error distortion) *Consider a continuous memoryless source $\{Z_i\}$ with a pdf of support $\mathbb{R}$ and finite differential entropy under the additive absolute error distortion measure. Then, its rate–distortion function satisfies*

$$R(D) \geq h(Z) - \log_2(2eD).$$

Proof The proof is left as an exercise.   □

Observation 6.31 (*Shannon lower bound on the rate–distortion function: difference distortion*) The general form of the Shannon lower bound for a source $\{Z_i\}$ with finite differential entropy under the difference distortion measure $\rho(z, \hat{z}) = d(x - \hat{x})$, where $d(\cdot, \cdot)$ is a nonnegative function, is as follows:

$$R(D) \geq h(Z) - \sup_{X: E[d(X)] \leq D} h(X),$$

where the supremum is taken over all random variables X with a pdf satisfying $E[d(X)] \leq D$. It can be readily seen that this bound encompasses those in Theorems 6.27 and 6.30 as special cases.[3]

6.5   Lossy Joint Source–Channel Coding Theorem

By combining the rate–distortion theorem with the channel coding theorem, the optimality of separation between lossy source coding and channel coding can be established and Shannon's *lossy joint source–channel coding theorem* (also known as the *lossy information-transmission theorem*) can be shown for the communication of a source over a noisy channel and its reconstruction within a distortion threshold at the receiver. These results can be viewed as the "lossy" counterparts of the lossless joint source–channel coding theorem and the separation principle discussed in Sect. 4.6.

Definition 6.32 (*Lossy source–channel block code*) Given a discrete-time source $\{Z_i\}_{i=1}^{\infty}$ with alphabet $\mathcal{Z}$ and reproduction alphabet $\widehat{\mathcal{Z}}$ and a discrete-time channel with input and output alphabets $\mathcal{X}$ and $\mathcal{Y}$, respectively, an m-to-n lossy source–channel block code with rate $\frac{m}{n}$ source symbol/channel symbol is a pair of mappings $(f^{(sc)}, g^{(sc)})$, where[4]

$$f^{(sc)} : \mathcal{Z}^m \to \mathcal{X}^n \quad \text{and} \quad g^{(sc)} : \mathcal{Y}^n \to \widehat{\mathcal{Z}}^m.$$

---

[3]The asymptotic tightness of this bound as D approaches zero is studied in [249].

[4]Note that as pointed out in Sect. 4.6, n, $f^{(sc)}$, and $g^{(sc)}$ are all a function of m.

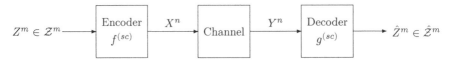

Fig. 6.3 An m-to-n block lossy source–channel coding system

The code's operation is illustrated in Fig. 6.3. The source m-tuple Z^m is encoded via the encoding function $f^{(sc)}$, yielding the codeword $X^n = f^{(sc)}(Z^m)$ as the channel input. The channel output Y^n, which is dependent on Z^m only via X^n (i.e., we have the Markov chain $Z^m \to X^n \to Y^n$), is decoded via $g^{(sc)}$ to obtain the source tuple estimate $\hat{Z}^m = g^{(sc)}(Y^n)$.

Given an additive distortion measure $\rho_m = \sum_{i=1}^{m} \rho(z_i, \hat{z}_i)$, where ρ is a distortion function on $\mathcal{Z} \times \hat{\mathcal{Z}}$, we say that the m-to-n lossy source–channel block code $(f^{(sc)}, g^{(sc)})$ *satisfies the average distortion fidelity criterion D*, where $D \geq 0$, if

$$\frac{1}{m} E[\rho_m(Z^m, \hat{Z}^m)] \leq D.$$

Theorem 6.33 (Lossy joint source–channel coding theorem) *Consider a discrete-time stationary ergodic source $\{Z_i\}_{i=1}^{\infty}$ with finite alphabet $\mathcal{Z}$, finite reproduction alphabet $\hat{\mathcal{Z}}$, bounded additive distortion measure $\rho_m(\cdot, \cdot)$ and rate–distortion function $R(D)$,[5] and consider a discrete-time memoryless channel with input alphabet $\mathcal{X}$, output alphabet $\mathcal{Y}$, and capacity C.[6] Assuming that both $R(D)$ and C are measured in the same units, the following hold:*

- *Forward part (achievability): For any $D > 0$, there exists a sequence of m-to-n_m lossy source–channel codes $(f^{(sc)}, g^{(sc)})$ satisfying the average distortion fidelity criterion D for sufficiently large m if*

$$\left(\limsup_{m \to \infty} \frac{m}{n_m} \right) \cdot R(D) < C.$$

- *Converse part: On the other hand, for any sequence of m-to-n_m lossy source–channel codes $(f^{(sc)}, g^{(sc)})$ satisfying the average distortion fidelity criterion D, we have*

$$\left(\frac{m}{n_m} \right) \cdot R(D) \leq C.$$

---

[5]Note that $\mathcal{Z}$ and $\hat{\mathcal{Z}}$ can also be continuous alphabets with an unbounded distortion function. In this case, the theorem still holds under appropriate conditions (e.g., [42, Problem 7.5], [135, Theorem 9.6.3]) that can accommodate, for example, the important class of Gaussian sources under the squared error distortion function (e.g., [135, p. 479]).

[6]The channel can have either finite or continuous alphabets. For example, it can be the memoryless Gaussian (i.e., AWGN) channel with input power P; in this case, $C = C(P)$.

Proof The proof uses both the channel coding theorem (i.e., Theorem 4.11) and the rate–distortion theorem (i.e., Theorem 6.21) and follows similar arguments as the proof of the lossless joint source–channel coding theorem presented in Sect. 4.6. We leave the proof as an exercise.

Observation 6.34 (*Lossy joint source–channel coding theorem with signaling rates*) The above theorem also admits another form when the source and channel are described in terms of "signaling rates" (e.g., [51]). More specifically, let T_s and T_c represent the durations (in seconds) per source letter and per channel input symbol, respectively.[7] In this case, $\frac{T_c}{T_s}$ represents the source–channel transmission rate measured in source symbols per channel use (or input symbol). Thus, again assuming that both $R(D)$ and C are measured in the same units, the theorem becomes as follows:

- The source can be reproduced at the output of the channel with distortion less than D (i.e., there exist lossy source–channel codes asymptotically satisfying the average distortion fidelity criterion D) if

$$\left(\frac{T_c}{T_s}\right) \cdot R(D) < C.$$

- Conversely, for any lossy source–channel codes satisfying the average distortion fidelity criterion D, we have

$$\left(\frac{T_c}{T_s}\right) \cdot R(D) \leq C.$$

6.6 Shannon Limit of Communication Systems

We close this chapter by applying Theorem 6.33 to a few useful examples of communication systems. Specifically, we obtain a bound on the end-to-end distortion of any communication system using the fact that if a source with rate–distortion function $R(D)$ can be transmitted over a channel with capacity C via a source–channel block code of rate $R_{sc} > 0$ (in source symbols/channel use) and reproduced at the destination with an average distortion no larger than D, then we must have that

$$R_{sc} \cdot R(D) \leq C$$

or equivalently,

$$R(D) \leq \frac{1}{R_{sc}} C. \tag{6.6.1}$$

---

[7]In other words, the source emits symbols at a rate of $1/T_s$ source symbols per second and the channel accepts inputs at a rate of $1/T_c$ channel symbols per second.

Solving for the smallest D, say D_{SL}, satisfying (6.6.1) with equality[8] yields a *lower bound*, called the *Shannon limit*,[9] on the distortion of all realizable lossy source–channel codes for the system with rate R_{sc}.

In the following examples, we calculate the Shannon limit for some source–channel configurations. The Shannon limit is not necessarily achievable in general, although this is the case in the first two examples.

Example 6.35 (Shannon limit for a binary uniform DMS over a BSC)[10]   Let $\mathcal{Z} = \widehat{\mathcal{Z}} = \{0, 1\}$ and consider a binary uniformly distributed DMS $\{Z_i\}$ (i.e., a Bernoulli(1/2) source) using the additive Hamming distortion measure. Note that in this case, $E[\rho(Z, \widehat{Z})] = P(Z \neq \widehat{Z}) := P_b$; in other words, the expected distortion is nothing but the source's *bit error probability* P_b. We desire to transmit the source over a BSC with crossover probability $\epsilon < 1/2$.

From Theorem 6.23, we know that for $0 \leq D \leq 1/2$, the source's rate–distortion function is given by

$$R(D) = 1 - h_b(D),$$

where $h_b(\cdot)$ is the binary entropy function. Also from (4.5.5), the channel's capacity is given by

$$C = 1 - h_b(\epsilon).$$

The Shannon limit D_{SL} for this system with source–channel transmission rate R_{sc} is determined by solving (6.6.1) with equality:

$$1 - h_b(D_{SL}) = \frac{1}{R_{sc}}[1 - h_b(\epsilon)].$$

This yields that

$$D_{SL} = h_b^{-1}\left(1 - \frac{1 - h_b(\epsilon)}{R_{sc}}\right) \tag{6.6.2}$$

for $D_{SL} \leq 1/2$, where for any $t \geq 0$,

$$h_b^{-1}(t) := \begin{cases} \inf\{x : t = h_b(x)\}, & \text{if } 0 \leq t \leq 1 \\ 0, & \text{if } t > 1 \end{cases} \tag{6.6.3}$$

is the inverse of the binary entropy function on the interval $[0, 1/2]$. Thus, D_{SL} given in (6.6.2) gives a lower bound on the bit error probability P_b of any rate-R_{sc} source–channel code used for this system. In particular, if $R_{sc} = 1$ source symbol/channel

---

[8]If the strict inequality $R(D) < \frac{1}{R_{sc}} C$ always holds, then in this case, the Shannon limit is $D_{SL} = D_{min} := E\left[\min_{\hat{z} \in \hat{\mathcal{Z}}} \rho(Z, \hat{z})\right]$.

[9]Other similar quantities used in the literature are the *optimal performance theoretically achievable (OPTA)* [42] and the *limit of the minimum transmission ratio (LMTR)* [87].

[10]This example appears in various sources including [205, Sect. 11.8], [87, Problem 2.2.16], and [266, Problem 5.7].

use, we directly obtain from (6.6.2) that[11]

$$D_{SL} = \epsilon. \tag{6.6.4}$$

Shannon limit over an equivalent binary-input hard-decision demodulated AWGN channel: It is well known that a BSC with crossover probability ϵ represents a binary-input AWGN channel used with antipodal (BPSK) signaling and hard-decision coherent demodulation (e.g., see [248]). More specifically, if the channel has noise power $\sigma_N^2 = N_0/2$ (i.e., the channel's underlying continuous-time white noise has power spectral density $N_0/2$) and uses antipodal signaling with average energy P per signal, then the BSCs crossover probability can be expressed in terms of P and N_0 as follows:

$$\epsilon = Q\left(\sqrt{\frac{P}{\sigma_N^2}}\right) = Q\left(\sqrt{\frac{2P}{N_0}}\right), \tag{6.6.5}$$

where

$$Q(x) = \frac{1}{\sqrt{2\pi}} \int_x^\infty e^{-\frac{t^2}{2}} \, dt$$

is the Gaussian Q-function. Furthermore, if the channel is used with a source–channel code of rate R_{sc} source (or information) bits/channel use, then $2P/N_0$ can be expressed in terms of a so-called *SNR per source (or information) bit* denoted by $\gamma_b := E_b/N_0$, where E_b is the *average energy per source bit*. Indeed, we have that $P = R_{sc}E_b$ and thus using (6.6.5), we have

$$\epsilon = Q\left(\sqrt{\frac{2R_{sc}E_b}{N_0}}\right) = Q\left(\sqrt{2R_{sc}\gamma_b}\right). \tag{6.6.6}$$

Thus, in light of (6.6.2), the Shannon limit for sending a uniform binary source over an AWGN channel used with antipodal modulation and hard-decision decoding satisfies the following in terms of the SNR per source bit γ_b:

$$R_{sc}\left(1 - h_b(D_{SL})\right) = 1 - h_b\left(Q\left(\sqrt{2R_{sc}\gamma_b}\right)\right) \tag{6.6.7}$$

for $D_{SL} \leq 1/2$. In Table 6.1, we use (6.6.7) to present the optimal (minimal) values of γ_b (in dB) for a given target value of D_{SL} and a given source–channel code rate $R_{sc} < 1$. The table indicates, for example, that if we desire to achieve an end-to-end bit error probability of no larger than 10^{-5} at a rate of 1/2, then the system's SNR per source bit can be no smaller than 1.772 dB. The Shannon limit values can similarly be computed for rates $R_{sc} > 1$.

---

[11] Source–channel systems with rate $R_{sc} = 1$ are typically referred to as systems with *matched* source and channel bandwidths (or signaling rates). Also, when $R_{sc} < 1$ (resp., > 1), the system is said to have *bandwidth compression* (resp., *bandwidth expansion*); e.g., cf. [274, 314, 358].

Table 6.1 Shannon limit values $\gamma_b = E_b/N_0$ (dB) for sending a binary uniform source over a BPSK-modulated AWGN used with hard-decision decoding

Rate R_{sc}	$D_{SL} = 0$	$D_{SL} = 10^{-5}$	$D_{SL} = 10^{-4}$	$D_{SL} = 10^{-3}$	$D_{SL} = 10^{-3}$
1/3	1.212	1.210	1.202	1.150	0.077
1/2	1.775	1.772	1.763	1.703	1.258
2/3	2.516	2.513	2.503	2.423	1.882
4/5	3.369	3.367	3.354	3.250	2.547

Optimality of uncoded communication: Note that for $R_{sc} = 1$, the Shannon limit in (6.6.4) can surprisingly be achieved by a simple *uncoded* scheme[12]: just directly transmit the source over the channel (i.e., set the blocklength $m = 1$ and the channel input $X_i = Z_i$ for any time instant $i = 1, 2, \ldots$) and declare the channel output as the reproduced source symbol (i.e., set $\widehat{Z}_i = Y_i$ for any i).[13]

In this case, the expected distortion (i.e., bit error probability) of this uncoded rate-one source–channel scheme is indeed given as follows:

$$
\begin{aligned}
E[\rho(Z, \widehat{Z})] &= P_b \\
&= P(X \neq Y) \\
&= P(Y \neq X | X = 1)(1/2) + P(Y \neq X | X = 0)(1/2) \\
&= \epsilon \\
&= D_{SL}.
\end{aligned}
$$

We conclude that this rate-one uncoded source–channel scheme achieves the Shannon limit and is hence *optimal*. Furthermore, this scheme, which has no encoding/decoding delay and no complexity, is clearly more desirable than using a separate source–channel coding scheme,[14] which would impose large encoding and decoding delays and would demand significant computational/storage resources.

Note that for rates $R_{sc} \neq 1$ and/or nonuniform sources, the uncoded scheme is not optimal and hence more complicated joint or separate source–channel codes would be required to yield a bit error probability arbitrarily close to (but strictly larger than) the Shannon limit D_{SL}. Finally, we refer the reader to [140], where necessary and sufficient conditions are established for source–channel pairs under which uncoded schemes are optimal.

Observation 6.36 The following two systems are extensions of the system considered in the above example.

---

[12]Uncoded transmission schemes are also referred to as scalar or single-letter codes.

[13]In other words, the code's encoding and decoding functions, $f^{(sc)}$ and $g^{(sc)}$, respectively, are both equal to the identity mapping.

[14]Note that in this system, since the source is incompressible, no source coding is actually required. Still the separate coding scheme will consist of a near-capacity achieving channel code.

- **Binary nonuniform DMS over a BSC:** The system is identical to that of Example 6.35 with the exception that the binary DMS is nonuniformly distributed with $P(Z = 0) = p$. Using the expression of $R(D)$ from Theorem 6.23, it can be readily shown that this system's Shannon limit is given by

$$D_{SL} = h_b^{-1}\left(h_b(p) - \frac{1 - h_b(\epsilon)}{R_{sc}}\right) \qquad (6.6.8)$$

for $D_{SL} \leq \min\{p, 1 - p\}$, where $h_b^{-1}(\cdot)$ is the inverse of the binary entropy function on the interval $[0, 1/2]$ defined in (6.6.3). Setting $p = 1/2$ in (6.6.8) directly results in the Shannon limit given in (6.6.2), as expected.

- **Nonbinary uniform DMS over a nonbinary symmetric channel:** Given integer $q \geq 2$, consider a q-ary uniformly distributed DMS with identical alphabet and reproduction alphabet $\mathcal{Z} = \widehat{\mathcal{Z}} = \{0, 1, \ldots, q - 1\}$ using the additive Hamming distortion measure and the q-ary symmetric DMC (with q-ary input and output alphabets and symbol error rate ϵ) described in (4.2.11). Thus using the expressions for the source's rate–distortion function in (6.4.2) and the channel's capacity in Example 4.19, we obtain that the Shannon limit of the system using rate-R_{sc} source–channel codes satisfies

$$\log_2(q) - D_{SL}\log_2(q - 1) - h_b(D_{SL}) = \frac{1}{R_{sc}}\left[\log_2(q) - \epsilon\log_2(q - 1) - h_b(\epsilon)\right] \qquad (6.6.9)$$

for $D_{SL} \leq \frac{q-1}{q}$. Setting $q = 2$ renders the source a Bernoulli(1/2) source and the channel a BSC with crossover probability ϵ, thus reducing (6.6.9) to (6.6.2).

Example 6.37 (Shannon limit for a memoryless Gaussian source over an AWGN channel [147]) Let $\mathcal{Z} = \widehat{\mathcal{Z}} = \mathbb{R}$ and consider a memoryless Gaussian source $\{Z_i\}$ of mean zero and variance σ^2 and the squared error distortion function. The objective is to transmit the source over an AWGN channel with input power constraint P and noise variance σ_N^2 and recover it with distortion fidelity no larger than D, for a given threshold $D > 0$.

By Theorem 6.26, the source's rate–distortion function is given by

$$R(D) = \frac{1}{2}\log_2\frac{\sigma^2}{D} \qquad \text{for} \qquad 0 < D < \sigma^2.$$

Furthermore, the capacity (or capacity-cost function) of the AWGN channel is given in (5.4.13) as

$$C(P) = \frac{1}{2}\log_2\left(1 + \frac{P}{\sigma_N^2}\right).$$

The Shannon limit D_{SL} for this system with rate R_{sc} is obtained by solving

$$R(D_{SL}) = \frac{C(P)}{R_{sc}}$$

or equivalently,

$$\frac{1}{2} \log_2 \frac{\sigma^2}{D_{SL}} = \frac{1}{2R_{sc}} \log_2 \left(1 + \frac{P}{\sigma_N^2}\right)$$

for $0 < D_{SL} < \sigma^2$, which gives

$$D_{SL} = \frac{\sigma^2}{\left(1 + \frac{P}{\sigma_N^2}\right)^{\frac{1}{R_{sc}}}}. \tag{6.6.10}$$

In particular, for a system with rate $R_{sc} = 1$, the Shannon limit in (6.6.10) becomes

$$D_{SL} = \frac{\sigma^2 \sigma_N^2}{P + \sigma_N^2}. \tag{6.6.11}$$

Optimality of a simple rate-one scalar source–channel coding scheme: The following simple (uncoded) source–channel coding scheme with rate $R_{sc} = 1$ can achieve the Shannon limit in (6.6.11). The code's encoding and decoding functions are scalar (with $m = 1$). More specifically, at any time instant i, the channel input (with power constraint P) is given by

$$X_i = f^{(sc)}(Z_i) = \sqrt{\frac{P}{\sigma^2}} Z_i,$$

and is sent over the AWGN channel. At the receiver, the corresponding channel output $Y_i = X_i + N_i$, where N_i is the additive Gaussian noise (which is independent of Z_i), is decoded via a scalar (MMSE) detector to yield the following reconstructed source symbol $\widehat{Z}_i$

$$\widehat{Z}_i = g^{(sc)}(Y_i) = \frac{\sqrt{P\sigma^2}}{P + \sigma_N^2} Y_i.$$

A simple calculation reveals that this code's expected distortion is given by

$$E[(Z - \widehat{Z})^2] = E\left[\left(Z - \frac{\sqrt{P\sigma^2}}{P + \sigma_N^2} \left(\sqrt{\frac{P}{\sigma^2}} Z + N\right)\right)^2\right]$$

$$= \frac{\sigma^2 \sigma_N^2}{P + \sigma_N^2}$$

$$= D_{SL},$$

which proves the optimality of this simple (delayless) source–channel code.

Extensions on the optimality of similar uncoded (scalar) schemes in Gaussian sensor networks can be found in [38, 141].

Example 6.38 (Shannon limit for a memoryless Gaussian source over a fading channel) Consider the same system as the above example except that now the channel is a memoryless fading channel as described in Observation 5.35 with input power constraint P and noise variance σ_N^2. We determine the Shannon limit of this system with rate R_{sc} for two cases: (1) the fading coefficients are known at the receiver, and (2) the fading coefficients are known at both the receiver and the transmitter.

1. *Shannon limit with decoder side information (DSI):* Using (5.4.17) for the channel capacity with DSI, we obtain that the Shannon limit with DSI is given by

$$D_{SL}^{(DSI)} = \frac{\sigma^2}{2^{\left\{ E_A \left[\log_2 \left(1 + \frac{A^2 P}{\sigma_N^2} \right)^{\frac{1}{R_{sc}}} \right] \right\}}} \tag{6.6.12}$$

   for $0 < D_{SL}^{(DSI)} < \sigma^2$. Making the fading process deterministic by setting $A = 1$ (almost surely) reduces (6.6.12) to (6.6.10), as expected.

2. *Shannon limit with full side information (FSI):* Similarly, using (5.4.19) for the fading channel capacity with FSI, we obtain the following Shannon limit:

$$D_{SL}^{(FSI)} = \frac{\sigma^2}{2^{\left\{ E_A \left[\log_2 \left(1 + \frac{A^2 p^*(A)}{\sigma_N^2} \right)^{\frac{1}{R_{sc}}} \right] \right\}}} \tag{6.6.13}$$

   for $0 < D_{SL}^{(FSI)} < \sigma^2$, where $p^*(\cdot)$ in (6.6.13) is given by

$$p^*(a) = \max \left(0, \frac{1}{\lambda} - \frac{\sigma^2}{a^2} \right)$$

   and λ is chosen to satisfy $E_A[p(A)] = P$.

Example 6.39 (Shannon limit for a binary uniform DMS over a binary-input AWGN channel) Consider the same binary uniform source as in Example 6.35 under the Hamming distortion measure to be sent via a source–channel code over a binary-input AWGN channel used with antipodal (BPSK) signaling of power P and noise variance $\sigma_N^2 = N_0/2$. Again, here the expected distortion is nothing but the source's bit error probability P_b. The source's rate–distortion function is given by Theorem 6.23 as presented in Example 6.35.

However, the channel capacity $C(P)$ of the AWGN whose input takes on two possible values $+\sqrt{P}$ or $-\sqrt{P}$, whose output is real-valued (unquantized), and whose noise variance is $\sigma_N^2 = \frac{N_0}{2}$, is given by evaluating the mutual information between the channel input and output under the input distribution $P_X(+\sqrt{P}) = P_X(-\sqrt{P}) = 1/2$ (e.g., see [63]):

$$C(P) = \frac{P}{\sigma_N^2} \log_2(e) - \frac{1}{\sqrt{2\pi}} \int_{-\infty}^{\infty} e^{-y^2/2} \log_2 \left[\cosh \left(\frac{P}{\sigma_N^2} + y\sqrt{\frac{P}{\sigma_N^2}} \right) \right] dy$$

$$= \frac{R_{sc}E_b}{N_0/2} \log_2(e) - \frac{1}{\sqrt{2\pi}} \int_{-\infty}^{\infty} e^{-y^2/2} \log_2 \left[\cosh \left(\frac{R_{sc}E_b}{N_0/2} + y\sqrt{\frac{R_{sc}E_b}{N_0/2}} \right) \right] dy$$

$$= 2R_{sc}\gamma_b \log_2(e) - \frac{1}{\sqrt{2\pi}} \int_{-\infty}^{\infty} e^{-y^2/2} \log_2 [\cosh(2R_{sc}\gamma_b + y\sqrt{2R_{sc}\gamma_b})] dy,$$

where $P = R_{sc}E_b$ is the channel signal power, E_b is the average energy per source bit, R_{sc} is the rate in source bit/channel use of the system's source–channel code, and $\gamma_b = E_b/N_0$ is the SNR per source bit. The system's Shannon limit satisfies

$$R_{sc}(1 - h_b(D_{SL})) = \Big[2R_{sc}\gamma_b \log_2(e)$$
$$- \frac{1}{\sqrt{2\pi}} \int_{-\infty}^{\infty} e^{-y^2/2} \log_2 [\cosh(2R_{sc}\gamma_b + y\sqrt{2R_{sc}\gamma_b})] dy \Big],$$

or equivalently,

$$h_b(D_{SL}) = 1 - 2\gamma_b \log_2(e) + \frac{1}{R_{sc}\sqrt{2\pi}} \int_{-\infty}^{\infty} e^{-y^2/2} \log_2 [\cosh(2R_{sc}\gamma_b + y\sqrt{2R_{sc}\gamma_b})] dy$$
$$(6.6.14)$$

for $D_{SL} \leq 1/2$. In Fig. 6.4, we use (6.6.14) to plot the Shannon limit versus γ_b (in dB) for codes with rates 1/2 and 1/3. We also provide in Table 6.2 the optimal values of γ_b for target values of D_{SL} and R_{sc}.

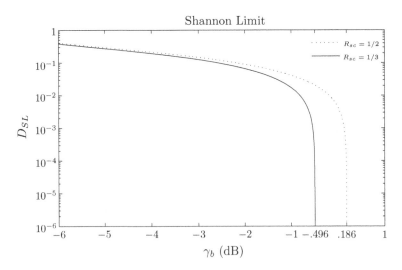

Fig. 6.4 The Shannon limit for sending a binary uniform source over a BPSK-modulated AWGN channel with unquantized output; rates $R_{sc} = 1/2$ and 1/3

Table 6.2 Shannon limit values $\gamma_b = E_b/N_0$ (dB) for sending a binary uniform source over a BPSK-modulated AWGN with unquantized output

Rate R_{sc}	$D_{SL} = 0$	$D_{SL} = 10^{-5}$	$D_{SL} = 10^{-4}$	$D_{SL} = 10^{-3}$	$D_{SL} = 10^{-3}$
1/3	−0.496	−0.496	−0.504	−0.559	−0.960
1/2	0.186	0.186	0.177	0.111	−0.357
2/3	1.060	1.057	1.047	0.963	0.382
4/5	2.040	2.038	2.023	1.909	1.152

The Shannon limits calculated above are pertinent due to the invention of near-capacity achieving channel codes, such as turbo [44, 45] or LDPC [133, 134, 251, 252] codes. For example, the rate-1/2 turbo coding system proposed in [44, 45] can approach a bit error rate of 10^{-5} at $\gamma_b = 0.9$ dB, which is only 0.714 dB away from the Shannon limit of 0.186 dB. This implies that a near-optimal channel code has been constructed, since in principle, no codes can perform better than the Shannon limit. Source–channel turbo codes for sending nonuniform memoryless and Markov binary sources over the BPSK-modulated AWGN channel are studied in [426–428].

Example 6.40 (Shannon limit for a binary uniform DMS over a binary-input Rayleigh fading channel) Consider the same system as the one in the above example, except that the channel is a unit-power BPSK-modulated Rayleigh fading channel (with unquantized output). The channel is described by (5.4.16), where the input can take on one of the two values, -1 or $+1$ (i.e., its input power is $P = 1 = R_{sc}E_b$), the noise variance is $\sigma_N^2 = N_0/2$, and the fading distribution is Rayleigh:

$$f_A(a) = 2ae^{-a^2}, \qquad a > 0.$$

Assume also that the receiver knows the fading amplitude (i.e., the case of decoder side information). Then, the channel capacity is given by evaluating $I(X; Y|A)$ under the uniform input distribution $P_X(-1) = P_X(+1) = 1/2$, yielding the following expression in terms of the SNR per source bit $\gamma_b = E_b/N_0$:

$$C_{DSI}(\gamma_b) = 1 - \sqrt{\frac{R_{sc}\gamma_b}{\pi}} \int_0^{+\infty} \int_{-\infty}^{+\infty} f_A(a)\, e^{-R_{sc}\gamma_b(y+a)^2} \log_2\left(1 + e^{4R_{sc}\gamma_b ya}\right) dy\, da.$$

Now, setting $R_{sc}R(D_{SL}) = C_{DSI}(\gamma_b)$ implies that the Shannon limit satisfies

$$
\begin{aligned}
h_b(D_{SL}) = 1 - \frac{1}{R_{sc}} + \sqrt{\frac{\gamma_b}{R_{sc}\pi}} \int_0^{+\infty} \int_{-\infty}^{+\infty} f_A(a) \\
\times e^{-R_{sc}\gamma_b(y+a)^2} \log_2\left(1 + e^{4R_{sc}\gamma_b ya}\right) dy\, da
\end{aligned}
\tag{6.6.15}
$$

for $D_{SL} \leq 1/2$. In Table 6.3, we present some Shannon limit values calculated from (6.6.15).

Table 6.3 Shannon limit values $\gamma_b = E_b/N_0$ (dB) for sending a binary uniform source over a BPSK-modulated Rayleigh fading channel with decoder side information

Rate R_{sc}	$D_{SL} = 0$	$D_{SL} = 10^{-5}$	$D_{SL} = 10^{-4}$	$D_{SL} = 10^{-3}$	$D_{SL} = 10^{-3}$
1/3	0.489	0.487	0.479	0.412	−0.066
1/2	1.830	1.829	1.817	1.729	1.107
2/3	3.667	3.664	3.647	3.516	2.627
4/5	5.936	5.932	5.904	5.690	4.331

Example 6.41 (Shannon limit for a binary uniform DMS over an AWGN channel) As in the above example, we consider a memoryless binary uniform source but we assume that the channel is an AWGN channel (with real inputs and outputs) with power constraint P and noise variance $\sigma_N^2 = N_0/2$. Recalling that the channel capacity is given by

$$C(P) = \frac{1}{2} \log_2 \left(1 + \frac{P}{\sigma_N^2} \right)$$
$$= \frac{1}{2} \log_2 \left(1 + 2R_{sc}\gamma_b \right),$$

we obtain that the system's Shannon limit satisfies

$$h_b(D_{SL}) = 1 - \frac{1}{2R_{sc}} \log_2 \left(1 + 2R_{sc}\gamma_b \right)$$

for $D_{SL} \le 1/2$. In Fig. 6.5, we plot the above Shannon limit versus γ_b for systems with $R_{sc} = 1/2$ and $1/3$.

Other examples of determining the Shannon limit for sending sources with memory over memoryless channels, such as discrete Markov sources under the Hamming distortion function[15] or Gauss–Markov sources under the squared error distortion measure (e.g., see [98]) can be similarly considered. Finally, we refer the reader to the end of Sect. 4.6 for a discussion of relevant works on lossy joint source–channel coding.

Problems

1. Prove Observation 6.15.
2. *Binary source with infinite distortion*: Let $\{Z_i\}$ be a DMS with binary source and reproduction alphabets $\mathcal{Z} = \hat{\mathcal{Z}} = \{0, 1\}$, distribution $P_Z(1) = p = 1 - P_Z(0)$, where $0 < p \le 1/2$, and the following distortion measure:

---

[15]For example, if the Markov source is binary symmetric, then its rate–distortion function is given by (6.3.9) for $D \le D_c$ and the Shannon limit for sending this source over say a BSC or an AWGN channel can be calculated. If the distortion region $D > D_c$ is of interest, then (6.3.8) or the right side of (6.3.9) can be used as lower bounds on $R(D)$; in this case, a lower bound on the Shannon limit can be obtained.

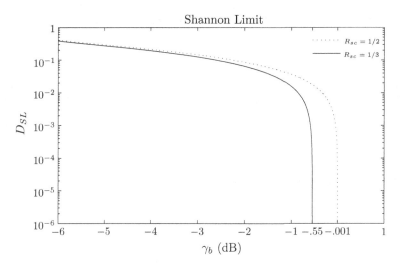

Fig. 6.5 The Shannon limits for sending a binary uniform source over a continuous-input AWGN channel; rates $R_{sc} = 1/2$ and $1/3$

$$p(z, \hat{z}) = \begin{cases} 0 & \text{if } \hat{z} = z \\ 1 & \text{if } z = 1 \text{ and } \hat{z} = 0 \\ \infty & \text{if } z = 0 \text{ and } \hat{z} = 1. \end{cases}$$

(a) Determine the source's rate–distortion function $R(D)$ (in your calculations, use the convention that $0 \cdot \infty = 0$).
(b) Specialize $R(D)$ to the case of $p = 1/2$ (uniform source).

3. *Binary uniform source with erasure and infinite distortion*: Consider a uniformly distributed DMS $\{Z_i\}$ with alphabet $\mathcal{Z} = \{0, 1\}$ and reproduction alphabet $\widehat{\mathcal{Z}} = \{0, 1, E\}$, where E represents an erasure. Let the source's distortion function be given as follows:

$$p(z, \hat{z}) = \begin{cases} 0 & \text{if } \hat{z} = z \\ 1 & \text{if } \hat{z} = E \\ \infty & \text{otherwise.} \end{cases}$$

Find the source's rate–distortion function.
4. For the binary source and distortion measure considered in Problem 6.3, describe a simple data compression scheme whose rate achieves the rate–distortion function $R(D)$ for any given distortion threshold D.
5. *Nonbinary uniform source with erasure and infinite distortion*: Consider a simple generalization of Problem 6.3 above, where $\{Z_i\}$ is a uniformly distributed nonbinary DMS with alphabet $\mathcal{Z} = \{0, 1, \ldots, q - 1\}$, reproduction alphabet $\widehat{\mathcal{Z}} = \{0, 1, \ldots, q - 1, E\}$ and the same distortion function as above, where

$q \geq 2$ is an integer. Find the source's rate–distortion function and verify that it reduces to the one derived in Problem 6.3 when $q = 2$.

6. *Translated distortion*: Consider a DMS $\{Z_i\}$ with alphabet $\mathcal{Z}$ and reproduction alphabet $\hat{\mathcal{Z}}$. Let $R(D)$ denote the source's rate–distortion function under the distortion function $\rho(\cdot, \cdot)$. Consider a new distortion function $\hat{\rho}(\cdot, \cdot)$ obtained by adding to $\rho(\cdot, \cdot)$ a constant that depends on the source symbols. More specifically, let

$$\hat{\rho}(z, \hat{z}) = \rho(z, \hat{z}) + c_z$$

where $z \in \mathcal{Z}$, $\hat{z} \in \hat{\mathcal{Z}}$, and c_z is a constant that depends on source symbol z. Show that the source's rate–distortion function $\widehat{R}(D)$ associated with the new distortion function $\hat{\rho}(\cdot, \cdot)$ can be expressed as follows in terms of $R(D)$:

$$\widehat{R}(D) = R(D - \bar{c})$$

for $D \geq \bar{c}$, where $\bar{c} = \sum_{z \in \mathcal{Z}} P_Z(z) c_z$.
Note: This result was originally shown by Pinkston [302].

7. *Scaled distortion*: Consider a DMS $\{Z_i\}$ with alphabet $\mathcal{Z}$, reproduction alphabet $\hat{\mathcal{Z}}$, distortion function $\rho(\cdot, \cdot)$, and rate–distortion function $R(D)$. Let $\hat{\rho}(\cdot, \cdot)$ be a new distortion function obtained by scaling $\rho(\cdot, \cdot)$ via a positive constant a:

$$\hat{\rho}(z, \hat{z}) = a\rho(z, \hat{z})$$

for $z \in \mathcal{Z}$ and $\hat{z} \in \hat{\mathcal{Z}}$. Determine the source's rate–distortion function $\hat{R}(D)$ associated with the new distortion function $\hat{\rho}(\cdot, \cdot)$ in terms of $R(D)$.

8. *Source symbols with zero distortion*: Consider a DMS $\{Z_i\}$ with alphabet $\mathcal{Z}$, reproduction alphabet $\hat{\mathcal{Z}}$, distortion function $\rho(\cdot, \cdot)$, and rate–distortion function $R(D)$. Assume that one source symbol, say z_1, in $\mathcal{Z}$ has zero distortion: $\rho(z_1, \hat{z}) = 0$ for all $\hat{z} \in \hat{\mathcal{Z}}$. Show that

$$R(D) = (1 - P_Z(z_1)) \widetilde{R} \left(\frac{D}{1 - P_Z(z_1)} \right)$$

where $\widetilde{R}(D)$ is the rate–distortion function of source $\{\widetilde{Z}_i\}$ with alphabet $\widetilde{\mathcal{Z}} = \mathcal{Z} \setminus \{z_1\}$ and distribution

$$P_{\widetilde{Z}}(z) = \frac{P_Z(z)}{1 - P_Z(z_1)}, \quad z \in \widetilde{\mathcal{Z}},$$

and with the same reproduction alphabet $\hat{\mathcal{Z}}$ and distortion function $\rho(\cdot, \cdot)$.
Note: This result first appeared in [302].

9. Consider a DMS $\{Z_i\}$ with quaternary source and reproduction alphabets $\mathcal{Z} = \hat{\mathcal{Z}} = \{0, 1, 2, 3\}$, probability distribution vector

$$(P_Z(0), P_Z(1), P_Z(2), P_Z(3)) = (p/3, p/3, 1 - p, p/3)$$

for fixed $0 < p < 1$, and distortion measure given by the following matrix:

$$[\rho(z, \hat{z})] = [\rho_{z\hat{z}}] = \begin{bmatrix} 0 & \infty & 1 & \infty \\ \infty & 0 & 1 & \infty \\ 0 & 0 & 0 & 0 \\ \infty & \infty & 1 & 0 \end{bmatrix}.$$

Determine the source's rate–distortion function.

10. Consider a binary DMS $\{Z_i\}$ with $\mathcal{Z} = \widehat{\mathcal{Z}} = \{0, 1\}$, distribution $P_Z(0) = p$, where $0 < p \le 1/2$, and the following distortion matrix

$$[\rho(z, \hat{z})] = [\rho_{z\hat{z}}] = \begin{bmatrix} b_1 & a + b_1 \\ a + b_2 & b_2 \end{bmatrix},$$

where $a > 0$, b_1, and b_2 are constants.

(a) Find $R(D)$ in terms of a, b_1, and b_2. (*Hint*: Use Problems 6.6 and 6.7.)
(b) What is $R(D)$ when $a = 1$ and $b_1 = b_2 = 0$?
(c) What is $R(D)$ when $a = b_2 = 1$ and $b_1 = 0$?

11. Prove Theorem 6.24.
12. Prove Theorem 6.30.
13. *Memory decreases the rate distortion function*: Give an example of a discrete source with memory whose rate–distortion function is strictly less (at least in some range of the distortion threshold) than the rate–distortion function of a memoryless source with identical marginal distribution.
14. *Lossy joint source–channel coding theorem—Forward part*: Prove the forward part of Theorem 6.33. For simplicity, assume that the source is memoryless.
15. *Lossy joint source–channel coding theorem—Converse part*: Prove the converse part of Theorem 6.33.
16. *Gap between the Laplacian rate–distortion function and the Shannon bound*: Consider a continuous memoryless source $\{Z_i\}$ with a pdf of support $\mathbb{R}$, zero mean, and $E[|Z|] = \lambda$, where $\lambda > 0$, under the absolute error distortion measure. Show that for any $0 < D \le \lambda$,

$$R_L(D) - R_{SD}^L(D) = D(Z\|Z_L),$$

where $D(Z\|Z_L)$ is the divergence between Z and a Laplacian random variable Z_L of mean zero and parameter λ, $R_L(D)$ denotes the rate–distortion function of source Z_L (see Theorem 6.29) and $R_{SD}^L(D)$ denotes the Shannon lower bound under the absolute error distortion measure (see Theorem 6.30).

17. *q-ary uniform DMS over the q-ary symmetric DMC*: Given integer $q \ge 2$, consider a q-ary DMS $\{Z_n\}$ that is uniformly distributed over its alphabet $\mathcal{Z} = \{0, 1, \ldots, q-1\}$, with reproduction alphabet $\widehat{\mathcal{Z}} = \mathcal{Z}$ and the Hamming distortion

measure. Consider also the q-ary symmetric DMC described in (4.2.11) with q-ary input and output alphabets and symbol error rate $\epsilon \leq \frac{q-1}{q}$.
Determine whether or not an uncoded source–channel transmission scheme of rate $R_{sc} = 1$ source symbol/channel use (i.e., a source–channel code whose encoder and decoder are both given by the identity function) is optimal for this communication system.

18. *Shannon limit of the erasure source–channel system*: Given integer $q \geq 2$, consider the q-ary uniform DMS together with the distortion measure of Problem 6.5 above and the q-ary erasure channel described in (4.2.12), see also Problem 4.13.

    (a) Find the system's Shannon limit under a transmission rate of R_{sc} source symbol/channel use.
    (b) Describe an uncoded source–channel transmission scheme for the system with rate $R_{sc} = 1$ and assess its optimality.

19. *Shannon limit for a Laplacian source over an AWGN channel*: Determine the Shannon limit under the absolute error distortion criterion and a transmission rate of R_{sc} source symbols/channel use for a communication system consisting of a memoryless zero-mean Laplacian source with parameter λ and an AWGN channel with input power constraint P and noise variance σ_N^2.

20. *Shannon limit for a nonuniform DMS over different channels*: Find the Shannon limit under the Hamming distortion criterion for each of the systems of Examples 6.39–6.41, where the source is a binary nonuniform DMS with $P_Z(0) = p$, where $0 \leq p \leq 1$.

Appendix A
Overview on Suprema and Limits

We herein review basic results on suprema and limits which are useful for the development of information theoretic coding theorems; they can be found in standard real analysis texts (e.g., see [262, 398]).

A.1 Supremum and Maximum

Throughout, we work on subsets of $\mathbb{R}$, the set of real numbers.

Definition A.1 (*Upper bound of a set*) A real number u is called an *upper bound* of a non-empty subset $\mathcal{A}$ of $\mathbb{R}$ if every element of $\mathcal{A}$ is less than or equal to u; we say that $\mathcal{A}$ is *bounded above*. Symbolically, the definition becomes

$$\mathcal{A} \subset \mathbb{R} \text{ is bounded above} \iff (\exists\, u \in \mathbb{R}) \text{ such that } (\forall\, a \in \mathcal{A}),\, a \leq u.$$

Definition A.2 (*Least upper bound or supremum*) Suppose $\mathcal{A}$ is a non-empty subset of $\mathbb{R}$. Then, we say that a real number s is a *least upper bound* or *supremum* of $\mathcal{A}$ if s is an upper bound of the set $\mathcal{A}$ and if $s \leq s'$ for each upper bound s' of $\mathcal{A}$. In this case, we write $s = \sup \mathcal{A}$; other notations are $s = \sup_{x \in \mathcal{A}} x$ and $s = \sup\{x : x \in \mathcal{A}\}$.

Completeness Axiom: (**Least upper bound property**) Let $\mathcal{A}$ be a non-empty subset of $\mathbb{R}$ that is bounded above. Then $\mathcal{A}$ has a least upper bound.

It follows directly that if a non-empty set in $\mathbb{R}$ has a supremum, and then this supremum is unique. Furthermore, note that the empty set ($\emptyset$) and any set not bounded above do not admit a supremum in $\mathbb{R}$. However, when working in the set of extended real numbers given by $\mathbb{R} \cup \{-\infty, \infty\}$, we can define the supremum of the empty set as $-\infty$ and that of a set not bounded above as ∞. These extended definitions will be adopted in the text.

We now distinguish between two situations: (i) the supremum of a set $\mathcal{A}$ belongs to $\mathcal{A}$, and (ii) the supremum of a set $\mathcal{A}$ does not belong to $\mathcal{A}$. It is quite easy to create examples for both situations. A quick example for (i) involves the set $(0, 1]$, while

© Springer Nature Singapore Pte Ltd. 2018   263
F. Alajaji and P.-N. Chen, *An Introduction to Single-User Information Theory*,
Springer Undergraduate Texts in Mathematics and Technology,
https://doi.org/10.1007/978-981-10-8001-2

the set $(0, 1)$ can be used for (ii). In both examples, the supremum is equal to 1; however, in the former case, the supremum belongs to the set, while in the latter case it does not. When a set contains its supremum, we call the supremum the *maximum* of the set.

Definition A.3 (*Maximum*) If $\sup \mathcal{A} \in \mathcal{A}$, then $\sup \mathcal{A}$ is also called the *maximum* of $\mathcal{A}$ and is denoted by $\max \mathcal{A}$. However, if $\sup \mathcal{A} \notin \mathcal{A}$, then we say that the maximum of $\mathcal{A}$ does not exist.

Property A.4 (Properties of the supremum)

1. *The supremum of any set in $\mathbb{R} \cup \{-\infty, \infty\}$ always exits.*
2. $(\forall a \in \mathcal{A}) \, a \leq \sup \mathcal{A}$.
3. *If $-\infty < \sup \mathcal{A} < \infty$, then $(\forall \varepsilon > 0)(\exists a_0 \in \mathcal{A}) \, a_0 > \sup \mathcal{A} - \varepsilon$.*
   (The existence of $a_0 \in (\sup \mathcal{A} - \varepsilon, \sup \mathcal{A}]$ for any $\varepsilon > 0$ under the condition of $|\sup \mathcal{A}| < \infty$ is called the approximation property for the supremum.)
4. *If $\sup \mathcal{A} = \infty$, then $(\forall L \in \mathbb{R})(\exists B_0 \in \mathcal{A}) \, B_0 > L$.*
5. *If $\sup \mathcal{A} = -\infty$, then $\mathcal{A}$ is empty.*

Observation A.5 (**Supremum of a set and channel coding theorems**) In information theory, a typical channel coding theorem establishes that a (finite) real number α is the supremum of a set $\mathcal{A}$. Thus, to prove such a theorem, one must show that α satisfies both Properties 3 and 2 above, i.e.,

$$(\forall \varepsilon > 0)(\exists a_0 \in \mathcal{A}) \, a_0 > \alpha - \varepsilon \qquad (A.1.1)$$

and

$$(\forall a \in \mathcal{A}) \, a \leq \alpha, \qquad (A.1.2)$$

where (A.1.1) and (A.1.2) are called the *achievability* (or *forward*) part and the *converse* part, respectively, of the theorem. Specifically, (A.1.2) states that α is an upper bound of $\mathcal{A}$, and (A.1.1) states that no number less than α can be an upper bound for $\mathcal{A}$.

Property A.6 (Properties of the maximum)

1. $(\forall a \in \mathcal{A}) \, a \leq \max \mathcal{A}$, *if* $\max \mathcal{A}$ *exists in* $\mathbb{R} \cup \{-\infty, \infty\}$.
2. $\max \mathcal{A} \in \mathcal{A}$.

From the above property, in order to obtain $\alpha = \max \mathcal{A}$, one needs to show that α satisfies both

$$(\forall a \in \mathcal{A}) \, a \leq \alpha \quad \text{and} \quad \alpha \in \mathcal{A}.$$

A.2 Infimum and Minimum

The concepts of infimum and minimum are dual to those of supremum and maximum.

Definition A.7 (*Lower bound of a set*) A real number ℓ is called a *lower bound* of a non-empty subset $\mathcal{A}$ in $\mathbb{R}$ if every element of $\mathcal{A}$ is greater than or equal to ℓ; we say that $\mathcal{A}$ is *bounded below*. Symbolically, the definition becomes

$$\mathcal{A} \subset \mathbb{R} \text{ is bounded below} \quad \Longleftrightarrow \quad (\exists \, \ell \in \mathbb{R}) \text{ such that } (\forall \, a \in \mathcal{A}) \, a \geq \ell.$$

Definition A.8 (*Greatest lower bound or infimum*) Suppose $\mathcal{A}$ is a non-empty subset of $\mathbb{R}$. Then, we say that a real number ℓ is a *greatest lower bound* or *infimum* of $\mathcal{A}$ if ℓ is a lower bound of $\mathcal{A}$ and if $\ell \geq \ell'$ for each lower bound ℓ' of $\mathcal{A}$. In this case, we write $\ell = \inf \mathcal{A}$; other notations are $\ell = \inf_{x \in \mathcal{A}} x$ and $\ell = \inf\{x : x \in \mathcal{A}\}$.

Completeness Axiom: (**Greatest lower bound property**) Let $\mathcal{A}$ be a non-empty subset of $\mathbb{R}$ that is bounded below. Then, $\mathcal{A}$ has a greatest lower bound.

As for the case of the supremum, it directly follows that if a non-empty set in $\mathbb{R}$ has an infimum, and then this infimum is unique. Furthermore, working in the set of extended real numbers, the infimum of the empty set is defined as ∞ and that of a set not bounded below as $-\infty$.

Definition A.9 (*Minimum*) If $\inf \mathcal{A} \in \mathcal{A}$, then $\inf \mathcal{A}$ is also called the *minimum* of $\mathcal{A}$ and is denoted by $\min \mathcal{A}$. However, if $\inf \mathcal{A} \notin \mathcal{A}$, we say that the minimum of $\mathcal{A}$ does not exist.

Property A.10 (Properties of the infimum)

1. *The infimum of any set in $\mathbb{R} \cup \{-\infty, \infty\}$ always exists.*
2. $(\forall \, a \in \mathcal{A}) \, a \geq \inf \mathcal{A}.$
3. *If $\infty > \inf \mathcal{A} > -\infty$, then $(\forall \, \varepsilon > 0)(\exists \, a_0 \in \mathcal{A}) \, a_0 < \inf \mathcal{A} + \varepsilon$.*
   (The existence of $a_0 \in [\inf \mathcal{A}, \inf \mathcal{A} + \varepsilon)$ for any $\varepsilon > 0$ under the assumption of $|\inf \mathcal{A}| < \infty$ is called the approximation property for the infimum.)
4. *If $\inf \mathcal{A} = -\infty$, then $(\forall A \in \mathbb{R})(\exists \, B_0 \in \mathcal{A}) B_0 < L$.*
5. *If $\inf \mathcal{A} = \infty$, then $\mathcal{A}$ is empty.*

Observation A.11 (**Infimum of a set and source coding theorems**) Analogously to Observation A.5, a typical source coding theorem in information theory establishes that a (finite) real number α is the infimum of a set $\mathcal{A}$. Thus, to prove such a theorem, one must show that α satisfies both Properties 3 and 2 above, i.e.,

$$(\forall \, \varepsilon > 0)(\exists \, a_0 \in \mathcal{A}) \, a_0 < \alpha + \varepsilon \tag{A.2.1}$$

and

$$(\forall \, a \in \mathcal{A}) \, a \geq \alpha. \tag{A.2.2}$$

Here, (A.2.1) is called the *achievability* or *forward* part of the coding theorem; it specifies that no number greater than α can be a lower bound for $\mathcal{A}$. Also, (A.2.2) is called the *converse* part of the theorem; it states that α is a lower bound of $\mathcal{A}$.

Property A.12 (Properties of the minimum)

1. $(\forall\, a \in \mathcal{A})\ a \geq \min \mathcal{A}$, *if* $\min \mathcal{A}$ *exists in* $\mathbb{R} \cup \{-\infty, \infty\}$.
2. $\min \mathcal{A} \in \mathcal{A}$.

A.3 Boundedness and Suprema Operations

Definition A.13 (*Boundedness*) A subset $\mathcal{A}$ of $\mathbb{R}$ is said to be *bounded* if it is both bounded above and bounded below; otherwise, it is called *unbounded*.

Lemma A.14 (Condition for boundedness) *A subset $\mathcal{A}$ of $\mathbb{R}$ is bounded iff* $(\exists\, k \in \mathbb{R})$ *such that* $(\forall\, a \in \mathcal{A})\ |a| \leq k$.

Lemma A.15 (Monotone property) *Suppose that $\mathcal{A}$ and $\mathcal{B}$ are non-empty subsets of $\mathbb{R}$ such that $\mathcal{A} \subset \mathcal{B}$. Then*

1. $\sup \mathcal{A} \leq \sup \mathcal{B}$.
2. $\inf \mathcal{A} \geq \inf \mathcal{B}$.

Lemma A.16 (Supremum for set operations) *Define the "addition" of two sets $\mathcal{A}$ and $\mathcal{B}$ as*

$$\mathcal{A} + \mathcal{B} := \{c \in \mathbb{R} : c = a + b \text{ for some } a \in \mathcal{A} \text{ and } b \in \mathcal{B}\}.$$

Define the "scalar multiplication" of a set $\mathcal{A}$ by a scalar $k \in \mathbb{R}$ as

$$k \cdot \mathcal{A} := \{c \in \mathbb{R} : c = k \cdot a \text{ for some } a \in \mathcal{A}\}.$$

Finally, define the "negation" of a set $\mathcal{A}$ as

$$-\mathcal{A} := \{c \in \mathbb{R} : c = -a \text{ for some } a \in \mathcal{A}\}.$$

Then, the following hold:

1. *If $\mathcal{A}$ and $\mathcal{B}$ are both bounded above, then $\mathcal{A} + \mathcal{B}$ is also bounded above and* $\sup(\mathcal{A} + \mathcal{B}) = \sup \mathcal{A} + \sup \mathcal{B}$.
2. *If $0 < k < \infty$ and $\mathcal{A}$ is bounded above, then $k \cdot \mathcal{A}$ is also bounded above and* $\sup(k \cdot \mathcal{A}) = k \cdot \sup \mathcal{A}$.
3. $\sup \mathcal{A} = -\inf(-\mathcal{A})$ and $\inf \mathcal{A} = -\sup(-\mathcal{A})$.

Property 1 does not hold for the "product" of two sets, where the "product" of sets $\mathcal{A}$ and $\mathcal{B}$ is defined as

$$\mathcal{A} \cdot \mathcal{B} := \{c \in \mathbb{R}: c = ab \text{ for some } a \in \mathcal{A} \text{ and } b \in \mathcal{B}\}.$$

In this case, both of these two situations can occur

$$\sup(\mathcal{A} \cdot \mathcal{B}) > (\sup \mathcal{A}) \cdot (\sup \mathcal{B})$$
$$\sup(\mathcal{A} \cdot \mathcal{B}) = (\sup \mathcal{A}) \cdot (\sup \mathcal{B}).$$

Lemma A.17 (Supremum/infimum for monotone functions)

1. *If $f : \mathbb{R} \to \mathbb{R}$ is a nondecreasing function, then*

$$\sup\{x \in \mathbb{R}: f(x) < \varepsilon\} = \inf\{x \in \mathbb{R}: f(x) \geq \varepsilon\}$$

and

$$\sup\{x \in \mathbb{R}: f(x) \leq \varepsilon\} = \inf\{x \in \mathbb{R}: f(x) > \varepsilon\}.$$

2. *If $f : \mathbb{R} \to \mathbb{R}$ is a nonincreasing function, then*

$$\sup\{x \in \mathbb{R}: f(x) > \varepsilon\} = \inf\{x \in \mathbb{R}: f(x) \leq \varepsilon\}$$

and

$$\sup\{x \in \mathbb{R}: f(x) \geq \varepsilon\} = \inf\{x \in \mathbb{R}: f(x) < \varepsilon\}.$$

The above lemma is illustrated in Fig. A.1.

A.4   Sequences and Their Limits

Let $\mathbb{N}$ denote the set of "natural numbers" (positive integers) 1, 2, 3, ... A sequence drawn from a real-valued function is denoted by

$$f : \mathbb{N} \to \mathbb{R}.$$

In other words, $f(n)$ is a real number for each $n = 1, 2, 3, \ldots$. It is usual to write $f(n) = a_n$, and we often indicate the sequence by any one of these notations

$$\{a_1, a_2, a_3, \ldots, a_n, \ldots\} \quad \text{or} \quad \{a_n\}_{n=1}^{\infty}.$$

One important question that arises with a sequence is what happens when n gets large. To be precise, we want to know that when n is large enough, whether or not every a_n is close to some fixed number L (which is the *limit* of a_n).

Fig. A.1 Illustration of
Lemma A.17

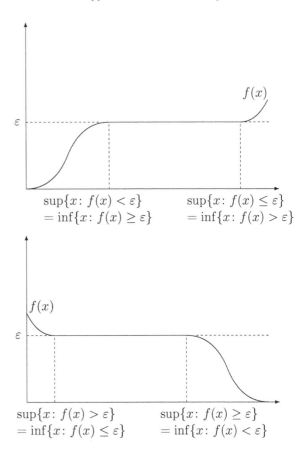

Definition A.18 (*Limit*) The *limit* of $\{a_n\}_{n=1}^{\infty}$ is the real number L satisfying: $(\forall\, \varepsilon > 0)(\exists\, N)$ such that $(\forall\, n > N)$

$$|a_n - L| < \varepsilon.$$

In this case, we write $L = \lim_{n\to\infty} a_n$. If no such L satisfies the above statement, we say that the limit of $\{a_n\}_{n=1}^{\infty}$ does not exist.

Property A.19 *If $\{a_n\}_{n=1}^{\infty}$ and $\{b_n\}_{n=1}^{\infty}$ both have a limit in $\mathbb{R}$, then the following hold:*

1. $\lim_{n\to\infty}(a_n + b_n) = \lim_{n\to\infty} a_n + \lim_{n\to\infty} b_n$.
2. $\lim_{n\to\infty}(\alpha \cdot a_n) = \alpha \cdot \lim_{n\to\infty} a_n$.
3. $\lim_{n\to\infty}(a_n b_n) = (\lim_{n\to\infty} a_n)(\lim_{n\to\infty} b_n)$.

Note that in the above definition, $-\infty$ and ∞ cannot be a legitimate limit for any sequence. In fact, if $(\forall\, L)(\exists\, N)$ such that $(\forall\, n > N)\, a_n > L$, then we say that a_n

diverges to ∞ and write $a_n \to \infty$. A similar argument applies to a_n diverging to $-\infty$. For convenience, we will work in the set of extended real numbers and thus state that a sequence $\{a_n\}_{n=1}^{\infty}$ that diverges to either ∞ or $-\infty$ has a limit in $\mathbb{R} \cup \{-\infty, \infty\}$.

Lemma A.20 (Convergence of monotone sequences) *If $\{a_n\}_{n=1}^{\infty}$ is nondecreasing in n, then $\lim_{n\to\infty} a_n$ exists in $\mathbb{R} \cup \{-\infty, \infty\}$. If $\{a_n\}_{n=1}^{\infty}$ is also bounded from above— i.e., $a_n \leq L$ $\forall n$ for some L in $\mathbb{R}$—then $\lim_{n\to\infty} a_n$ exists in $\mathbb{R}$.*

Likewise, if $\{a_n\}_{n=1}^{\infty}$ is nonincreasing in n, then $\lim_{n\to\infty} a_n$ exists in $\mathbb{R} \cup \{-\infty, \infty\}$. If $\{a_n\}_{n=1}^{\infty}$ is also bounded from below—i.e., $a_n \geq L$ $\forall n$ for some L in $\mathbb{R}$—then $\lim_{n\to\infty} a_n$ exists in $\mathbb{R}$.

As stated above, the limit of a sequence may not exist. For example, $a_n = (-1)^n$. Then, a_n will be close to either -1 or 1 for n large. Hence, more generalized definitions that can describe the general limiting behavior of a sequence is required.

Definition A.21 (*limsup and liminf*) The *limit supremum* of $\{a_n\}_{n=1}^{\infty}$ is the extended real number in $\mathbb{R} \cup \{-\infty, \infty\}$ defined by

$$\limsup_{n\to\infty} a_n := \lim_{n\to\infty} (\sup_{k\geq n} a_k),$$

and the *limit infimum* of $\{a_n\}_{n=1}^{\infty}$ is the extended real number defined by

$$\liminf_{n\to\infty} a_n := \lim_{n\to\infty} (\inf_{k\geq n} a_k).$$

Some also use the notations $\overline{\lim}$ and $\underline{\lim}$ to denote limsup and liminf, respectively.

Note that the limit supremum and the limit infimum of a sequence are always defined in $\mathbb{R} \cup \{-\infty, \infty\}$, since the sequences $\sup_{k\geq n} a_k = \sup\{a_k : k \geq n\}$ and $\inf_{k\geq n} a_k = \inf\{a_k : k \geq n\}$ are monotone in n (cf. Lemma A.20). An immediate result follows from the definitions of limsup and liminf.

Lemma A.22 (Limit) *For a sequence $\{a_n\}_{n=1}^{\infty}$,*

$$\lim_{n\to\infty} a_n = L \iff \limsup_{n\to\infty} a_n = \liminf_{n\to\infty} a_n = L.$$

Some properties regarding the limsup and liminf of sequences (which are parallel to Properties A.4 and A.10) are listed below.

Property A.23 (Properties of the limit supremum)

1. *The limit supremum always exists in $\mathbb{R} \cup \{-\infty, \infty\}$.*
2. *If $|\limsup_{m\to\infty} a_m| < \infty$, then $(\forall \varepsilon > 0)(\exists N)$ such that $(\forall n > N)$ $a_n < \limsup_{m\to\infty} a_m + \varepsilon$. (Note that this holds for every $n > N$.)*

3. *If* $|\lim\sup_{m\to\infty} a_m| < \infty$, *then* $(\forall\, \varepsilon > 0$ *and integer* $K)(\exists\, N > K)$ *such that* $a_N > \lim\sup_{m\to\infty} a_m - \varepsilon$. *(Note that this holds only for one* N, *which is larger than* K.)

Property A.24 (Properties of the limit infimum)

1. *The limit infimum always exists in* $\mathbb{R} \cup \{-\infty, \infty\}$.
2. *If* $|\lim\inf_{m\to\infty} a_m| < \infty$, *then* $(\forall\, \varepsilon > 0$ *and* $K)(\exists\, N > K)$ *such that* $a_N < \lim\inf_{m\to\infty} a_m + \varepsilon$. *(Note that this holds only for one* N, *which is larger than* K.)
3. *If* $|\lim\inf_{m\to\infty} a_m| < \infty$, *then* $(\forall\, \varepsilon > 0)(\exists\, N)$ *such that* $(\forall\, n > N)\, a_n > \lim\inf_{m\to\infty} a_m - \varepsilon$. *(Note that this holds for every* $n > N$.)

The last two items in Properties A.23 and A.24 can be stated using the terminology of *sufficiently large* and *infinitely often*, which is often adopted in information theory.

Definition A.25 (*Sufficiently large*) We say that a property holds for a sequence $\{a_n\}_{n=1}^\infty$ *almost always* or for *all sufficiently large* n if the property holds for every $n > N$ for some N.

Definition A.26 (*Infinitely often*) We say that a property holds for a sequence $\{a_n\}_{n=1}^\infty$ *infinitely often* or for *infinitely many* n if for every K, the property holds for *one* (specific) N with $N > K$.

Then, Properties 2 and 3 of Property A.23 can be, respectively, rephrased as if $|\lim\sup_{m\to\infty} a_m| < \infty$, then $(\forall\, \varepsilon > 0)$

$$a_n < \lim_{m\to\infty}\sup a_m + \varepsilon \quad \text{for all sufficiently large } n$$

and

$$a_n > \lim_{m\to\infty}\sup a_m - \varepsilon \quad \text{for infinitely many } n.$$

Similarly, Properties 2 and 3 of Property A.24 becomes: if $|\lim\inf_{m\to\infty} a_m| < \infty$, then $(\forall\, \varepsilon > 0)$

$$a_n < \lim_{m\to\infty}\inf a_m + \varepsilon \quad \text{for infinitely many } n$$

and

$$a_n > \lim_{m\to\infty}\inf a_m - \varepsilon \quad \text{for all sufficiently large } n.$$

Lemma A.27

1. $\lim\inf_{n\to\infty} a_n \leq \lim\sup_{n\to\infty} a_n$.
2. *If* $a_n \leq b_n$ *for all sufficiently large* n, *then*

$$\lim_{n\to\infty}\inf a_n \leq \lim_{n\to\infty}\inf b_n \quad \text{and} \quad \lim_{n\to\infty}\sup a_n \leq \lim_{n\to\infty}\sup b_n.$$

3. $\limsup_{n\to\infty} a_n < r \implies a_n < r$ for all sufficiently large n.
4. $\limsup_{n\to\infty} a_n > r \implies a_n > r$ for infinitely many n.
5.

$$\liminf_{n\to\infty} a_n + \liminf_{n\to\infty} b_n \leq \liminf_{n\to\infty}(a_n + b_n)$$
$$\leq \limsup_{n\to\infty} a_n + \liminf_{n\to\infty} b_n$$
$$\leq \limsup_{n\to\infty}(a_n + b_n)$$
$$\leq \limsup_{n\to\infty} a_n + \limsup_{n\to\infty} b_n.$$

6. *If $\lim_{n\to\infty} a_n$ exists, then*

$$\liminf_{n\to\infty}(a_n + b_n) = \lim_{n\to\infty} a_n + \liminf_{n\to\infty} b_n$$

and

$$\limsup_{n\to\infty}(a_n + b_n) = \lim_{n\to\infty} a_n + \limsup_{n\to\infty} b_n.$$

Finally, one can also interpret the limit supremum and limit infimum in terms of the concept of *clustering points*. A clustering point is a point that a sequence $\{a_n\}_{n=1}^{\infty}$ approaches (i.e., belonging to a ball with arbitrarily small radius and that point as center) infinitely many times. For example, if $a_n = \sin(n\pi/2)$, then $\{a_n\}_{n=1}^{\infty} = \{1, 0, -1, 0, 1, 0, -1, 0, \ldots\}$. Hence, there are three clustering points in this sequence, which are -1, 0 and 1. Then, the limit supremum of the sequence is nothing but its *largest clustering point*, and its limit infimum is exactly its *smallest clustering point*. Specifically, $\limsup_{n\to\infty} a_n = 1$ and $\liminf_{n\to\infty} a_n = -1$. This approach can sometimes be useful to determine the limsup and liminf quantities.

A.5 Equivalence

We close this appendix by providing some equivalent statements that are often used to simplify proofs. For example, instead of directly showing that quantity x is less than or equal to quantity y, one can take an arbitrary constant $\varepsilon > 0$ and prove that $x < y + \varepsilon$. Since $y + \varepsilon$ is a larger quantity than y, in some cases it might be easier to show $x < y + \varepsilon$ than proving $x \leq y$. By the next theorem, any proof that concludes that "$x < y + \varepsilon$ for all $\varepsilon > 0$" immediately gives the desired result of $x \leq y$.

Theorem A.28 *For any x, y and a in $\mathbb{R}$,*

1. $x < y + \varepsilon$ *for all $\varepsilon > 0$ iff $x \leq y$;*
2. $x < y - \varepsilon$ *for some $\varepsilon > 0$ iff $x < y$;*
3. $x > y - \varepsilon$ *for all $\varepsilon > 0$ iff $x \geq y$;*
4. $x > y + \varepsilon$ *for some $\varepsilon > 0$ iff $x > y$;*
5. $|a| < \varepsilon$ *for all $\varepsilon > 0$ iff $a = 0$.*

Appendix B
Overview in Probability and Random Processes

This appendix presents a quick overview of important concepts from probability theory and the theory of random processes. We assume that the reader has a basic knowledge of these subjects; for a thorough study, comprehensive texts such as [30, 47, 104, 162, 170] should be consulted. We close the appendix with a brief discussion of Jensen's inequality and the Lagrange multipliers optimization technique [46, 56].

B.1 Probability Space

Definition B.1 (σ-*Fields*) Let $\mathcal{F}$ be a collection of subsets of a non-empty set Ω. Then, $\mathcal{F}$ is called a σ-*field* (or σ-*algebra*) if the following hold:

1. $\Omega \in \mathcal{F}$.
2. $\mathcal{F}$ *is closed under complementation:* If $A \in \mathcal{F}$, then $A^c \in \mathcal{F}$, where $A^c = \{\omega \in \Omega : \omega \notin A\}$.
3. $\mathcal{F}$ *is closed under countable unions:* If $A_i \in \mathcal{F}$ for $i = 1, 2, 3, \ldots$, then $\bigcup_{i=1}^{\infty} A_i \in \mathcal{F}$.

It directly follows that the empty set $\emptyset$ is also an element of $\mathcal{F}$ (as $\Omega^c = \emptyset$) and that $\mathcal{F}$ is closed under countable intersection since

$$\bigcap_{i=1}^{\infty} A_i^c = \left(\bigcup_{i=1}^{\infty} A_i\right)^c .$$

The largest σ-field of subsets of a given set Ω is the collection of all subsets of Ω (i.e., its powerset), while the smallest σ-field is given by $\{\Omega, \emptyset\}$. Also, if A is a proper (strict) non-empty subset of Ω, then the smallest σ-field containing A is given by $\{\Omega, \emptyset, A, A^c\}$.

© Springer Nature Singapore Pte Ltd. 2018
F. Alajaji and P.-N. Chen, *An Introduction to Single-User Information Theory*,
Springer Undergraduate Texts in Mathematics and Technology,
https://doi.org/10.1007/978-981-10-8001-2

Definition B.2 (*Probability space*) A *probability space* is a triple $(\Omega, \mathcal{F}, P)$, where Ω is a given set called *sample space* containing all possible outcomes (usually observed from an experiment), $\mathcal{F}$ is a σ-field of subsets of Ω and P is a probability measure $P : \mathcal{F} \rightarrow [0, 1]$ on the σ-field satisfying the following:

1. $0 \leq P(A) \leq 1$ for all $A \in \mathcal{F}$.
2. $P(\Omega) = 1$.
3. *Countable additivity*: If $A_1, A_2, \ldots$ is a sequence of disjoint sets (i.e., $A_i \cap A_j = \emptyset$ for all $i \neq j$) in $\mathcal{F}$, then

$$P\left(\bigcup_{k=1}^{\infty} A_k\right) = \sum_{k=1}^{\infty} P(A_k).$$

It directly follows from Properties 1–3 of the above definition that $P(\emptyset) = 0$. Usually, the σ-field $\mathcal{F}$ is called the *event space* and its elements (which are subsets of Ω satisfying the properties of Definition B.1) are called *events*.

B.2   Random Variables and Random Processes

A random variable X defined over the probability space $(\Omega, \mathcal{F}, P)$ is a real-valued function $X : \Omega \rightarrow \mathbb{R}$ that is *measurable* (or $\mathcal{F}$-*measurable*), i.e., satisfying the property that

$$X^{-1}((-\infty, t]) := \{\omega \in \Omega : X(\omega) \leq t\} \in \mathcal{F}$$

for each real t.[1]

The Borel σ-field of $\mathbb{R}$, denoted by $\mathscr{B}(\mathbb{R})$, is the smallest σ-field of subsets of $\mathbb{R}$ containing all open intervals in $\mathbb{R}$. The elements of $\mathscr{B}(\mathbb{R})$ are called Borel sets. For any random variable X, we use P_X to denote the probability distribution on $\mathscr{B}(\mathbb{R})$ induced by X, given by

$$P_X(B) := \Pr[X \in B] = P(w \in \Omega : X(w) \in B), \qquad B \in \mathscr{B}(\mathbb{R}).$$

Note that the quantities $P_X(B)$, $B \in \mathscr{B}(\mathbb{R})$, fully characterize the random variable X as they determine the probabilities of all events that concern X.

---

[1] One may question why bother defining random variables based on some abstract probability space. One may continue that "a random variable X can simply be defined based on its probability distribution," which is indeed true (cf. Observation B.3). A perhaps easier way to understand the abstract definition of a random variable is that the underlying probability space $(\Omega, \mathcal{F}, P)$ on which it is defined is what truly occurs *internally*, but it is *possibly non-observable*. In order to infer which of the *non-observable* ω occurs, an experiment is performed resulting in an observable x that is a function of ω. Such experiment yields the random variable X whose probability is defined over the probability space $(\Omega, \mathcal{F}, P)$.

Observation B.3 (**Distribution function versus probability space**) In many applications, we are perhaps more interested in the distribution functions of random variables than the underlying probability space on which they are defined. It can be proved [47, Theorem 14.1] that given a real-valued nonnegative function $F(\cdot)$ that is nondecreasing and right-continuous and satisfies $\lim_{x\downarrow-\infty} F(x) = 0$ and $\lim_{x\uparrow\infty} F(x) = 1$, there exist a random variable X and an underlying probability space such that the cumulative distribution function (cdf) of the random variable, $\Pr[X \leq x] = P_X((-\infty, x])$, defined over the probability space is equal to $F(\cdot)$. This result releases us from the burden of referring to a probability space before defining the random variable. In other words, we can define a random variable X directly by its cdf, $F_X(x) = \Pr[X \leq x]$, without bothering to refer to its underlying probability space. Nevertheless, it is important to keep in mind that, formally, random variables are defined over underlying probability spaces.

The n-tuple of random variables $X^n := (X_1, X_2, \ldots, X_n)$ is called a random vector of length n. In other words, given a probability space $(\Omega, \mathcal{F}, P)$, X^n is a measurable function from Ω to $\mathbb{R}^n$, where $\mathbb{R}^n$ denotes the n-fold Cartesian product of $\mathbb{R}$ with itself: $\mathbb{R}^n := \mathbb{R} \times \mathbb{R} \times \cdots \times \mathbb{R}$. Also, the probability distribution P_{X^n} induced by X^n is given by

$$P_{X^n}(B) = P\left(w \in \Omega : X^n(w) \in B\right), \qquad B \in \mathscr{B}(\mathbb{R}^n),$$

where $\mathscr{B}(\mathbb{R}^n)$ is the Borel σ-field of $\mathbb{R}^n$; i.e., the smallest σ-field of subsets of $\mathbb{R}^n$ containing all open sets in $\mathbb{R}^n$.

The joint cdf F_{X^n} of X^n is the function from $\mathbb{R}^n$ to $[0, 1]$ given by

$$F_{X^n}(x^n) = P_X((-\infty, x_i), i = 1, \ldots, n) = P(\omega \in \Omega : X_i(w) \leq x_i, i = 1, \ldots, n)$$

for $x^n = (x_1, \ldots, x_n) \in \mathbb{R}^n$.

A random process (or random source) is a collection of random variables that arise from the same probability space. It can be mathematically represented by the collection

$$\{X_t, t \in I\},$$

where X_t denotes the tth random variable in the process, and the index t runs over an index set I which is arbitrary. The index set I can be uncountably infinite (e.g., $I = \mathbb{R}$), in which case we are dealing with a continuous-time process. We will, however, exclude such a case in this appendix for the sake of simplicity.

In this text, we focus mostly on *discrete-time* sources; i.e., sources with the countable index set $I = \{1, 2, \ldots\}$. Each such source is denoted by

$$X := \{X_n\}_{n=1}^{\infty} = \{X_1, X_2, \ldots\},$$

as an infinite sequence of random variables, where all the random variables take on values from a common *generic alphabet* $\mathcal{X} \subseteq \mathbb{R}$. The elements in $\mathcal{X}$ are usually called letters (or symbols or values). When $\mathcal{X}$ is a finite set, the letters of $\mathcal{X}$ can be conveniently expressed via the elements of any appropriately chosen finite set (i.e., the letters of $\mathcal{X}$ need not be real numbers).[2]

The source X is completely characterized by the sequence of joint cdf's $\{F_{X^n}\}_{n=1}^{\infty}$. When the alphabet $\mathcal{X}$ is finite, the source can be equivalently described by the sequence of joint probability mass functions (pmf's):

$$P_{X^n}(a^n) = \Pr[X_1 = a_1, X_2 = a_2, \ldots, X_n = a_n]$$

for all $a^n = (a_1, a_2, \ldots, a_n) \in \mathcal{X}^n, n = 1, 2, \ldots$.

B.3 Statistical Properties of Random Sources

For a random process $X = \{X_n\}_{n=1}^{\infty}$ with alphabet $\mathcal{X}$ (i.e., $\mathcal{X} \subseteq \mathbb{R}$ is the range of each X_i) defined over probability space $(\Omega, \mathcal{F}, P)$, consider $\mathcal{X}^{\infty}$, the set of all sequences $x := (x_1, x_2, x_3, \ldots)$ of real numbers in $\mathcal{X}$. An event E in $\mathcal{F}_X$, the smallest σ-field generated by all open sets of $\mathcal{X}^{\infty}$ (i.e., the Borel σ-field of $\mathcal{X}^{\infty}$), is said to be $\mathbb{T}$-*invariant* with respect to the left-shift (or shift transformation) $\mathbb{T}: \mathcal{X}^{\infty} \to \mathcal{X}^{\infty}$ if

$$\mathbb{T}E \subseteq E,$$

where

$$\mathbb{T}E := \{\mathbb{T}x : x \in E\} \quad \text{and} \quad \mathbb{T}x := \mathbb{T}(x_1, x_2, x_3, \ldots) = (x_2, x_3, \ldots).$$

In other words, $\mathbb{T}$ is equivalent to "chopping the first component." For example, applying $\mathbb{T}$ onto an event E_1 defined below:

$$E_1 := \{(x_1 = 1, x_2 = 1, x_3 = 1, x_4 = 1, \ldots), (x_1 = 0, x_2 = 1, x_3 = 1, x_4 = 1, \ldots),$$
$$(x_1 = 0, x_2 = 0, x_3 = 1, x_4 = 1, \ldots)\}, \tag{B.3.1}$$

yields

---

[2]More formally, the definition of a random variable X can be generalized by allowing it to take values that are not real numbers: a random variable over the probability space $(\Omega, \mathcal{F}, P)$ is a function $X : \Omega \to \mathcal{X}$ satisfying the property that for every $F \in \mathcal{F}_X$,

$$X^{-1}(F) := \{w \in \Omega : X(w) \in F\} \in \mathcal{F},$$

where the alphabet $\mathcal{X}$ is a general set and $\mathcal{F}_X$ is a σ-field of subsets of $\mathcal{X}$ [159, 349]. Note that this definition allows $\mathcal{X}$ to be an arbitrary set (including being an arbitrary finite set). Furthermore, if we set $\mathcal{X} = \mathbb{R}$, then we revert to the earlier (standard) definition of a random variable.

$$\mathbb{T}E_1 = \{(x_1 = 1, x_2 = 1, x_3 = 1, \ldots), (x_1 = 1, x_2 = 1, x_3 = 1 \ldots),$$
$$(x_1 = 0, x_2 = 1, x_3 = 1, \ldots)\}$$
$$= \{(x_1 = 1, x_2 = 1, x_3 = 1, \ldots), (x_1 = 0, x_2 = 1, x_3 = 1, \ldots)\}.$$

We then have $\mathbb{T}E_1 \subseteq E_1$, and hence E_1 is $\mathbb{T}$-invariant.

It can be proved[3] that if $\mathbb{T}E \subseteq E$, then $\mathbb{T}^2 E \subseteq \mathbb{T}E$. By induction, we can further obtain

$$\cdots \subseteq \mathbb{T}^3 E \subseteq \mathbb{T}^2 E \subseteq \mathbb{T}E \subseteq E.$$

Thus, if an element say $(1, 0, 0, 1, 0, 0, \ldots)$ is in a $\mathbb{T}$-invariant set E, then all its left-shift counterparts (i.e., $(0, 0, 1, 0, 0, 1 \ldots)$ and $(0, 1, 0, 0, 1, 0, \ldots)$) should be contained in E. As a result, for a $\mathbb{T}$-invariant set E, an element and all its left-shift counterparts are either all in E or all outside E, but cannot be partially inside E. Hence, a "$\mathbb{T}$-invariant group" such as one containing $(1, 0, 0, 1, 0, 0, \ldots)$, $(0, 0, 1, 0, 0, 1 \ldots)$ and $(0, 1, 0, 0, 1, 0, \ldots)$ should be treated as an indecomposable group in $\mathbb{T}$-invariant sets.

Although we are in particular interested in these "$\mathbb{T}$-invariant indecomposable groups" (especially when defining an ergodic random process), it is possible that some single "transient" element, such as $(0, 0, 1, 1, \ldots)$ in (B.3.1), is included in a $\mathbb{T}$-invariant set, and will be excluded after applying left-shift operation $\mathbb{T}$. This, however, can be resolved by introducing the inverse operation $\mathbb{T}^{-1}$. Note that $\mathbb{T}$ is a many-to-one mapping, so its inverse operation does not exist in general. Similar to taking the closure of an open set, the definition adopted below [349, p. 3] allows us to "enlarge" the $\mathbb{T}$-invariant set such that all right-shift counterparts of the single "transient" element are included

$$\mathbb{T}^{-1} E := \left\{ x \in \mathcal{X}^\infty : \mathbb{T}x \in E \right\}.$$

We then notice from the above definition that if

$$\mathbb{T}^{-1} E = E, \tag{B.3.2}$$

then[4]

$$\mathbb{T}E = \mathbb{T}(\mathbb{T}^{-1} E) = E,$$

and hence E is constituted only by the $\mathbb{T}$-invariant groups because

$$\cdots = \mathbb{T}^{-2} E = \mathbb{T}^{-1} E = E = \mathbb{T}E = \mathbb{T}^2 E = \cdots.$$

---

[3]If $A \subseteq B$, then $\mathbb{T}A \subseteq \mathbb{T}B$. Thus $\mathbb{T}^2 E \subseteq \mathbb{T}E$ holds whenever $\mathbb{T}E \subseteq E$.

[4]The proof of $\mathbb{T}(\mathbb{T}^{-1} E) = E$ is as follows. If $y \in \mathbb{T}(\mathbb{T}^{-1} E) = \mathbb{T}(\{x \in \mathcal{X}^\infty : \mathbb{T}x \in E\})$, then there must exist an element $x \in \{x \in \mathcal{X}^\infty : \mathbb{T}x \in E\}$ such that $y = \mathbb{T}x$. Since $\mathbb{T}x \in E$, we have $y \in E$ and $\mathbb{T}(\mathbb{T}^{-1} E) \subseteq E$.

On the contrary, if $y \in E$, all x's satisfying $\mathbb{T}x = y$ belong to $\mathbb{T}^{-1} E$. Thus, $y \in \mathbb{T}(\mathbb{T}^{-1} E)$, which implies $E \subseteq \mathbb{T}(\mathbb{T}^{-1} E)$.

The sets that satisfy (B.3.2) are sometimes referred to as *ergodic sets* because as time goes by (the left-shift operator $\mathbb{T}$ can be regarded as a shift to a future time), the set always stays in the state that it has been before. A quick example of an ergodic set for $\mathcal{X} = \{0, 1\}$ is one that consists of all binary sequences that contain finitely many 0's.[5]

We now classify several useful statistical properties of random process $X = \{X_1, X_2, \ldots\}$.

- *Memoryless process:*The process or source X is said to be *memoryless* if its random variables are *independent and identically distributed* (i.i.d.). Here by independence, we mean that any finite sequence $X_{i_1}, X_{i_2}, \ldots, X_{i_n}$ of random variables satisfies

$$\Pr[X_{i_1} = x_1, X_{i_2} = x_2, \ldots, X_{i_n} = x_n] = \prod_{l=i}^{n} \Pr[X_{i_l} = x_l]$$

for all $x_l \in \mathcal{X}, l = 1, \ldots, n$; we also say that these random variables are mutually independent. Furthermore, the notion of identical distribution means that

$$\Pr[X_i = x] = \Pr[X_1 = x]$$

for any $x \in \mathcal{X}$ and $i = 1, 2, \ldots$; i.e., all the source's random variables are governed by the same marginal distribution.
- *Stationary process* The process X is said to be *stationary* (or *strictly stationary*) if the probability of every sequence or event is unchanged by a left (time) shift, or equivalently, if any $j = 1, 2, \ldots$, the joint distribution of $(X_1, X_2, \ldots, X_n)$ satisfies

$$\Pr[X_1 = x_1, X_2 = x_2, \ldots, X_n = x_n]$$
$$= \Pr[X_{j+1} = x_1, X_{j+2} = x_2, \ldots, X_{j+n} = x_n]$$

for all $x_l \in \mathcal{X}, l = 1, \ldots, n$.

It is direct to verify that a memoryless source is stationary. Also, for a stationary source, its random variables are identically distributed.
- *Ergodic process:* The process X is said to be *ergodic* if any ergodic set (satisfying (B.3.2)) in $\mathcal{F}_X$ has probability either 1 or 0. This definition is not very intuitive, but some interpretations and examples may shed some light.

---

[5]As the textbook only deals with one-sided random processes, the discussion on $\mathbb{T}$-invariance only focuses on sets of one-sided sequences. When a two-sided random process $\ldots, X_{-2}, X_{-1}, X_0, X_1, X_2, \ldots$ is considered, the left-shift operation $\mathbb{T}$ of a two-sided sequence actually has a unique inverse. Hence, $\mathbb{T}E \subseteq E$ implies $\mathbb{T}E = E$. Also, $\mathbb{T}E = E$ iff $\mathbb{T}^{-1}E = E$. Ergodicity for two-sided sequences can therefore be directly defined using $\mathbb{T}E = E$.

Observe that the definition has nothing to do with stationarity. It simply states that events that are unaffected by time-shifting (both left- and right-shifting) must have probability either zero or one.

Ergodicity implies that all convergent sample averages[6] converge to a constant (but not necessarily to the ensemble average or statistical expectation), and stationarity assures that the time average converges to a random variable; hence, it is reasonable to expect that they jointly imply the ultimate time average equals the ensemble average. This is validated by the well-known *ergodic theorem* by Birkhoff and Khinchin.

Theorem B.4 (Pointwise ergodic theorem) *Consider a discrete-time stationary random process,* $X = \{X_n\}_{n=1}^{\infty}$. *For real-valued function* $f(\cdot)$ *on* $\mathbb{R}$ *with finite mean (i.e.,* $|E[f(X_n)]| < \infty$), *there exists a random variable* Y *such that*

$$\lim_{n \to \infty} \frac{1}{n} \sum_{k=1}^{n} f(X_k) = Y \quad \text{with probability 1.}$$

If, in addition to stationarity, the process is also ergodic, then

$$\lim_{n \to \infty} \frac{1}{n} \sum_{k=1}^{n} f(X_k) = E[f(X_1)] \quad \text{with probability 1.}$$

Example B.5 Consider the process $\{X_i\}_{i=1}^{\infty}$ consisting of a family of i.i.d. binary random variables (obviously, it is stationary and ergodic). Define the function $f(\cdot)$ by $f(0) = 0$ and $f(1) = 1$. Hence,[7]

$$E[f(X_n)] = P_{X_n}(0)f(0) + P_{X_n}(1)f(1) = P_{X_n}(1)$$

is finite. By the pointwise ergodic theorem, we have

$$\lim_{n \to \infty} \frac{f(X_1) + f(X_2) + \cdots + f(X_n)}{n} = \lim_{n \to \infty} \frac{X_1 + X_2 + \cdots + X_n}{n}$$
$$= P_X(1).$$

As seen in the above example, one of the important consequences that the pointwise ergodic theorem indicates is that the time average can ultimately replace the statistical average, which is a useful result. Hence, with stationarity and ergodicity, one, who observes

---

[6]Two alternative names for *sample average* are *time average* and *Cesàro mean*. In this book, these names will be used interchangeably.

[7]As specified in Sect. B.2, $P_{X_n}(0) = \Pr[X_n = 0]$. These two representations will be used alternatively throughout the book.

$$X_1^{30} = 154326543334225632425644234443$$

from the experiment of rolling a dice, can draw the conclusion that the true distribution of rolling the dice can be well approximated by

$$\Pr\{X_i = 1\} \approx \frac{1}{30} \qquad \Pr\{X_i = 2\} \approx \frac{6}{30} \qquad \Pr\{X_i = 3\} \approx \frac{7}{30}$$
$$\Pr\{X_i = 4\} \approx \frac{9}{30} \qquad \Pr\{X_i = 5\} \approx \frac{4}{30} \qquad \Pr\{X_i = 6\} \approx \frac{3}{30}$$

Such result is also known by the *law of large numbers*. The relation between ergodicity and the law of large numbers will be further explored in Sect. B.5.

We close the discussion on ergodicity by remarking that in communications theory, one may assume that *the source is stationary* or *the source is stationary ergodic*. But it is not common to see the assumption of *the source being ergodic but nonstationary*. This is perhaps because an ergodic but nonstationary source does not in general facilitate the analytical study of communications problems. This, to some extent, justifies that the *ergodicity* assumption usually comes after the *stationarity* assumption. A specific example is the pointwise ergodic theorem, where the random processes considered is presumed to be stationary.

- *Markov chain for three random variables*: Three random variables X, Y, and Z are said to form a Markov chain if

$$P_{X,Y,Z}(x, y, z) = P_X(x) \cdot P_{Y|X}(y|x) \cdot P_{Z|Y}(z|y); \qquad (B.3.3)$$

i.e., $P_{Z|X,Y}(z|x, y) = P_{Z|Y}(z|y)$. This is usually denoted by $X \to Y \to Z$.

$X \to Y \to Z$ is sometimes read as "X and Z are conditionally independent given Y" because it can be shown that (B.3.3) is equivalent to

$$P_{X,Z|Y}(x, z|y) = P_{X|Y}(x|y) \cdot P_{Z|Y}(z|y).$$

Therefore, $X \to Y \to Z$ is equivalent to $Z \to Y \to X$. Accordingly, the Markovian notation is sometimes expressed as $X \leftrightarrow Y \leftrightarrow Z$.

- *kth order Markov sources*: The sequence of random variables $\{X_n\}_{n=1}^{\infty} = X_1, X_2, X_3, \ldots$ with common finite-alphabet $\mathcal{X}$ is said to form a kth order Markov chain (or kth order Markov source or process) if for all $n > k$, $x_i \in \mathcal{X}$, $i = 1, \ldots, n$,

$$\Pr[X_n = x_n | X_{n-1} = x_{n-1}, \ldots, X_1 = x_1]$$
$$= \Pr[X_n = x_n | X_{n-1} = x_{n-1}, \ldots, X_{n-k} = x_{n-k}]. \qquad (B.3.4)$$

Each $x_{n-k}^{n-1} := (x_{n-k}, x_{n-k+1}, \ldots, x_{n-1}) \in \mathcal{X}^k$ is called the *state* of the Markov chain at time n.

When $k = 1$, then $\{X_n\}_{n=1}^{\infty}$ is called a first-order Markov source (or just a Markov source or chain). In light of (B.3.4), for any $n > 1$, the random variables $X_1, X_2, \ldots, X_n$ directly satisfy the conditional independence property

$$\Pr[X_i = x_i | X^{i-1} = x^{i-1}] = \Pr[X_i = x_i | X_{i-1} = x_{i-1}] \qquad (\text{B.3.5})$$

for all $x_i \in \mathcal{X}$, $i = 1, \ldots, n$; this property is denoted as in (B.3.3) by

$$X_1 \to X_2 \to \cdots \to X_n$$

for $n > 2$. The same property applies to any finite number of random variables from the source ordered in terms of increasing time indices.

We next summarize important concepts and facts about Markov sources (e.g., see [137, 162]).

- A kth order Markov chain is *irreducible* if with some probability, we can go from any state in $\mathcal{X}^k$ to another state in a finite number of steps, i.e., for all $x^k, y^k \in \mathcal{X}^k$ there exists an integer $j \geq 1$ such that

$$\Pr\left\{ X_j^{k+j-1} = x^k \,\middle|\, X_1^k = y^k \right\} > 0.$$

- A kth order Markov chain is said to be *time-invariant* or *homogeneous*, if for every $n > k$,

$$\Pr[X_n = x_n | X_{n-1} = x_{n-1}, \ldots, X_{n-k} = x_{n-k}]$$
$$= \Pr[X_{k+1} = x_{k+1} | X_k = x_k, \ldots, X_1 = x_1].$$

Therefore, a homogeneous first-order Markov chain can be defined through its transition probability:

$$\left[\Pr\{X_2 = x_2 | X_1 = x_1\} \right]_{|\mathcal{X}| \times |\mathcal{X}|},$$

and its initial state distribution $P_{X_1}(x)$.

- In a first-order Markov chain, the *period* $d(x)$ of state $x \in \mathcal{X}$ is defined by

$$d(x) := \gcd \{n \in \{1, 2, 3, \ldots\}: \Pr\{X_{n+1} = x | X_1 = x\} > 0\},$$

where gcd denotes the greatest common divisor; in other words, if the Markov chain starts in state x, then the chain cannot return to state x at any time that is not a multiple of $d(x)$. If $\Pr\{X_{n+1} = x | X_1 = x\} = 0$ for all n, we say that state x has an infinite period and write $d(x) = \infty$. We also say that *state x is aperiodic* if $d(x) = 1$ and *periodic* if $d(x) > 1$. Furthermore, the first-order Markov chain is called *aperiodic* if all its states are aperiodic. In other words, the first-order Markov chain is aperiodic if

$$\gcd\{n \in \{1, 2, 3, \ldots\}: \Pr\{X_{n+1} = x | X_1 = x\} > 0\} = 1 \quad \forall x \in \mathcal{X}.$$

- In an irreducible first-order Markov chain, all states have the same period. Hence, if one state in such a chain is aperiodic, then the entire Markov chain is aperiodic.
- A distribution $\pi(\cdot)$ on $\mathcal{X}$ is said to be a *stationary* distribution for a homogeneous first-order Markov chain, if for every $y \in \mathcal{X}$,

$$\pi(y) = \sum_{x \in \mathcal{X}} \pi(x) \Pr\{X_2 = y | X_1 = x\}.$$

For a finite-alphabet homogeneous first-order Markov chain, $\pi(\cdot)$ always exists; furthermore, $\pi(\cdot)$ is unique if the Markov chain is irreducible. For a finite-alphabet homogeneous first-order Markov chain that is both irreducible and aperiodic,

$$\lim_{n \to \infty} \Pr\{X_{n+1} = y | X_1 = x\} = \pi(y)$$

for all states x and y in $\mathcal{X}$. If the initial state distribution is equal to a stationary distribution, then the homogeneous first-order Markov chain becomes a stationary process.
- A finite-alphabet stationary Markov source is an ergodic process (and hence satisfies the pointwise ergodic theorem) iff it is irreducible; see [30, p. 371] and [349, Prop. I.2.9].

The general relations among i.i.d. sources, Markov sources, stationary sources, and ergodic sources are depicted in Fig. B.1

B.4   Convergence of Sequences of Random Variables

In this section, we will discuss modes in which a random process $X_1, X_2, \ldots$ converges to a limiting random variable X. Recall that a random variable is a real-valued measurable function from Ω to $\mathbb{R}$, where Ω the sample space of the probability space over which the random variable is defined. So the following two expressions will be used interchangeably: $X_1(\omega), X_2(\omega), X_3(\omega), \ldots \equiv X_1, X_2, X_3, \ldots$, for $\omega \in \Omega$. Note that the random variables in a random process are defined over the same probability space $(\Omega, \mathcal{F}, P)$,

Fig. B.1 General relations
of random processes

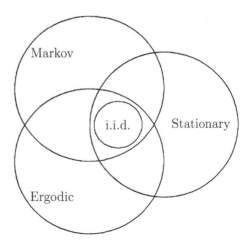

Definition B.6 (*Convergence modes for random sequences*)

1. *Pointwise convergence on* Ω.[8]

   $\{X_n\}_{n=1}^{\infty}$ is said to converge to X *pointwise on* Ω if

   $$\lim_{n\to\infty} X_n(\omega) = X(\omega) \quad \text{for all } \omega \in \Omega.$$

   This notion of convergence, which is familiar to us from real analysis, is denoted by $X_n \xrightarrow{p.w.} X$.

2. *Almost sure convergence* or *convergence with probability* 1.

   $\{X_n\}_{n=1}^{\infty}$ is said to converge to X *with probability* 1, if

   $$P\{\omega \in \Omega : \lim_{n\to\infty} X_n(\omega) = X(\omega)\} = 1.$$

   Almost sure convergence is denoted by $X_n \xrightarrow{a.s.} X$; note that it is nothing but a probabilistic version of pointwise convergence.

3. *Convergence in probability*.

   $\{X_n\}_{n=1}^{\infty}$ is said to converge to X *in probability*, if for any $\varepsilon > 0$,

   $$\lim_{n\to\infty} P\{\omega \in \Omega : |X_n(\omega) - X(\omega)| > \varepsilon\} = \lim_{n\to\infty} \Pr\{|X_n - X| > \varepsilon\} = 0.$$

   This mode of convergence is denoted by $X_n \xrightarrow{p} X$.

---

[8] Although such mode of convergence is not used in probability theory, we introduce it herein to contrast it with the almost sure convergence mode (see Example B.7).

4. *Convergence in rth mean.*

   $\{X_n\}_{n=1}^{\infty}$ is said to converge to X in rth *mean*, if

   $$\lim_{n \to \infty} E[|X - X_n|^r] = 0.$$

   This is denoted by $X_n \xrightarrow{\mathcal{L}_r} X$.

5. *Convergence in distribution.*

   $\{X_n\}_{n=1}^{\infty}$ is said to converge to X *in distribution*, if

   $$\lim_{n \to \infty} F_{X_n}(x) = F_X(x)$$

   for every continuity point of $F(x)$, where

   $$F_{X_n}(x) = \Pr\{X_n \le x\} \quad \text{and} \quad F_X(x) = \Pr\{X \le x\}.$$

   We denote this notion of convergence by $X_n \xrightarrow{d} X$.

An example that facilitates the understanding of pointwise convergence and almost sure convergence is as follows.

Example B.7 Consider a probability space $(\Omega, 2^{\Omega}, P)$, where $\Omega = \{0, 1, 2, 3\}$, 2^{Ω} is the power set of Ω and $P(0) = P(1) = P(2) = 1/3$ and $P(3) = 0$. Define a random variable as

$$X_n(\omega) = \frac{\omega}{n}.$$

Then

$$\Pr\{X_n = 0\} = \Pr\left\{X_n = \frac{1}{n}\right\} = \Pr\left\{X_n = \frac{2}{n}\right\} = \frac{1}{3}.$$

It is clear that for every ω in Ω, $X_n(\omega)$ converges to $X(\omega)$, where $X(\omega) = 0$ for every $\omega \in \Omega$; so

$$X_n \xrightarrow{p.w.} X.$$

Now let $\tilde{X}(\omega) = 0$ for $\omega = 0, 1, 2$ and $\tilde{X}(\omega) = 1$ for $\omega = 3$. Then, both of the following statements are true:

$$X_n \xrightarrow{a.s.} X \quad \text{and} \quad X_n \xrightarrow{a.s.} \tilde{X},$$

since

$$\Pr\left\{\lim_{n \to \infty} X_n = \tilde{X}\right\} = \sum_{\omega=0}^{3} P(\omega) \cdot \mathbf{1}\left\{\lim_{n \to \infty} X_n(\omega) = \tilde{X}(\omega)\right\} = 1,$$

where $\mathbf{1}\{\cdot\}$ represents the set indicator function. However, X_n does not converge to $\tilde{X}$ pointwise because

$$\lim_{n \to \infty} X_n(3) \neq \tilde{X}(3).$$

In other words, pointwise convergence requires "equality" even for samples without probability mass; however, these samples are ignored under almost sure convergence.

Observation B.8 (Uniqueness of convergence)

1. If $X_n \xrightarrow{p.w.} X$ and $X_n \xrightarrow{p.w.} Y$, then $X = Y$ pointwise. That is, $(\forall \, \omega \in \Omega)$ $X(\omega) = Y(\omega)$.
2. If $X_n \xrightarrow{a.s.} X$ and $X_n \xrightarrow{a.s.} Y$ (or $X_n \xrightarrow{p} X$ and $X_n \xrightarrow{p} Y$) (or $X_n \xrightarrow{\mathcal{L}_r} X$ and $X_n \xrightarrow{\mathcal{L}_r} Y$), then $X = Y$ with probability 1. That is, $\Pr\{X = Y\} = 1$.
3. $X_n \xrightarrow{d} X$ and $X_n \xrightarrow{d} Y$, then $F_X(x) = F_Y(x)$ for all x.

For ease of understanding, the relations of the five modes of convergence can be depicted as follows. As usual, a double arrow denotes implication.

$$
\begin{array}{c}
X_n \xrightarrow{p.w.} X \\
\Downarrow \\
X_n \xrightarrow{a.s.} X \dashrightarrow[\text{Thm. B.9}]{\text{Thm. B.10}} X_n \xrightarrow{\mathcal{L}_r} X \ (r \geq 1) \\
X_n \xrightarrow{p} X \\
\Downarrow \\
X_n \xrightarrow{d} X
\end{array}
$$

There are some other relations among these five convergence modes that are also depicted in the above graph (via the dotted line); they are stated below.

Theorem B.9 (Monotone convergence theorem [47])

$$X_n \xrightarrow{a.s.} X, \ (\forall \, n) Y \leq X_n \leq X_{n+1}, \text{ and } E[|Y|] < \infty \implies X_n \xrightarrow{\mathcal{L}_1} X$$
$$\implies E[X_n] \to E[X].$$

Theorem B.10 (Dominated convergence theorem [47])

$$X_n \xrightarrow{a.s.} X, \ (\forall \, n)|X_n| \leq Y, \text{ and } E[|Y|] < \infty \implies X_n \xrightarrow{\mathcal{L}_1} X$$
$$\implies E[X_n] \to E[X].$$

The implication of $X_n \xrightarrow{\mathcal{L}_1} X$ to $E[X_n] \rightarrow E[X]$ can be easily seen from

$$|E[X_n] - E[X]| = |E[X_n - X]| \leq E[|X_n - X|].$$

B.5 Ergodicity and Laws of Large Numbers

B.5.1 Laws of Large Numbers

Consider a random process $X_1, X_2, \ldots$ with common marginal mean μ. Suppose that we wish to estimate μ on the basis of the observed sequence $x_1, x_2, x_3, \ldots$. The *weak* and *strong* laws of large numbers ensure that such inference is possible (with reasonable accuracy), provided that the dependencies between X_n's are suitably restricted: e.g., the weak law is valid for uncorrelated X_n's, while the strong law is valid for independent X_n's. Since independence is a more restrictive condition than the absence of correlation, one expects the strong law to be more powerful than the weak law. This is indeed the case, as the weak law states that the sample average

$$\frac{X_1 + \cdots + X_n}{n}$$

converges to μ in probability, while the strong law asserts that this convergence takes place with probability 1.

The following two inequalities will be useful in the discussion of this subject.

Lemma B.11 (Markov's inequality) *For any integer $k > 0$, real number $\alpha > 0$, and any random variable X,*

$$\Pr[|X| \geq \alpha] \leq \frac{1}{\alpha^k} E[|X|^k].$$

Proof Let $F_X(\cdot)$ be the cdf of random variable X. Then,

$$
\begin{aligned}
E[|X|^k] &= \int_{-\infty}^{\infty} |x|^k d F_X(x) \\
&\geq \int_{\{x \in \mathbb{R}: |x| \geq \alpha\}} |x|^k d F_X(x) \\
&\geq \int_{\{x \in \mathbb{R}: |x| \geq \alpha\}} \alpha^k d F_X(x) \\
&= \alpha^k \int_{\{x \in \mathbb{R}: |x| \geq \alpha\}} d F_X(x) \\
&= \alpha^k \Pr[|X| \geq \alpha].
\end{aligned}
$$

Equality holds iff

$$\int_{\{x\in\mathbb{R}:\, |x|<\alpha\}} |x|^k dF_X(x) = 0 \quad \text{and} \quad \int_{\{x\in\mathbb{R}:\, |x|>\alpha\}} |x|^k dF_X(x) = 0,$$

namely,

$$\Pr[X = 0] + \Pr[|X| = \alpha] = 1.$$

□

In the proof of Markov's inequality, we use the general representation for integration with respect to a (cumulative) distribution function $F_X(\cdot)$, i.e.,

$$\int_{\mathcal{X}} \cdot\, dF_X(x), \tag{B.5.1}$$

which is named the *Lebesgue–Stieltjes integral*. Such a representation can be applied for both discrete and continuous supports as well as the case that the probability density function does not exist. We use this notational convention to remove the burden of differentiating discrete random variables from continuous ones.

Lemma B.12 (Chebyshev's inequality) *For any random variable X with variance $Var[X]$ and real number $\alpha > 0$,*

$$\Pr[|X - E[X]| \geq \alpha] \leq \frac{1}{\alpha^2} Var[X].$$

Proof By Markov's inequality with $k = 2$, we have

$$\Pr[|X - E[X]| \geq \alpha] \leq \frac{1}{\alpha^2} E[|X - E[X]|^2].$$

Equality holds iff

$$\Pr[|X - E[X]| = 0] + \Pr\left[|X - E[X]| = \alpha\right] = 1,$$

equivalently, there exists $p \in [0, 1]$ such that

$$\Pr\left[X = E[X] + \alpha\right] = \Pr\left[X = E[X] - \alpha\right] = p \quad \text{and} \quad \Pr[X = E[X]] = 1 - 2p.$$

□

In the proofs of the above two lemmas, we also provide the condition under which equality holds. These conditions indicate that equality usually cannot be fulfilled. Hence in most cases, the two inequalities are strict.

Theorem B.13 (Weak law of large numbers) *Let $\{X_n\}_{n=1}^{\infty}$ be a sequence of uncorrelated random variables with common mean $E[X_i] = \mu$. If the variables also have common variance, or more generally,*

$$\lim_{n\to\infty} \frac{1}{n^2} \sum_{i=1}^{n} \text{Var}[X_i] = 0, \quad \left(\text{equivalently}, \frac{X_1 + \cdots + X_n}{n} \xrightarrow{\mathcal{L}_2} \cdot\right)$$

then the sample average

$$\frac{X_1 + \cdots + X_n}{n}$$

converges to the mean μ in probability.

Proof By Chebyshev's inequality,

$$\Pr\left\{ \left| \frac{1}{n} \sum_{i=1}^{n} X_i - \mu \right| \geq \varepsilon \right\} \leq \frac{1}{n^2 \varepsilon^2} \sum_{i=1}^{n} \text{Var}[X_i].$$

$\square$

Note that the right-hand side of the above Chebyshev's inequality is just the second moment of the difference between the n-sample average and the mean μ. Thus, the variance constraint is equivalent to the statement that $X_n \xrightarrow{\mathcal{L}_2} \mu$ implies $X_n \xrightarrow{p} \mu$.

Theorem B.14 (Kolmogorov's strong law of large numbers) *Let $\{X_n\}_{n=1}^{\infty}$ be a sequence of independent random variables with common mean $E[X_n] = \mu$. If either*

1. *X_n's are identically distributed; or*
2. *X_n's are square-integrable[9] with variances satisfying*

$$\sum_{i=1}^{\infty} \frac{\text{Var}[X_i]}{i^2} < \infty,$$

then

$$\frac{X_1 + \cdots + X_n}{n} \xrightarrow{a.s.} \mu.$$

Note that the above i.i.d. assumption does not exclude the possibility of $\mu = \infty$ (or $\mu = -\infty$), in which case the sample average converges to ∞ (or $-\infty$) with probability 1. Also note that there are cases of sequences of independent random variables for which the weak law applies, but the strong law does not. This is due to the fact that

---

[9] A random variable X is said to be *square-integrable* if $E[|X|^2] < \infty$.

$$\sum_{i=1}^{n} \frac{\text{Var}[X_i]}{i^2} \geq \frac{1}{n^2} \sum_{i=1}^{n} \text{Var}[X_i].$$

The final remark is that Kolmogorov's strong law of large number can be extended to a function of a sequence of independent random variables:

$$\frac{g(X_1) + \cdots + g(X_n)}{n} \xrightarrow{a.s.} E[g(X_1)].$$

But such extension cannot be applied to the weak law of large numbers, since $g(Y_i)$ and $g(Y_j)$ can be correlated even if Y_i and Y_j are not.

B.5.2 Ergodicity Versus Law of Large Numbers

After the introduction of Kolmogorov's strong law of large numbers, one may find that the pointwise ergodic theorem (Theorem B.4) actually indicates a similar result. In fact, the pointwise ergodic theorem can be viewed as another version of the strong law of large numbers, which states that *for stationary and ergodic processes, time averages converge with probability 1 to the ensemble expectation.*

The notion of ergodicity is often misinterpreted, since the definition is not very intuitive. Some texts may provide a definition that a stationary process satisfying the ergodic theorem is also ergodic.[10] However, the ergodic theorem is indeed a consequence of the original mathematical definition of ergodicity in terms of the shift-invariant property (see Sect. B.3 and the discussion in [160, pp. 174–175]).

Let us try to clarify the notion of ergodicity by the following remarks:

- The concept of ergodicity does not require stationarity. In other words, a nonstationary process can be ergodic.
- Many perfectly good models of physical processes are not ergodic, yet they obey some form of law of large numbers. In other words, non-ergodic processes can be perfectly good and useful models.

---

[10]Here is one example. *A stationary random process $\{X_n\}_{n=1}^{\infty}$ is called ergodic if for arbitrary integer k and function $f(\cdot)$ on $\mathcal{X}^k$ of finite mean,*

$$\frac{1}{n} \sum_{i=1}^{n} f(X_{i+1}, \ldots, X_{i+k}) \xrightarrow{a.s.} E[f(X_1, \ldots, X_k)].$$

As a result of this definition, a stationary ergodic source is the most general dependent random process for which the strong law of large numbers holds. This definition somehow implies that if a process is not stationary ergodic, then the strong law of large numbers is violated (or the time average does not converge with probability 1 to its ensemble expectation). But this is not true. One can weaken the conditions of stationarity and ergodicity from its original mathematical definitions to *asymptotic* stationarity and ergodicity, and still make the strong law of large numbers hold. (Cf. the last remark in this section and also Fig. B.2.)

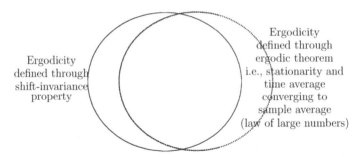

Fig. B.2 Relation of ergodic random processes, respectively, defined through time-shift invariance and ergodic theorem

- There is no finite-dimensional equivalent definition of ergodicity as there is for stationarity. This fact makes it more difficult to describe and interpret ergodicity.
- I.i.d. processes are ergodic; hence, ergodicity can be thought of as a (kind of) generalization of i.i.d.
- As mentioned earlier, stationarity and ergodicity imply the time average converges with probability 1 to the ensemble mean. Now if a process is stationary but not ergodic, then the time average still converges, but possibly not to the ensemble mean.

For example, let $\{A_n\}_{n=1}^{\infty}$ and $\{B_n\}_{n=1}^{\infty}$ be two sequences of i.i.d. binary-valued random variables with $\Pr\{A_n = 0\} = \Pr\{B_n = 1\} = 1/4$. Suppose that $X_n = A_n$ if $U = 1$, and $X_n = B_n$ if $U = 0$, where U is an equiprobable binary random variable, and $\{A_n\}_{n=1}^{\infty}$, $\{B_n\}_{n=1}^{\infty}$ and U are independent. Then $\{X_n\}_{n=1}^{\infty}$ is stationary. Is the process ergodic? The answer is negative. If the stationary process were ergodic, then from the pointwise ergodic theorem (Theorem B.4), its relative frequency would converge to

$$\Pr(X_n = 1) = \Pr(U = 1)\Pr(X_n = 1|U = 1)$$
$$+ \Pr(U = 0)\Pr(X_n = 1|U = 0)$$
$$= \Pr(U = 1)\Pr(A_n = 1) + \Pr(U = 0)\Pr(B_n = 1) = \frac{1}{2}.$$

However, one should observe that the outputs of $(X_1, \ldots, X_n)$ form a Bernoulli process with relative frequency of 1's being either 3/4 or 1/4, depending on the value of U. Therefore,

$$\lim_{n \to \infty} \frac{1}{n} \sum_{i=1}^{n} X_n \xrightarrow{a.s.} Y,$$

where $\Pr(Y = 1/4) = \Pr(Y = 3/4) = 1/2$, which contradicts the ergodic theorem.

From the above example, the pointwise ergodic theorem can actually be made useful in such a stationary but non-ergodic case, since an "apparent" stationary ergodic process (either $\{A_n\}_{n=1}^{\infty}$ or $\{B_n\}_{n=1}^{\infty}$) is actually being observed when measuring the relative frequency ($3/4$ or $1/4$). This renders a surprising fundamental result for random processes—the *ergodic decomposition theorem:* under fairly general assumptions, any (not necessarily ergodic) stationary process is in fact a mixture of stationary ergodic processes, and hence one always observes a stationary ergodic outcome (e.g., see [159, 349]). As in the above example, one always observe either $A_1, A_2, A_3, \ldots$ or $B_1, B_2, B_3, \ldots$, depending on the value of U, for which both sequences are stationary ergodic (i.e., the time-stationary observation X_n satisfies $X_n = U \cdot A_n + (1 - U) \cdot B_n$).

- The previous remark implies that ergodicity is not required for the strong law of large numbers to be useful. The next question is whether or not stationarity is required. Again, the answer is negative. In fact, the main concern of the law of large numbers is the convergence of sample averages to its ensemble expectation. It should be reasonable to expect that random processes could exhibit transient behaviors that violate the stationarity definition, with their sample average still converging. One can then introduce the notion of *asymptotically mean stationary* to achieve the law of large numbers [159]. For example, a finite-alphabet time-invariant (but not necessarily stationary) irreducible Markov chain satisfies the law of large numbers. Thus, the stationarity and/or ergodicity properties of a process can be weakened with the process still admitting laws of large numbers (i.e., time averages and relative frequencies have desired and well-defined limits).

B.6   Central Limit Theorem

Theorem B.15 (Central limit theorem) *If $\{X_n\}_{n=1}^{\infty}$ is a sequence of i.i.d. random variables with finite common marginal mean μ and variance σ^2, then*

$$\frac{1}{\sqrt{n}} \sum_{i=1}^{n} (X_i - \mu) \xrightarrow{d} Z \sim \mathcal{N}(0, \sigma^2),$$

where the convergence is in distribution (as $n \to \infty$) and $Z \sim \mathcal{N}(0, \sigma^2)$ is a Gaussian distributed random variable with mean 0 and variance σ^2.

B.7   Convexity, Concavity, and Jensen's inequality

Jensen's inequality provides a useful bound for the expectation of convex (or concave) functions.

Definition B.16 (*Convexity*) Consider a convex set[11] $\mathcal{O} \in \mathbb{R}^m$, where m is a fixed positive integer. Then, a function $f : \mathcal{O} \to \mathbb{R}$ is said to be *convex* over $\mathcal{O}$ if for every $\underline{x}$, $\underline{y}$ in $\mathcal{O}$ and $0 \leq \lambda \leq 1$,

$$f\left(\lambda \underline{x} + (1 - \lambda)\underline{y}\right) \leq \lambda f(\underline{x}) + (1 - \lambda)f(\underline{y}).$$

Furthermore, a function f is said to be *strictly convex* if equality holds only when $\lambda = 0$ or $\lambda = 1$.

Note that different from the usual notations $x^n = (x_1, x_2, \ldots, x_n)$ or $\boldsymbol{x} = (x_1, x_2, \ldots)$ throughout this book, we use $\underline{x}$ to denote a column vector in this section.

Definition B.17 (*Concavity*) A function f is *concave* if $-f$ is convex.

Note that when $\mathcal{O} = (a, b)$ is an interval in $\mathbb{R}$ and function $f : \mathcal{O} \to \mathbb{R}$ has a nonnegative (resp. positive) second derivative over $\mathcal{O}$, then the function is convex (resp. strictly convex). This can be shown via the Taylor series expansion of the function.

Theorem B.18 (Jensen's inequality) *If* $f : \mathcal{O} \to \mathbb{R}$ *is convex over a convex set* $\mathcal{O} \subset \mathbb{R}^m$, *and* $\underline{X} = (X_1, X_2, \ldots, X_m)^T$ *is an m-dimensional random vector with alphabet* $\mathcal{X} \subset \mathcal{O}$, *then*

$$E[f(\underline{X})] \geq f(E[\underline{X}]).$$

Moreover, if f is strictly convex, then equality in the above inequality immediately implies $\underline{X} = E[\underline{X}]$ with probability 1.

Note: $\mathcal{O}$ is a convex set; hence, $\mathcal{X} \subset \mathcal{O}$ implies $E[\underline{X}] \in \mathcal{O}$. This guarantees that $f(E[\underline{X}])$ is defined. Similarly, if f is concave, then

$$E[f(\underline{X})] \leq f(E[\underline{X}]).$$

Furthermore, if f is strictly concave, then equality in the above inequality immediately implies that $\underline{X} = E[\underline{X}]$ with probability 1.

Proof Let $y = \underline{a}^T \underline{x} + b$ be a "support hyperplane" for f with "slope" vector $\underline{a}^T$ and affine parameter b that passes through the point $(E[\underline{X}], f(E[\underline{X}]))$, where a support hyperplane[12] for function f at $\underline{x}'$ is by definition a hyperplane passing through

---

[11]A set $\mathcal{O} \in \mathbb{R}^m$ is said to be *convex* if for every $\underline{x} = (x_1, x_2, \ldots, x_m)^T$ and $\underline{y} = (y_1, y_2, \ldots, y_m)^T$ in $\mathcal{O}$ (where T denotes transposition), and every $0 \leq \lambda \leq 1$, $\lambda \underline{x} + (1 - \lambda)\underline{y} \in \mathcal{O}$; in other words, the "convex combination" of any two "points" $\underline{x}$ and $\underline{y}$ in $\mathcal{O}$ also belongs to $\mathcal{O}$.

[12]A hyperplane $y = \underline{a}^T \underline{x} + b$ is said to be a support hyperplane for a function f with "slope" vector $\underline{a}^T \in \mathbb{R}^m$ and affine parameter $b \in \mathbb{R}$ if among all hyperplanes of the same slope vector $\underline{a}^T$, it is the largest one satisfying $\underline{a}^T \underline{x} + b \leq f(\underline{x})$ for every $\underline{x} \in \mathcal{O}$. A support hyperplane may not necessarily be made to pass through the desired point $(\underline{x}', f(\underline{x}'))$. Here, since we only consider convex functions, the validity of the support hyperplane passing $(\underline{x}', f(\underline{x}'))$ is therefore guaranteed. Note that when $\underline{x}$ is one-dimensional (i.e., $m = 1$), a support hyperplane is simply referred to as a support line.

Fig. B.3 The support line $y = ax + b$ of the convex function $f(x)$

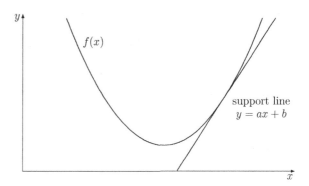

the point $(\underline{x}', f(\underline{x}'))$ and lying entirely below the graph of f (see Fig. B.3 for an illustration of a support line for a convex function over $\mathbb{R}$). Thus,

$$(\forall \underline{x} \in \mathcal{X}) \quad \underline{a}^T \underline{x} + b \le f(\underline{x}).$$

By taking the expectation of both sides, we obtain

$$\underline{a}^T E[\underline{X}] + b \le E[f(\underline{X})],$$

but we know that $\underline{a}^T E[\underline{X}] + b = f(E[\underline{X}])$. Consequently,

$$f(E[\underline{X}]) \le E[f(\underline{X})].$$

□

B.8 Lagrange Multipliers Technique and Karush–Kuhn–Tucker (KKT) Conditions

Optimization of a function $f(x)$ over $x = (x_1, \ldots, x_n) \in \mathcal{X} \subseteq \mathbb{R}^n$ subject to some inequality constraints $g_i(x) \le 0$ for $1 \le i \le m$ and equality constraints $h_j(x) = 0$ for $1 \le j \le \ell$ is a central technique to problems in information theory. An immediate example is to maximize the mutual information subject to an "inequality" power constraint and an "equality" probability unity-sum constraint in order to determine the channel capacity. We can formulate such an optimization problem [56, Eq. (5.1)] mathematically as[13]

$$\min_{x \in Q} f(x), \tag{B.8.1}$$

---

[13]Since maximization of $f(\cdot)$ is equivalent to minimization of $-f(\cdot)$, it suffices to discuss the KKT conditions for the minimization problem defined in (B.8.1).

294 Appendix B: Overview in Probability and Random Processes

where

$$Q := \{x \in \mathcal{X} : g_i(x) \leq 0 \text{ for } 1 \leq i \leq m \text{ and } h_i(x) = 0 \text{ for } 1 \leq j \leq \ell\}.$$

In most cases, solving the constrained optimization problem defined in (B.8.1) is hard. Instead, one may introduce a dual optimization problem without constraints as

$$L(\boldsymbol{\lambda}, \boldsymbol{\nu}) := \min_{x \in \mathcal{X}} \left(f(x) + \sum_{i=1}^{m} \lambda_i g_i(x) + \sum_{j=1}^{\ell} \nu_j h_j(x) \right). \tag{B.8.2}$$

In the literature, $\boldsymbol{\lambda} = (\lambda_1, \ldots, \lambda_m)$ and $\boldsymbol{\nu} = (\nu_1, \ldots, \nu_\ell)$ are usually referred to as Lagrange multipliers, and $L(\boldsymbol{\lambda}, \boldsymbol{\nu})$ is called the Lagrange dual function. Note that $L(\boldsymbol{\lambda}, \boldsymbol{\nu})$ is a concave function of $\boldsymbol{\lambda}$ and $\boldsymbol{\nu}$ since it is the minimization of affine functions of $\boldsymbol{\lambda}$ and $\boldsymbol{\nu}$.

It can be verified that when $\lambda_i \geq 0$ for $1 \leq i \leq m$,

$$L(\boldsymbol{\lambda}, \boldsymbol{\nu}) \leq \min_{x \in Q} \left(f(x) + \sum_{i=1}^{m} \lambda_i g_i(x) + \sum_{j=1}^{\ell} \nu_j h_j(x) \right) \leq \min_{x \in Q} f(x). \tag{B.8.3}$$

We are however interested in when the above inequality becomes equality (i.e., when the so-called *strong duality* holds) because if there exist nonnegative $\tilde{\boldsymbol{\lambda}}$ and $\tilde{\boldsymbol{\nu}}$ that equate (B.8.3), then

$$f(x^*) = \min_{x \in Q} f(x) = L(\tilde{\boldsymbol{\lambda}}, \tilde{\boldsymbol{\nu}})$$

$$= \min_{x \in \mathcal{X}} \left[f(x) + \sum_{i=1}^{m} \tilde{\lambda}_i g_i(x) + \sum_{j=1}^{\ell} \tilde{\nu}_j h_j(x) \right]$$

$$\leq f(x^*) + \sum_{i=1}^{m} \tilde{\lambda}_i g_i(x^*) + \sum_{j=1}^{\ell} \tilde{\nu}_j h_j(x^*)$$

$$\leq f(x^*), \tag{B.8.4}$$

where (B.8.4) follows because the minimizer x^* of (B.8.1) lies in Q. Hence, if the strong duality holds, the same x^* achieves both $\min_{x \in Q} f(x)$ and $L(\tilde{\boldsymbol{\lambda}}, \tilde{\boldsymbol{\nu}})$, and $\tilde{\lambda}_i g_i(x^*) = 0$ for $1 \leq i \leq m$.[14]

The strong duality does not in general hold. A situation that guarantees the validity of the strong duality has been determined by William Karush [212], and separately Harold W. Kuhn and Albert W. Tucker [235]. In particular, when $f(\cdot)$ and $\{g_i(\cdot)\}_{i=1}^{m}$ are both convex, and $\{h_j(\cdot)\}_{j=1}^{\ell}$ are affine, and these functions are all differentiable,

---

[14]Equating (B.8.4) implies $\sum_{i=1}^{m} \tilde{\lambda}_i g_i(x^*) = 0$. It can then be easily verified from $\tilde{\lambda}_i g_i(x^*) \leq 0$ for every $1 \leq i \leq m$ that $\tilde{\lambda}_i g_i(x^*) = 0$ for $1 \leq i \leq m$.

they found that the strong duality holds iff the KKT conditions are satisfied [56, p. 258].

Definition B.19 (*Karush–Kuhn–Tucker (KKT) conditions*) Point $x = (x_1, \ldots, x_n)$ and multipliers $\lambda = (\lambda_1, \ldots, \lambda_m)$ and $\nu = (\nu_1, \ldots, \nu_\ell)$ are said to satisfy the KKT conditions if

$$\begin{cases} g_i(x) \leq 0, \quad \lambda_i \geq 0, \quad \lambda_i g_i(x) = 0 & i = 1, \ldots, m \\[2mm] h_j(x) = 0 & j = 1, \ldots, \ell \\[2mm] \frac{\partial f}{\partial x_k}(x) + \sum_{i=1}^{m} \lambda_i \frac{\partial g_i}{\partial x_k}(x) + \sum_{j=1}^{\ell} \nu_j \frac{\partial h_j}{\partial x_k}(x) = 0 & k = 1, \ldots, n \end{cases}$$

Note that when $f(\cdot)$ and constraints $\{g_i(\cdot)\}_{i=1}^m$ and $\{h_j(\cdot)\}_{j=1}^\ell$ are arbitrary functions, the KKT conditions are only necessary for the validity of the strong duality. In other words, for a non-convex optimization problem, we can only claim that if the strong duality holds, then the KKT conditions are satisfied but not vice versa.

A case that is particularly useful in information theory is when x is restricted to be a probability distribution. In such case, apart from other problem-specific constraints, we have additionally n inequality constraints $g_{m+i}(x) = -x_i \leq 0$ for $1 \leq i \leq n$ and one equality constraint $h_{\ell+1}(x) = \sum_{k=1}^n x_k - 1 = 0$. Hence, the KKT conditions become

$$\begin{cases} g_i(x) \leq 0, \quad \lambda_i \geq 0, \quad \lambda_i g_i(x) = 0 & i = 1, \ldots, m \\[2mm] g_{m+k}(x) = -x_k \leq 0, \quad \lambda_{m+k} \geq 0, \quad \lambda_{m+k} x_k = 0 & k = 1, \ldots, n \\[2mm] h_j(x) = 0 & j = 1, \ldots, \ell \\[2mm] h_{\ell+1}(x) = \sum_{k=1}^n x_k - 1 = 0 \\[2mm] \frac{\partial f}{\partial x_k}(x) + \sum_{i=1}^{m} \lambda_i \frac{\partial g_i}{\partial x_k}(x) - \lambda_{m+k} + \sum_{j=1}^{\ell} \nu_j \frac{\partial h_j}{\partial x_k}(x) + \nu_{\ell+1} = 0 & k = 1, \ldots, n \end{cases}$$

From $\lambda_{m+k} \geq 0$ and $\lambda_{m+k} x_k = 0$, we can obtain the well-known relation below.

$$\lambda_{m+k} = \begin{cases} \frac{\partial f}{\partial x_k}(x) + \sum_{i=1}^{m} \lambda_i \frac{\partial g_i}{\partial x_k}(x) + \sum_{j=1}^{\ell} \nu_j \frac{\partial h_j}{\partial x_k}(x) + \nu_{\ell+1} = 0 & \text{if } x_k > 0 \\[2mm] \frac{\partial f}{\partial x_k}(x) + \sum_{i=1}^{m} \lambda_i \frac{\partial g_i}{\partial x_k}(x) + \sum_{j=1}^{\ell} \nu_j \frac{\partial h_j}{\partial x_k}(x) + \nu_{\ell+1} \geq 0 & \text{if } x_k = 0. \end{cases}$$

The above relation is the most seen form of the KKT conditions when it is used in problems in information theory.

Example B.20 Suppose for nonnegative $\{q_{i,j}\}_{1 \leq i \leq n, 1 \leq j \leq n'}$ with $\sum_{j=1}^{n'} q_{i,j} = 1$,

$$
\begin{cases}
f(x) = -\sum_{i=1}^{n} \sum_{j=1}^{n'} x_i q_{i,j} \log \dfrac{q_{i,j}}{\sum_{i'=1}^{n} x_{i'} q_{i',j}} \\[2ex]
g_i(x) = -x_i \le 0 & i = 1, \ldots, n \\[2ex]
h(x) = \sum_{i=1}^{n} x_i - 1 = 0
\end{cases}
$$

Then, the KKT conditions imply

$$
\begin{cases}
x_i \ge 0, \quad \lambda_i \ge 0, \quad \lambda_i x_i = 0 & i = 1, \ldots, n \\[2ex]
\sum_{i=1}^{n} x_i = 1 \\[2ex]
-\sum_{j=1}^{n'} q_{k,j} \log \dfrac{q_{k,j}}{\sum_{i'=1}^{n} x_{i'} q_{i',j}} + 1 - \lambda_k + \nu = 0 & k = 1, \ldots, n
\end{cases}
$$

which further implies that

$$
\lambda_k =
\begin{cases}
-\sum_{j=1}^{n'} q_{k,j} \log \dfrac{q_{k,j}}{\sum_{i'=1}^{n} x_{i'} q_{i',j}} + 1 + \nu = 0 & x_k > 0 \\[3ex]
-\sum_{j=1}^{n'} q_{k,j} \log \dfrac{q_{k,j}}{\sum_{i'=1}^{n} x_{i'} q_{i',j}} + 1 + \nu \ge 0 & x_k = 0
\end{cases}
$$

By this, the input distributions that achieve the channel capacities of some channels such as BSC and BEC can be identified. $\qquad\square$

The next example shows the analogy of determining the channel capacity to the problem of optimal power allocation.

Example B.21 (**Water-filling**) Suppose with $\sigma_i^2 > 0$ for $1 \le i \le n$ and $P > 0$,

$$
\begin{cases}
f(x) = -\sum_{i=1}^{n} \log \left(1 + \dfrac{x_i}{\sigma_i^2} \right) \\[2ex]
g_i(x) = -x_i \le 0 & i = 1, \ldots, n \\[2ex]
h(x) = \sum_{i=1}^{n} x_i - P = 0
\end{cases}
$$

Then, the KKT conditions imply

$$\begin{cases} x_i \geq 0, \quad \lambda_i \geq 0, \quad \lambda_i x_i = 0 \quad i = 1, \dots, n \\[2mm] \sum_{i=1}^{n} x_i = P \\[2mm] -\frac{1}{\sigma_i^2 + x_i} - \lambda_i + \nu = 0 \qquad\qquad i = 1, \dots, n \end{cases}$$

which further implies that

$$\lambda_i = \begin{cases} -\frac{1}{\sigma_i^2 + x_i} + \nu = 0 & x_i > 0 \\[2mm] -\frac{1}{\sigma_i^2 + x_i} + \nu \geq 0 & x_i = 0 \end{cases} \text{ equivalently } x_i = \begin{cases} \frac{1}{\nu} - \sigma_i^2 & \sigma_i^2 < \frac{1}{\nu} \\[2mm] 0 & \sigma_i^2 \geq \frac{1}{\nu} \end{cases}$$

This then gives the water-filling solution for the power allocation over parallel continuous-input AWGN channels. $\square$

References

1. J. Aczél, Z. Daróczy, *On Measures of Information and Their Characterization* (Academic Press, New York, 1975)
2. C. Adam, Information theory in molecular biology. Phys. Life Rev. **1**, 3–22 (2004)
3. R. Ahlswede, Certain results in coding theory for compound channels I, in *Proceedings of Bolyai Colloquium on Information Theory*, Debrecen, Hungary, 1967, pp. 35–60
4. R. Ahlswede, The weak capacity of averaged channels. Z. Wahrscheinlichkeitstheorie Verw. Gebiete **11**, 61–73 (1968)
5. R. Ahlswede, J. Körner, Source coding with side information and a converse for degraded broadcast channels. IEEE Trans. Inf. Theory **21**(6), 629–637 (1975)
6. R. Ahlswede, I. Csiszár, Common randomness in information theory and cryptography. I. Secret sharing. IEEE Trans. Inf. Theory **39**(4), 1121–1132 (1993)
7. R. Ahlswede, I. Csiszár, Common randomness in information theory and cryptography. II. CR capacity. IEEE Trans. Inf. Theory **44**(1), 225–240 (1998)
8. E. Akyol, K.B. Viswanatha, K. Rose, T.A. Ramstad, On zero-delay source-channel coding. IEEE Trans. Inf. Theory **60**(12), 7473–7489 (2014)
9. F. Alajaji, Feedback does not increase the capacity of discrete channels with additive noise. IEEE Trans. Inf. Theory **41**, 546–549 (1995)
10. F. Alajaji, T. Fuja, The performance of focused error control codes. IEEE Trans. Commun. **42**(2/3/4), 272–280 (1994)
11. F. Alajaji, T. Fuja, Effect of feedback on the capacity of discrete additive channels with memory, in *Proceedings of IEEE International Symposium on Information Theory*, Trondheim, Norway (1994)
12. F. Alajaji, T. Fuja, A communication channel modeled on contagion. IEEE Trans. Inf. Theory **40**(6), 2035–2041 (1994)
13. F. Alajaji, N. Phamdo, N. Farvardin, T.E. Fuja, Detection of binary Markov sources over channels with additive Markov noise. IEEE Trans. Inf. Theory **42**(1), 230–239 (1996)
14. F. Alajaji, N. Phamdo, T.E. Fuja, Channel codes that exploit the residual redundancy in CELP-encoded speech. IEEE Trans. Speech Audio Process. **4**(5), 325–336 (1996)
15. F. Alajaji, N. Phamdo, Soft-decision COVQ for Rayleigh-fading channels. IEEE Commun. Lett. **2**, 162–164 (1998)

© Springer Nature Singapore Pte Ltd. 2018

F. Alajaji and P.-N. Chen, *An Introduction to Single-User Information Theory*,

Springer Undergraduate Texts in Mathematics and Technology,

https://doi.org/10.1007/978-981-10-8001-2

16. F. Alajaji, N. Whalen, The capacity-cost function of discrete additive noise channels with and without feedback. IEEE Trans. Inf. Theory **46**(3), 1131–1140 (2000)

17. F. Alajaji, P.-N. Chen, Z. Rached, Csiszár's cutoff rates for the general hypothesis testing problem. IEEE Trans. Inf. Theory **50**(4), 663–678 (2004)

18. V.R. Algazi, R.M. Lerner, *Binary detection in white non-Gaussian noise*, Technical Report DS-2138, M.I.T. Lincoln Lab, Lexington, MA (1964)

19. S.A. Al-Semari, F. Alajaji, T. Fuja, Sequence MAP decoding of trellis codes for Gaussian and Rayleigh channels. IEEE Trans. Veh. Technol. **48**(4), 1130–1140 (1999)

20. E. Arikan, An inequality on guessing and its application to sequential decoding. IEEE Trans. Inf. Theory **42**(1), 99–105 (1996)

21. E. Arikan, N. Merhav, Joint source-channel coding and guessing with application to sequential decoding. IEEE Trans. Inf. Theory **44**, 1756–1769 (1998)

22. E. Arikan, A performance comparison of polar codes and Reed-Muller codes. IEEE Commun. Lett. **12**(6), 447–449 (2008)

23. E. Arikan, Channel polarization: a method for constructing capacity-achieving codes for symmetric binary-input memoryless channels. IEEE Trans. Inf. Theory **55**(7), 3051–3073 (2009)

24. E. Arikan, Source polarization, in *Proceedings of International Symposium on Information Theory and Applications*, July 2010, pp. 899–903

25. E. Arikan, I.E. Telatar, On the rate of channel polarization, in *Proceedings of IEEE International Symposium on Information Theory*, Seoul, Korea, June–July 2009, pp. 1493–1495

26. E. Arikan, N. ul Hassan, M. Lentmaier, G. Montorsi, J. Sayir, Challenges and some new directions in channel coding. J. Commun. Netw. **17**(4), 328–338 (2015)

27. S. Arimoto, An algorithm for computing the capacity of arbitrary discrete memoryless channel. IEEE Trans. Inf. Theory **18**(1), 14–20 (1972)

28. S. Arimoto, Information measures and capacity of order α for discrete memoryless channels, *Topics in Information Theory, Proceedings of Colloquium Mathematical Society Janos Bolyai*, Keszthely, Hungary, 1977, pp. 41–52

29. R.B. Ash, *Information Theory* (Interscience, New York, 1965)

30. R.B. Ash, C.A. Doléans-Dade, *Probability and Measure Theory* (Academic Press, MA, 2000)

31. S. Asoodeh, M. Diaz, F. Alajaji, T. Linder, Information extraction under privacy constraints. Information **7**(1), 1–37 (2016)

32. E. Ayanoğlu, R. Gray, The design of joint source and channel trellis waveform coders. IEEE Trans. Inf. Theory **33**, 855–865 (1987)

33. J. Bakus, A.K. Khandani, Quantizer design for channel codes with soft-output decoding. IEEE Trans. Veh. Technol. **54**(2), 495–507 (2005)

34. V.B. Balakirsky, Joint source-channel coding with variable length codes. Probl. Inf. Transm. **1**(37), 10–23 (2001)

35. A. Banerjee, P. Burlina, F. Alajaji, Image segmentation and labeling using the Polya urn model. IEEE Trans. Image Process. **8**(9), 1243–1253 (1999)

36. M.B. Bassat, J. Raviv, Rényi's entropy and the probability of error. IEEE Trans. Inf. Theory **24**(3), 324–330 (1978)

37. F. Behnamfar, F. Alajaji, T. Linder, MAP decoding for multi-antenna systems with non-uniform sources: exact pairwise error probability and applications. IEEE Trans. Commun. **57**(1), 242–254 (2009)

38. H. Behroozi, F. Alajaji, T. Linder, On the optimal performance in asymmetric Gaussian wireless sensor networks with fading. IEEE Trans. Signal Process. **58**(4), 2436–2441 (2010)

39. S. Ben-Jamaa, C. Weidmann, M. Kieffer, Analytical tools for optimizing the error correction performance of arithmetic codes. IEEE Trans. Commun. **56**(9), 1458–1468 (2008)

40. C.H. Bennett, G. Brassard, Quantum cryptography: public key and coin tossing, in *Proceedings of International Conference on Computer Systems and Signal Processing*, Bangalore, India, Dec 1984, pp. 175–179

41. C.H. Bennett, G. Brassard, C. Crepeau, U.M. Maurer, Generalized privacy amplification. IEEE Trans. Inf. Theory **41**(6), 1915–1923 (1995)

42. T. Berger, *Rate Distortion Theory: A Mathematical Basis for Data Compression* (Prentice-Hall, New Jersey, 1971)
43. T. Berger, Explicit bounds to $R(D)$ for a binary symmetric Markov source. IEEE Trans. Inf. Theory **23**(1), 52–59 (1977)
44. C. Berrou, A. Glavieux, P. Thitimajshima, Near Shannon limit error-correcting coding and decoding: Turbo-codes(1), in *Proceedings of IEEE International Conference on Communications*, Geneva, Switzerland, May 1993, pp. 1064–1070
45. C. Berrou, A. Glavieux, Near optimum error correcting coding and decoding: turbo-codes. IEEE Trans. Commun. **44**(10), 1261–1271 (1996)
46. D.P. Bertsekas, with A. Nedić, A.E. Ozdagler, Convex Analysis and Optimization (Athena Scientific, Belmont, MA, 2003)
47. P. Billingsley, *Probability and Measure*, 2nd edn. (Wiley, New York, 1995)
48. C.M. Bishop, *Pattern Recognition and Machine Learning* (Springer, Berlin, 2006)
49. R.E. Blahut, Computation of channel capacity and rate-distortion functions. IEEE Trans. Inf. Theory **18**(4), 460–473 (1972)
50. R.E. Blahut, *Theory and Practice of Error Control Codes* (Addison-Wesley, MA, 1983)
51. R.E. Blahut, *Principles and Practice of Information Theory* (Addison Wesley, MA, 1988)
52. R.E. Blahut, *Algebraic Codes for Data Transmission* (Cambridge University Press, Cambridge, 2003)
53. M. Bloch, J. Barros, *Physical-Layer Security: From Information Theory to Security Engineering* (Cambridge University Press, Cambridge, 2011)
54. A.C. Blumer, R.J. McEliece, The Rényi redundancy of generalized Huffman codes. IEEE Trans. Inf. Theory **34**(5), 1242–1249 (1988)
55. L. Boltzmann, Uber die beziehung zwischen dem hauptsatze der mechanischen warmetheorie und der wahrscheinlicjkeitsrechnung respective den satzen uber das warmegleichgewicht. Wiener Berichte **76**, 373–435 (1877)
56. S. Boyd, L. Vandenberghe, *Convex Optimization* (Cambridge University Press, Cambridge, UK, 2003)
57. G. Brante, R. Souza, J. Garcia-Frias, Spatial diversity using analog joint source channel coding in wireless channels. IEEE Trans. Commun. **61**(1), 301–311 (2013)
58. L. Breiman, The individual ergodic theorems of information theory. Ann. Math. Stat. **28**, 809–811 (1957). (with acorrection made in vol. 31, pp. 809–810, 1960)
59. D.R. Brooks, E.O. Wiley, *Evolution as Entropy: Toward a Unified Theory of Biology* (University of Chicago Press, Chicago, 1988)
60. N. Brunel, J.P. Nadal, Mutual information, Fisher information, and population coding. Neural Comput. **10**(7), 1731–1757 (1998)
61. O. Bursalioglu, G. Caire, D. Divsalar, Joint source-channel coding for deep-space image transmission using rateless codes. IEEE Trans. Commun. **61**(8), 3448–3461 (2013)
62. V. Buttigieg, P.G. Farrell, Variable-length error-correcting codes. IEE Proc. Commun. **147**(4), 211–215 (2000)
63. S.A. Butman, R.J. McEliece, *The ultimate limits of binary coding for a wideband Gaussian channel*, DSN Progress Report 42–22, Jet Propulsion Lab, Pasadena, CA, Aug 1974, pp. 78–80
64. G. Caire, K. Narayanan, On the distortion SNR exponent of hybrid digital-analog space-time coding. IEEE Trans. Inf. Theory **53**, 2867–2878 (2007)
65. F.P. Calmon, *Information-Theoretic Metrics for Security and Privacy*, Ph.D. thesis, MIT, Sept 2015
66. F.P. Calmon, A. Makhdoumi, M. Médard, Fundamental limits of perfect privacy, in *Proceedings of IEEE International Symposium on Information Theory*, Hong Kong, pp. 1796–1800, June 2015
67. L.L. Campbell, A coding theorem and Rényi's entropy. Inf. Control **8**, 423–429 (1965)
68. L.L. Campbell, A block coding theorem and Rényi's entropy. Int. J. Math. Stat. Sci. **6**, 41–47 (1997)

69. A.T. Campo, G. Vazquez-Vilar, A. Guillen i Fabregas, T. Koch, A. Martinez, A derivation of the source-channel error exponent using non-identical product distributions. IEEE Trans. Inf. Theory **60**(6), 3209–3217 (2014)

70. C. Chang, Error exponents for joint source-channel coding with side information. IEEE Trans. Inf. Theory **57**(10), 6877–6889 (2011)

71. B. Chen, G. Wornell, Analog error-correcting codes based on chaotic dynamical systems. IEEE Trans. Commun. **46**(7), 881–890 (1998)

72. P.-N. Chen, F. Alajaji, Strong converse, feedback capacity and hypothesis testing, in *Proceedings of Conference on Information Sciences and Systems*, John Hopkins University, Baltimore (1995)

73. P.-N. Chen, F. Alajaji, Generalized source coding theorems and hypothesis testing (Parts I and II). J. Chin. Inst. Eng. **21**(3), 283–303 (1998)

74. P.-N. Chen, F. Alajaji, A rate-distortion theorem for arbitrary discrete sources. IEEE Trans. Inf. Theory **44**(4), 1666–1668 (1998)

75. P.-N. Chen, F. Alajaji, Optimistic Shannon coding theorems for arbitrary single-user systems. IEEE Trans. Inf. Theory **45**, 2623–2629 (1999)

76. P.-N. Chen, F. Alajaji, Csiszár's cutoff rates for arbitrary discrete sources. IEEE Trans. Inf. Theory **47**(1), 330–338 (2001)

77. X. Chen, E. Tuncel, Zero-delay joint source-channel coding using hybrid digital-analog schemes in the Wyner-Ziv setting. IEEE Trans. Commun. **62**(2), 726–735 (2014)

78. H. Chernoff, A measure of asymptotic efficiency for tests of a hypothesis based on the sum of observations. Ann. Math. Stat. **23**(4), 493–507 (1952)

79. S.Y. Chung, *On the Construction of some Capacity Approaching Coding Schemes*, Ph.D. thesis, MIT, 2000

80. R.H. Clarke, A statistical theory of mobile radio reception. Bell Syst. Tech. J. **47**, 957–1000 (1968)

81. T.M. Cover, A. El Gamal, M. Salehi, Multiple access channels with arbitrarily correlated sources. IEEE Trans. Inf. Theory **26**(6), 648–657 (1980)

82. T.M. Cover, An algorithm for maximizing expected log investment return. IEEE Trans. Inf. Theory **30**(2), 369–373 (1984)

83. T.M. Cover, J.A. Thomas, *Elements of Information Theory*, 2nd edn. (Wiley, New York, 2006)

84. I. Csiszár, Joint source-channel error exponent. Probl. Control Inf. Theory **9**, 315–328 (1980)

85. I. Csiszár, On the error exponent of source-channel transmission with a distortion threshold. IEEE Trans. Inf. Theory **28**(6), 823–828 (1982)

86. I. Csiszár, Generalized cutoff rates and Rényi's information measures. IEEE Trans. Inf. Theory **41**(1), 26–34 (1995)

87. I. Csiszár, J. Körner, *Information Theory: Coding Theorems for Discrete Memoryless Systems* (Academic, New York, 1981)

88. I. Csiszár, G. Tusnady, Information geometry and alternating minimization procedures, in *Statistics and Decision*, Supplement Issue, vol. 1 (1984), pp. 205–237

89. J.V. Davis, B. Kulis, P. Jain, S. Sra, I.S. Dhillon, Information-theoretic metric learning, in *Proceedings of 24th International Conference on Machine Learning*, June 2007, pp. 209–216

90. B. de Finetti, Funcione caratteristica di un fenomeno aleatorio, *Atti della R. Academia Nazionale dei Lincii Ser. 6, Memorie,classe di Scienze, Fisiche, Matamatiche e Naturali*, vol. 4 (1931), pp. 251–299

91. J. del Ser, P.M. Crespo, I. Esnaola, J. Garcia-Frias, Joint source-channel coding of sources with memory using Turbo codes and the Burrows-Wheeler transform. IEEE Trans. Commun. **58**(7), 1984–1992 (2010)

92. J. Devore, A note on the observation of a Markov source through a noisy channel. IEEE Trans. Inf. Theory **20**(6), 762–764 (1974)

93. A. Diallo, C. Weidmann, M. Kieffer, New free distance bounds and design techniques for joint source-channel variable-length codes. IEEE Trans. Commun. **60**(10), 3080–3090 (2012)

94. W. Diffie, M. Hellman, New directions in cryptography. IEEE Trans. Inf. Theory **22**(6), 644–654 (1976)

95. R.L. Dobrushin, Asymptotic bounds of the probability of error for the transmission of messages over a memoryless channel with a symmetric transition probability matrix (in Russian). Teor. Veroyatnost. i Primenen **7**(3), 283–311 (1962)

96. R.L. Dobrushin, General formulation of Shannon's basic theorems of information theory, *AMS Translations*, vol. 33, AMS, Providence, RI (1963), pp. 323–438

97. R.L. Dobrushin, M.S. Pinsker, Memory increases transmission capacity. Probl. Inf. Transm. **5**(1), 94–95 (1969)

98. S.J. Dolinar, F. Pollara, The theoretical limits of source and channel coding, *TDA Progress Report 42-102*, Jet Propulsion Lab, Pasadena, CA, Aug 1990, pp. 62–72

99. *Draft report of 3GPP TSG RAN WG1* #87 v0.2.0, The 3rd Generation Partnership Project (3GPP),Reno, Nevada, USA, Nov. 2016

100. P. Duhamel, M. Kieffer, *Joint Source-Channel Decoding: A Cross- Layer Perspective with Applications in Video Broadcasting over Mobile and Wireless Networks* (Academic Press, 2010)

101. S. Dumitrescu, X. Wu, On the complexity of joint source-channel decoding of Markov sequences over memoryless channels. IEEE Trans. Commun. **56**(6), 877–885 (2008)

102. S. Dumitrescu, Y. Wan, Bit-error resilient index assignment for multiple description scalar quantizers. IEEE Trans. Inf. Theory **61**(5), 2748–2763 (2015)

103. J.G. Dunham, R.M. Gray, Joint source and noisy channel trellis encoding. IEEE Trans. Inf. Theory **27**, 516–519 (1981)

104. R. Durrett, *Probability: Theory and Examples* (Cambridge University Press, Cambridge, 2015)

105. A.K. Ekert, Quantum cryptography based on Bell's theorem. Phys. Rev. Lett. **67**(6), 661–663 (1991)

106. A. El Gamal, Y.-H. Kim, *Network Information Theory* (Cambridge University Press, Cambridge, 2011)

107. P. Elias, Coding for noisy channels, *IRE Convention Record*, Part 4, pp. 37–46 (1955)

108. E.O. Elliott, Estimates of error rates for codes on burst-noise channel. Bell Syst. Tech. J. **42**, 1977–1997 (1963)

109. S. Emami, S.L. Miller, Nonsymmetric sources and optimum signal selection. IEEE Trans. Commun. **44**(4), 440–447 (1996)

110. M. Ergen, *Mobile Broadband: Including WiMAX and LTE* (Springer, Berlin, 2009)

111. F. Escolano, P. Suau, B. Bonev, *Information Theory in Computer Vision and Pattern Recognition* (Springer, Berlin, 2009)

112. I. Esnaola, A.M. Tulino, J. Garcia-Frias, Linear analog coding of correlated multivariate Gaussian sources. IEEE Trans. Commun. **61**(8), 3438–3447 (2013)

113. R.M. Fano, Class notes for "Transmission of Information," Course 6.574, MIT, 1952

114. R.M. Fano, *Transmission of Information: A Statistical Theory of Communication* (Wiley, New York, 1961)

115. B. Farbre, K. Zeger, Quantizers with uniform decoders and channel-optimized encoders. IEEE Trans. Inf. Theory **52**(2), 640–661 (2006)

116. N. Farvardin, A study of vector quantization for noisy channels. IEEE Trans. Inf. Theory **36**(4), 799–809 (1990)

117. N. Farvardin, V. Vaishampayan, On the performance and complexity of channel-optimized vector quantizers. IEEE Trans. Inf. Theory **37**(1), 155–159 (1991)

118. T. Fazel, T. Fuja, Robust transmission of MELP-compressed speech: an illustrative example of joint source-channel decoding. IEEE Trans. Commun. **51**(6), 973–982 (2003)

119. W. Feller, *An Introduction to Probability Theory and its Applications*, vol. I, 3rd edn. (Wiley, New York, 1970)

120. W. Feller, *An Introduction to Probability Theory and its Applications*, vol. II, 2nd edn. (Wiley, New York, 1971)

121. T. Fine, Properties of an optimum digital system and applications. IEEE Trans. Inf. Theory **10**, 443–457 (1964)

122. T. Fingscheidt, T. Hindelang, R.V. Cox, N. Seshadri, Joint source-channel (de)coding for mobile communications. IEEE Trans. Commun. **50**, 200–212 (2002)

123. F. Fleuret, Fast binary feature selection with conditional mutual information. J. Mach. Learn. Res. **5**, 1531–1555 (2004)

124. M. Fossorier, Z. Xiong, K. Zeger, Progressive source coding for a power constrained Gaussian channel. IEEE Trans. Commun. **49**(8), 1301–1306 (2001)

125. B.D. Fritchman, A binary channel characterization using partitioned Markov chains. IEEE Trans. Inf. Theory **13**(2), 221–227 (1967)

126. M. Fresia, G. Caire, A linear encoding approach to index assignment in lossy source-channel coding. IEEE Trans. Inf. Theory **56**(3), 1322–1344 (2010)

127. M. Fresia, F. Perez-Cruz, H.V. Poor, S. Verdú, Joint source-channel coding. IEEE Signal Process. Mag. **27**(6), 104–113 (2010)

128. S.H. Friedberg, A.J. Insel, L.E. Spence, *Linear Algebra*, 4th edn. (Prentice Hall, 2002)

129. T.E. Fuja, C. Heegard, Focused codes for channels with skewed errors. IEEE Trans. Inf. Theory **36**(9), 773–783 (1990)

130. A. Fuldseth, T.A. Ramstad, Bandwidth compression for continuous amplitude channels based onvector approximation to a continuous subset of the source signal space, in *Proceedings IEEE International Conference on Acoustics, Speech and Signal Processing*, Munich, Germany, Apr 1997, pp. 3093–3096

131. S. Gadkari, K. Rose, Robust vector quantizer design by noisy channel relaxation. IEEE Trans. Commun. **47**(8), 1113–1116 (1999)

132. S. Gadkari, K. Rose, Unequally protected multistage vector quantization for time-varying CDMA channels. IEEE Trans. Commun. **49**(6), 1045–1054 (2001)

133. R.G. Gallager, Low-density parity-check codes. IRE Trans. Inf. Theory **28**(1), 8–21 (1962)

134. R.G. Gallager, *Low-Density Parity-Check Codes* (MIT Press, 1963)

135. R.G. Gallager, *Information Theory and Reliable Communication* (Wiley, New York, 1968)

136. R.G. Gallager, Variations on a theme by Huffman. IEEE Trans. Inf. Theory **24**(6), 668–674 (1978)

137. R.G. Gallager, *Discrete Stochastic Processes* (Kluwer Academic, Boston, 1996)

138. Y. Gao, E. Tuncel, New hybrid digital/analog schemes for transmission of a Gaussian source over a Gaussian channel. IEEE Trans. Inf. Theory **56**(12), 6014–6019 (2010)

139. J. Garcia-Frias, J.D. Villasenor, Joint Turbo decoding and estimation of hidden Markov sources. IEEE J. Sel. Areas Commun. **19**, 1671–1679 (2001)

140. M. Gastpar, B. Rimoldi, M. Vetterli, To code, or not to code: lossy source-channel communication revisited. IEEE Trans. Inf. Theory **49**, 1147–1158 (2003)

141. M. Gastpar, Uncoded transmission is exactly optimal for a simple Gaussian sensor network. IEEE Trans. Inf. Theory **54**(11), 5247–5251 (2008)

142. A. Gersho, R.M. Gray, *Vector Quantization and Signal Compression* (Kluwer Academic Press/Springer, 1992)

143. J.D. Gibson, T.R. Fisher, Alphabet-constrained data compression. IEEE Trans. Inf. Theory **28**, 443–457 (1982)

144. M. Gil, F. Alajaji, T. Linder, Rényi divergence measures for commonly used univariate continuous distributions. Inf. Sci. **249**, 124–131 (2013)

145. E.N. Gilbert, Capacity of a burst-noise channel. Bell Syst. Tech. J. **39**, 1253–1266 (1960)

146. J. Gleick, *The Information: A History, a Theory and a Flood* (Pantheon Books, New York, 2011)

147. T. Goblick Jr., Theoretical limitations on the transmission of data from analog sources. IEEE Trans. Inf. Theory **11**(4), 558–567 (1965)

148. N. Goela, E. Abbe, M. Gastpar, Polar codes for broadcast channels. IEEE Trans. Inf. Theory **61**(2), 758–782 (2015)

149. N. Görtz, On the iterative approximation of optimal joint source-channel decoding. IEEE J. Sel. Areas Commun. **19**(9), 1662–1670 (2001)

150. N. Görtz, *Joint Source-Channel Coding of Discrete-Time Signals with Continuous Amplitudes* (Imperial College Press, London, UK, 2007)

151. A. Goldsmith, *Wireless Communications* (Cambridge University Press, UK, 2005)
152. A. Goldsmith, M. Effros, Joint design of fixed-rate source codes and multiresolution channel codes. IEEE Trans. Commun. **46**(10), 1301–1312 (1998)
153. A. Goldsmith, S.A. Jafar, N. Jindal, S. Vishwanath, Capacity limits of MIMO channels. IEEE J. Sel. Areas Commun. **21**(5), 684–702 (2003)
154. L. Golshani, E. Pasha, Rényi entropy rate for Gaussian processes. Inf. Sci. **180**(8), 1486–1491 (2010)
155. M. Grangetto, P. Cosman, G. Olmo, Joint source/channel coding and MAP decoding of arithmetic codes. IEEE Trans. Commun. **53**(6), 1007–1016 (2005)
156. R.M. Gray, Information rates for autoregressive processes. IEEE Trans. Inf. Theory **16**(4), 412–421 (1970)
157. R.M. Gray, *Entropy and Information Theory* (Springer, New York, 1990)
158. R.M. Gray, *Source Coding Theory* (Kluwer Academic Press/Springer, 1990)
159. R.M. Gray, *Probability, Random Processes, and Ergodic Properties* (Springer, Berlin, 1988), last revised 2010
160. R.M. Gray, L.D. Davisson, *Random Processes: A Mathematical Approach for Engineers* (Prentice-Hall, 1986)
161. R.M. Gray, D.S. Ornstein, Sliding-block joint source/noisy-channel coding theorems. IEEE Trans. Inf. Theory **22**, 682–690 (1976)
162. G.R. Grimmett, D.R. Stirzaker, *Probability and Random Processes*, 3rd edn. (Oxford University Press, New York, 2001)
163. S.F. Gull, J. Skilling, Maximum entropy method in image processing. IEE Proc. F (Commun. Radar Signal Process.) **131**(6), 646–659 (1984)
164. D. Gunduz, E. Erkip, Joint source-channel codes for MIMO block-fading channels. IEEE Trans. Inf. Theory **54**(1), 116–134 (2008)
165. D. Guo, S. Shamai, S. Verdú, Mutual information and minimum mean-square error in Gaussian channels. IEEE Trans. Inf. Theory **51**(4), 1261–1283 (2005)
166. A. Guyader, E. Fabre, C. Guillemot, M. Robert, Joint source-channel Turbo decoding of entropy-coded sources. IEEE J. Sel. Areas Commun. **19**(9), 1680–1696 (2001)
167. R. Hagen, P. Hedelin, Robust vector quantization by a linear mapping of a block code. IEEE Trans. Inf. Theory **45**(1), 200–218 (1999)
168. J. Hagenauer, R. Bauer, The Turbo principle in joint source channel decoding of variable length codes, in *Proceedings IEEE Information Theory Workshop*, Sept 2001, pp. 33–35
169. J. Hagenauer, Source-controlled channel decoding. IEEE Trans. Commun. **43**, 2449–2457 (1995)
170. B. Hajek, *Random Processes for Engineers* (Cambridge University Press, Cambridge, 2015)
171. R. Hamzaoui, V. Stankovic, Z. Xiong, Optimized error protection of scalable image bit streams. IEEE Signal Process. Mag. **22**(6), 91–107 (2005)
172. T.S. Han, *Information-Spectrum Methods in Information Theory* (Springer, Berlin, 2003)
173. T.S. Han, Multicasting multiple correlated sources to multiple sinks over a noisy channel network. IEEE Trans. Inf. Theory **57**(1), 4–13 (2011)
174. T.S. Han, M.H.M. Costa, Broadcast channels with arbitrarily correlated sources. IEEE Trans. Inf. Theory **33**(5), 641–650 (1987)
175. T.S. Han, S. Verdú, Approximation theory of output statistics. IEEE Trans. Inf. Theory **39**(3), 752–772 (1993)
176. G.H. Hardy, J.E. Littlewood, G. Polya, *Inequalities* (Cambridge University Press, Cambridge, 1934)
177. E.A. Haroutunian, Estimates of the error probability exponent for a semi-continuous memoryless channel (in Russian). Probl. Inf. Transm. **4**(4), 37–48 (1968)
178. E.A. Haroutunian, M.E. Haroutunian, A.N. Harutyunyan, Reliability criteria in information theory and in statistical hypothesis testing. Found. Trends Commun. Inf. Theory **4**(2–3), 97–263 (2008)
179. J. Harte, T. Zillio, E. Conlisk, A.B. Smith, Maximum entropy and the state-variable approach to macroecology. Ecology **89**(10), 2700–2711 (2008)

180. R.V.L. Hartley, Transmission of information. Bell Syst. Tech. J. **7**, 535 (1928)
181. B. Hayes, C. Wilson, A maximum entropy model of phonotactics and phonotactic learning. Linguist. Inq. **39**(3), 379–440 (2008)
182. M. Hayhoe, F. Alajaji, B. Gharesifard, A Polya urn-based model for epidemics on networks, in *Proceedings of American Control Conference*, Seattle, May 2017, pp. 358–363
183. A. Hedayat, A. Nosratinia, Performance analysis and design criteria for finite-alphabet source/channel codes. IEEE Trans. Commun. **52**(11), 1872–1879 (2004)
184. S. Heinen, P. Vary, Source-optimized channel coding for digital transmission channels. IEEE Trans. Commun. **53**(4), 592–600 (2005)
185. F. Hekland, P.A. Floor, T.A. Ramstad, Shannon-Kotelnikov mappings in joint source-channel coding. IEEE Trans. Commun. **57**(1), 94–105 (2009)
186. M. Hellman, J. Raviv, Probability of error, equivocation and the Chernoff bound. IEEE Trans. Inf. Theory **16**(4), 368–372 (1970)
187. M.E. Hellman, Convolutional source encoding. IEEE Trans. Inf. Theory **21**, 651–656 (1975)
188. M. Hirvensalo, *Quantum Computing* (Springer, Berlin, 2013)
189. B. Hochwald, K. Zeger, Tradeoff between source and channel coding. IEEE Trans. Inf. Theory **43**, 1412–1424 (1997)
190. T. Holliday, A. Goldsmith, H.V. Poor, Joint source and channel coding for MIMO systems: is it better to be robust or quick? IEEE Trans. Inf. Theory **54**(4), 1393–1405 (2008)
191. G.D. Hu, On Shannon theorem and its converse for sequence of communication schemes in the case of abstract random variables, *Transactions of 3rd Prague Conference on Information Theory, Statistical Decision Functions, Random Processes* (Czechoslovak Academy of Sciences, Prague, 1964), pp. 285–333
192. T.C. Hu, D.J. Kleitman, J.K. Tamaki, Binary trees optimum under various criteria. SIAM J. Appl. Math. **37**(2), 246–256 (1979)
193. Y. Hu, J. Garcia-Frias, M. Lamarca, Analog joint source-channel coding using non-linear curves and MMSE decoding. IEEE Trans. Commun. **59**(11), 3016–3026 (2011)
194. J. Huang, S. Meyn, M. Medard, Error exponents for channel coding with application to signal constellation design. IEEE J. Sel. Areas Commun. **24**(8), 1647–1661 (2006)
195. D.A. Huffman, A method for the construction of minimum redundancy codes. Proc. IRE **40**, 1098–1101 (1952)
196. S. Ihara, *Information Theory for Continuous Systems* (World-Scientific, Singapore, 1993)
197. I. Issa, S. Kamath, A.B. Wagner, An operational measure of information leakage, in *Proceedings of the Conference on Information Sciences and Systems*, Princeton University, Mar 2016, pp. 234–239
198. H. Jafarkhani, N. Farvardin, Design of channel-optimized vector quantizers in the presence of channel mismatch. IEEE Trans. Commun. **48**(1), 118–124 (2000)
199. K. Jacobs, Almost periodic channels, *Colloquium on Combinatorial Methods in Probability Theory*, Aarhus, 1962, pp. 118–126
200. X. Jaspar, C. Guillemot, L. Vandendorpe, Joint source-channel turbo techniques for discrete-valued sources: from theory to practice. Proc. IEEE **95**, 1345–1361 (2007)
201. E.T. Jaynes, Information theory and statistical mechanics. Phys. Rev. **106**(4), 620–630 (1957)
202. E.T. Jaynes, Information theory and statistical mechanics II. Phys. Rev. **108**(2), 171–190 (1957)
203. E.T. Jaynes, On the rationale of maximum-entropy methods. Proc. IEEE **70**(9), 939–952 (1982)
204. M. Jeanne, J.-C. Carlach, P. Siohan, Joint source-channel decoding of variable-length codes for convolutional codes and Turbo codes. IEEE Trans. Commun. **53**(1), 10–15 (2005)
205. F. Jelinek, *Probabilistic Information Theory* (McGraw Hill, 1968)
206. F. Jelinek, Buffer overflow in variable length coding of fixed rate sources. IEEE Trans. Inf. Theory **14**, 490–501 (1968)
207. V.D. Jerohin, ϵ-entropy of discrete random objects. Teor. Veroyatnost. i Primenen **3**, 103–107 (1958)
208. R. Johanesson, K. Zigangirov, *Fundamentals of Convolutional Coding* (IEEE, 1999)

209. O. Johnson, *Information Theory and the Central Limit Theorem* (Imperial College Press, London, 2004)
210. N.L. Johnson, S. Kotz, *Urn Models and Their Application: An Approach to Modern Discrete Probability Theory* (Wiley, New York, 1977)
211. L.N. Kanal, A.R.K. Sastry, Models for channels with memory and their applications to error control. Proc. IEEE **66**(7), 724–744 (1978)
212. W. Karush, *Minima of Functions of Several Variables with Inequalities as Side Constraints*, M.Sc. Dissertation, Department of Mathematics, University of Chicago, Chicago, Illinois, 1939
213. A. Khisti, G. Wornell, Secure transmission with multiple antennas I: The MISOME wiretap channel. IEEE Trans. Inf. Theory **56**(7), 3088–3104 (2010)
214. A. Khisti, G. Wornell, Secure transmission with multiple antennas II: the MIMOME wiretap channel. IEEE Trans. Inf. Theory **56**(11), 5515–5532 (2010)
215. Y.H. Kim, A coding theorem for a class of stationary channels with feedback. IEEE Trans. Inf. Theory **54**(4), 1488–1499 (2008)
216. Y.H. Kim, A. Sutivong, T.M. Cover, State amplification. IEEE Trans. Inf. Theory **54**(5), 1850–1859 (2008)
217. J. Kliewer, R. Thobaben, Iterative joint source-channel decoding of variable-length codes using residual source redundancy. IEEE Trans. Wireless Commun. **4**(3), 919–929 (2005)
218. P. Knagenhjelm, E. Agrell, The Hadamard transform—a tool for index assignment. IEEE Trans. Inf. Theory **42**(4), 1139–1151 (1996)
219. Y. Kochman, R. Zamir, Analog matching of colored sources to colored channels. IEEE Trans. Inf. Theory **57**(6), 3180–3195 (2011)
220. Y. Kochman, G. Wornell, On uncoded transmission and blocklength, in *Proceedings IEEE Information Theory Workshop*, Sept 2012, pp. 15–19
221. E. Koken, E. Tuncel, On robustness of hybrid digital/analog source-channel coding with bandwidth mismatch. IEEE Trans. Inf. Theory **61**(9), 4968–4983 (2015)
222. A.N. Kolmogorov, On the Shannon theory of information transmission in the case of continuous signals. IEEE Trans. Inf. Theory **2**(4), 102–108 (1956)
223. A.N. Kolmogorov, A new metric invariant of transient dynamical systems and automorphisms, *Lebesgue Spaces.18 Dokl. Akad. Nauk. SSSR*, 119.61-864 (1958)
224. A.N. Kolmogorov, S.V. Fomin, *Introductory Real Analysis* (Dover Publications, New York, 1970)
225. L.H. Koopmans, Asymptotic rate of discrimination for Markov processes. Ann. Math. Stat. **31**, 982–994 (1960)
226. S.B. Korada, *Polar Codes for Channel and Source Coding*, Ph.D. Dissertation, EPFL, Lausanne, Switzerland, 2009
227. S.B. Korada, R.L. Urbanke, Polar codes are optimal for lossy source coding. IEEE Trans. Inf. Theory **56**(4), 1751–1768 (2010)
228. S.B. Korada, E. Şaşoğlu, R. Urbanke, Polar codes: characterization of exponent, bounds, and constructions. IEEE Trans. Inf. Theory **56**(12), 6253–6264 (2010)
229. I. Korn, J.P. Fonseka, S. Xing, Optimal binary communication with nonequal probabilities. IEEE Trans. Commun. **51**(9), 1435–1438 (2003)
230. V.N. Koshelev, Direct sequential encoding and decoding for discrete sources. IEEE Trans. Inf. Theory **19**, 340–343 (1973)
231. V. Kostina, S. Verdú, Lossy joint source-channel coding in the finite blocklength regime. IEEE Trans. Inf. Theory **59**(5), 2545–2575 (2013)
232. V.A. Kotelnikov, *The Theory of Optimum Noise Immunity* (McGraw-Hill, New York, 1959)
233. G. Kramer, *Directed Information for Channels with Feedback*, Ph.D. Dissertation, ser. ETH Series in Information Processing. Konstanz, Switzerland: Hartung-Gorre Verlag, vol. 11 (1998)
234. J. Kroll, N. Phamdo, Analysis and design of trellis codes optimized for a binary symmetric Markov source with MAP detection. IEEE Trans. Inf. Theory **44**(7), 2977–2987 (1998)

235. H.W. Kuhn, A.W. Tucker, Nonlinear programming, in *Proceedings of 2nd Berkeley Symposium*, Berkeley, University of California Press, 1951, pp. 481–492

236. S. Kullback, R.A. Leibler, On information and sufficiency. Ann. Math. Stat. **22**(1), 79–86 (1951)

237. S. Kullback, *Information Theory and Statistics* (Wiley, New York, 1959)

238. H. Kumazawa, M. Kasahara, T. Namekawa, A construction of vector quantizers for noisy channels. Electron. Eng. Jpn. **67–B**(4), 39–47 (1984)

239. A. Kurtenbach, P. Wintz, Quantizing for noisy channels. IEEE Trans. Commun. Technol. **17**, 291–302 (1969)

240. F. Lahouti, A.K. Khandani, Efficient source decoding over memoryless noisy channels using higher order Markov models. IEEE Trans. Inf. Theory **50**(9), 2103–2118 (2004)

241. J.N. Laneman, E. Martinian, G. Wornell, J.G. Apostolopoulos, Source-channel diversity for parallel channels. IEEE Trans. Inf. Theory **51**(10), 3518–3539 (2005)

242. G.G. Langdon, An introduction to arithmetic coding. IBM J. Res. Dev. **28**, 135–149 (1984)

243. G.G. Langdon, J. Rissanen, A simple general binary source code. IEEE Trans. Inf. Theory **28**(5), 800–803 (1982)

244. K.H. Lee, D. Petersen, Optimal linear coding for vector channels. IEEE Trans. Commun. **24**(12), 1283–1290 (1976)

245. J.M. Lervik, A. Grovlen, T.A. Ramstad, Robust digital signal compression and modulation exploiting the advantages of analog communications, in *Proceedings of IEEEGLOBECOM*, Nov 1995, pp. 1044–1048

246. F. Liese, I. Vajda, *Convex Statistical Distances*, Treubner, 1987

247. J. Lim, D.L. Neuhoff, Joint and tandem source-channel coding with complexity and delay constraints. IEEE Trans. Commun. **51**(5), 757–766 (2003)

248. S. Lin, D.J. Costello, *Error Control Coding: Fundamentals and Applications*, 2nd edn. (Prentice Hall, Upper Saddle River, NJ, 2004)

249. T. Linder, R. Zamir, On the asymptotic tightness of the Shannon lower bound. IEEE Trans. Inf. Theory **40**(6), 2026–2031 (1994)

250. A. Lozano, A.M. Tulino, S. Verdú, Optimum power allocation for parallel Gaussian channels with arbitrary input distributions. IEEE Trans. Inf. Theory **52**(7), 3033–3051 (2006)

251. D.J.C. MacKay, R.M. Neal, Near Shannon limit performance of low density parity check codes. Electron. Lett. **33**(6) (1997)

252. D.J.C. MacKay, Good error correcting codes based on very sparse matrices. IEEE Trans. Inf. Theory **45**(2), 399–431 (1999)

253. D.J.C. MacKay, *Information Theory, Inference and Learning Algorithms* (Cambridge University Press, Cambridge, 2003)

254. F.J. MacWilliams, N.J.A. Sloane, *The Theory of Error Correcting Codes* (North-Holland Pub. Co., 1978)

255. U. Madhow, *Fundamentals of Digital Communication* (Cambridge University Press, Cambridge, 2008)

256. H. Mahdavifar, A. Vardy, Achieving the secrecy capacity of wiretap channels using polar codes. IEEE Trans. Inf. Theory **57**(10), 6428–6443 (2011)

257. H.M. Mahmoud, *Polya Urn Models* (Chapman and Hall/CRC, 2008)

258. S. Mallat, A theory for multiresolution signal decomposition: the wavelet representation. IEEE Trans. Pattern Anal. Mach. Intell. **11**(7), 674–693 (1989)

259. C.D. Manning, H. Schütze, *Foundations of Statistical Natural Language Processing* (MIT Press, Cambridge, MA, 1999)

260. W. Mao, *Modern Cryptography: Theory and Practice* (Prentice Hall Professional Technical Reference, 2003)

261. H. Marko, The bidirectional communication theory—a generalization of information theory. IEEE Trans. Commun. Theory **21**(12), 1335–1351 (1973)

262. J.E. Marsden, M.J. Hoffman, *Elementary Classical Analysis* (W.H. Freeman & Company, 1993)

263. J.L. Massey, Joint source and channel coding, in *Communications and Random Process Theory*, ed. by J.K. Skwirzynski (Sijthoff and Nordhoff, The Netherlands, 1978), pp. 279–293

264. J.L. Massey, Cryptography—a selective survey, in *Digital Communications*, ed. by E. Biglieri, G. Prati (Elsevier, 1986), pp. 3–21

265. J. Massey, Causality, feedback, and directed information, in *Proceedings of International Symposium on Information Theory and Applications*, 1990, pp. 303–305

266. R.J. McEliece, *The Theory of Information and Coding*, 2nd edn. (Cambridge University Press, Cambridge, 2002)

267. B. McMillan, The basic theorems of information theory. Ann. Math. Stat. **24**, 196–219 (1953)

268. A. Méhes, K. Zeger, Performance of quantizers on noisy channels using structured families of codes. IEEE Trans. Inf. Theory **46**(7), 2468–2476 (2000)

269. N. Merhav, Shannon's secrecy system with informed receivers and its application to systematic coding for wiretapped channels. IEEE Trans. Inf. Theory **54**(6), 2723–2734 (2008)

270. N. Merhav, E. Arikan, The Shannon cipher system with a guessing wiretapper. IEEE Trans. Inf. Theory **45**(6), 1860–1866 (1999)

271. N. Merhav, S. Shamai, On joint source-channel coding for the Wyner-Ziv source and the Gel'fand-Pinsker channel. IEEE Trans. Inf. Theory **49**(11), 2844–2855 (2003)

272. D. Miller, K. Rose, Combined source-channel vector quantization using deterministic annealing. IEEE Trans. Commun. **42**, 347–356 (1994)

273. U. Mittal, N. Phamdo, Duality theorems for joint source-channel coding. IEEE Trans. Inf. Theory **46**(4), 1263–1275 (2000)

274. U. Mittal, N. Phamdo, Hybrid digital-analog (HDA) joint source-channel codes for broadcasting and robust communications. IEEE Trans. Inf. Theory **48**(5), 1082–1102 (2002)

275. J.W. Modestino, D.G. Daut, Combined source-channel coding of images. IEEE Trans. Commun. **27**, 1644–1659 (1979)

276. B. Moore, G. Takahara, F. Alajaji, Pairwise optimization of modulation constellations for non-uniform sources. IEEE Can. J. Electr. Comput. Eng. **34**(4), 167–177 (2009)

277. M. Mushkin, I. Bar-David, Capacity and coding for the Gilbert-Elliott channel. IEEE Trans. Inf. Theory **35**(6), 1277–1290 (1989)

278. T. Nakano, A.M. Eckford, T. Haraguchi, *Molecular Communication* (Cambridge University Press, Cambridge, 2013)

279. T. Nemetz, On the α-divergence rate for Markov-dependent hypotheses. Probl. Control Inf. Theory **3**(2), 147–155 (1974)

280. T. Nemetz, *Information Type Measures and Their Applications to Finite Decision-Problems*, Carleton Mathematical Lecture Notes, no. 17, May 1977

281. J. Neyman, E.S. Pearson, On the problem of the most efficient tests of statistical hypotheses. Philos. Trans. R. Soc. Lond. A **231**, 289–337 (1933)

282. H. Nguyen, P. Duhamel, Iterative joint source-channel decoding of VLC exploiting source semantics over realistic radio-mobile channels. IEEE Trans. Commun. **57**(6), 1701–1711 (2009)

283. A. Nosratinia, J. Lu, B. Aazhang, Source-channel rate allocation for progressive transmission of images. IEEE Trans. Commun. **51**(2), 186–196 (2003)

284. J.M. Ooi, *Coding for Channels with Feedback* (Springer, Berlin, 1998)

285. E. Ordentlich, T. Weissman, On the optimality of symbol-by-symbol filtering and denoising. IEEE Trans. Inf. Theory **52**(1), 19–40 (2006)

286. X. Pan, A. Banihashemi, A. Cuhadar, Progressive transmission of images over fading channels using rate-compatible LDPC codes. IEEE Trans. Image Process. **15**(12), 3627–3635 (2006)

287. L. Paninski, Estimation of entropy and mutual information. Neural Comput. **15**(6), 1191–1253 (2003)

288. M. Park, D. Miller, Joint source-channel decoding for variable-length encoded data by exact and approximate MAP source estimation. IEEE Trans. Commun. **48**(1), 1–6 (2000)

289. R. Pemantle, A survey of random processes with reinforcement. Probab. Surv. **4**, 1–79 (2007)

290. W.B. Pennebaker, J.L. Mitchell, *JPEG: Still Image Data Compression Standard* (Kluwer Academic Press/Springer, 1992)

291. H. Permuter, T. Weissman, A.J. Goldsmith, Finite state channels with time-invariant deterministic feedback. IEEE Trans. Inform. Theory **55**, 644–662 (2009)

292. H. Permuter, H. Asnani, T. Weissman, Capacity of a POST channel with and without feedback. IEEE Trans. Inf. Theory **60**(10), 6041–6057 (2014)

293. N. Phamdo, N. Farvardin, T. Moriya, A unified approach to tree-structured and multistage vector quantization for noisy channels. IEEE Trans. Inf. Theory **39**(3), 835–850 (1993)

294. N. Phamdo, N. Farvardin, Optimal detection of discrete Markov sources over discrete memoryless channels—applications to combined source-channel coding. IEEE Trans. Inf. Theory **40**(1), 186–193 (1994)

295. N. Phamdo, F. Alajaji, N. Farvardin, Quantization of memoryless and Gauss-Markov sources over binary Markov channels. IEEE Trans. Commun. **45**(6), 668–675 (1997)

296. N. Phamdo, F. Alajaji, Soft-decision demodulation design for COVQ over white, colored, and ISI Gaussian channels. IEEE Trans. Commun. **46**(9), 1499–1506 (2000)

297. J.R. Pierce, *An Introduction to Information Theory: Symbols, Signals and Noise*, 2nd edn. (Dover Publications Inc., New York, 1980)

298. C. Pimentel, I.F. Blake, Modeling burst channels using partitioned Fritchman's Markov models. IEEE Trans. Veh. Technol. **47**(3), 885–899 (1998)

299. C. Pimentel, T.H. Falk, L. Lisbôa, Finite-state Markov modeling of correlated Rician-fading channels. IEEE Trans. Veh. Technol. **53**(5), 1491–1501 (2004)

300. C. Pimentel, F. Alajaji, Packet-based modeling of Reed-Solomon block coded correlated fading channels via a Markov finite queue model. IEEE Trans. Veh. Technol. **58**(7), 3124–3136 (2009)

301. C. Pimentel, F. Alajaji, P. Melo, A discrete queue-based model for capturing memory and soft-decision information in correlated fading channels. IEEE Trans. Commun. **60**(5), 1702–1711 (2012)

302. J.T. Pinkston, An application of rate-distortion theory to a converse to the coding theorem. IEEE Trans. Inf. Theory **15**(1), 66–71 (1969)

303. M.S. Pinsker, *Information and Information Stability of Random Variables and Processes* (Holden-Day, San Francisco, 1964)

304. G. Polya, F. Eggenberger, Über die Statistik Verketteter Vorgänge. Z. Angew. Math. Mech. **3**, 279–289 (1923)

305. G. Polya, F. Eggenberger, Sur l'Interpretation de Certaines Courbes de Fréquences. Comptes Rendus C.R. **187**, 870–872 (1928)

306. G. Polya, Sur Quelques Points de la Théorie des Probabilités. Ann. Inst. H. Poincarré **1**, 117–161 (1931)

307. J.G. Proakis, *Digital Communications* (McGraw Hill, 1983)

308. L. Pronzato, H.P. Wynn, A.A. Zhigljavsky, Using Rényi entropies to measure uncertainty in search problems. Lect. Appl. Math. **33**, 253–268 (1997)

309. Z. Rached, *Information Measures for Sources with Memory and their Application to Hypothesis Testing and Source Coding*, Doctoral dissertation, Queen's University, 2002

310. Z. Rached, F. Alajaji, L.L. Campbell, Rényi's entropy rate for discrete Markov sources, in *Proceedings on Conference of Information Sciences and Systems*, Baltimore, Mar 1999

311. Z. Rached, F. Alajaji, L.L. Campbell, Rényi's divergence and entropy rates for finite alphabet Markov sources. IEEE Trans. Inf. Theory **47**(4), 1553–1561 (2001)

312. Z. Rached, F. Alajaji, L.L. Campbell, The Kullback-Leibler divergence rate between Markov sources. IEEE Trans. Inf. Theory **50**(5), 917–921 (2004)

313. M. Raginsky, I. Sason, Concentration of measure inequalities in information theory, communications, and coding, Found. Trends Commun. Inf. Theory. **10**(1-2), 1–246, Now Publishers, Oct 2013

314. T.A. Ramstad, Shannon mappings for robust communication. Telektronikk **98**(1), 114–128 (2002)

315. R.C. Reininger, J.D. Gibson, Distributions of the two-dimensional DCT coefficients for images. IEEE Trans. Commun. **31**(6), 835–839 (1983)

316. A. Rényi, On the dimension and entropy of probability distributions. Acta Math. Acad. Sci. Hung. **10**, 193–215 (1959)
317. A. Rényi, On measures of entropy and information, in *Proceedings of the Fourth Berkeley Symposium on Mathematical Statistics Probability*, vol. 1 (University of California Press, Berkeley, 1961), pp. 547–561
318. A. Rényi, On the foundations of information theory. Rev. Inst. Int. Stat. **33**, 1–14 (1965)
319. M. Rezaeian, A. Grant, Computation of total capacity for discrete memoryless multiple-access channels. IEEE Trans. Inf. Theory **50**(11), 2779–2784 (2004)
320. Z. Reznic, M. Feder, R. Zamir, Distortion bounds for broadcasting with bandwidth expansion. IEEE Trans. Inf. Theory **52**(8), 3778–3788 (2006)
321. T.J. Richardson, R.L. Urbanke, *Modern Coding Theory* (Cambridge University Press, Cambridge, 2008)
322. J. Rissanen, Generalized Kraft inequality and arithmetic coding. IBM J. Res. Dev. **20**, 198–203 (1976)
323. H.L. Royden, *Real Analysis*, 3rd edn. (Macmillan Publishing Company, New York, 1988)
324. M. Rüngeler, J. Bunte, P. Vary, Design and evaluation of hybrid digital-analog transmission outperforming purely digital concepts. IEEE Trans. Commun. **62**(11), 3983–3996 (2014)
325. P. Sadeghi, R.A. Kennedy, P.B. Rapajic, R. Shams, Finite-state Markov modeling of fading channels. IEEE Signal Process. Mag. **25**(5), 57–80 (2008)
326. D. Salomon, *Data Compression: The Complete Reference*, 3rd edn. (Springer, Berlin, 2004)
327. L. Sankar, S.R. Rajagopalan, H.V. Poor, Utility-privacy tradeoffs in databases: an information-theoretic approach. IEEE Trans. Inf. Forensic Secur. **8**(6), 838–852 (2013)
328. L. Sankar, S.R. Rajagopalan, S. Mohajer, H.V. Poor, Smart meter privacy: a theoretical framework. IEEE Trans. Smart Grid **4**(2), 837–846 (2013)
329. E. Şaşoğlu, Polarization and polar codes. Found. Trends Commun. Inf. Theory **8**(4), 259–381 (2011)
330. K. Sayood, *Introduction to Data Compression*, 4th edn. (Morgan Kaufmann, 2012)
331. K. Sayood, J.C. Borkenhagen, Use of residual redundancy in the design of joint source/channel coders. IEEE Trans. Commun. **39**, 838–846 (1991)
332. L. Schmalen, M. Adrat, T. Clevorn, P. Vary, EXIT chart based system design for iterative source-channel decoding with fixed-length codes. IEEE Trans. Commun. **59**(9), 2406–2413 (2011)
333. N. Sen, F. Alajaji, S. Yüksel, Feedback capacity of a class of symmetric finite-state Markov channels. IEEE Trans. Inf. Theory **57**, 4110–4122 (2011)
334. S. Shahidi, F. Alajaji, T. Linder, MAP detection and robust lossy coding over soft-decision correlated fading channels. IEEE Trans. Veh. Technol. **62**(7), 3175–3187 (2013)
335. S. Shamai, S. Verdú, R. Zamir, Systematic lossy source/channel coding. IEEE Trans. Inf. Theory **44**, 564–579 (1998)
336. G.I. Shamir, K. Xie, Universal source controlled channel decoding with nonsystematic quick-look-in Turbo codes. IEEE Trans. Commun. **57**(4), 960–971 (2009)
337. C.E. Shannon, A symbolic analysis of relay and switching circuits. Trans. Am. Inst. Electr. Eng. **57**(12), 713–723 (1938)
338. C.E. Shannon, *A Symbolic Analysis of Relay and Switching Circuits*, M.Sc. Thesis, Department of Electrical Engineering, MIT, 1940
339. C.E. Shannon, *An Algebra for Theoretical Genetics*, Ph.D. Dissertation, Department of Mathematics, MIT, 1940
340. C.E. Shannon, A mathematical theory of communications. Bell Syst. Tech. J. **27**, 379–423 and 623–656 (1948)
341. C.E. Shannon, Communication in the presence of noise. Proc. IRE **37**, 10–21 (1949)
342. C.E. Shannon, Communication theory of secrecy systems. Bell Syst. Tech. J. **28**, 656–715 (1949)
343. C.E. Shannon, The zero-error capacity of a noisy channel. IRE Trans. Inf. Theory **2**, 8–19 (1956)

344. C.E. Shannon, Certain results in coding theory for noisy channels. Inf. Control **1**(1), 6–25 (1957)
345. C.E. Shannon, Coding theorems for a discrete source with a fidelity criterion. IRE Nat. Conv. Rec. **4**, 142–163 (1959)
346. C.E. Shannon, W.W. Weaver, *The Mathematical Theory of Communication* (University of Illinois Press, Urbana, IL, 1949)
347. C.E. Shannon, R.G. Gallager, E.R. Berlekamp, Lower bounds to error probability for coding in discrete memoryless channels I. Inf. Control **10**(1), 65–103 (1967)
348. C.E. Shannon, R.G. Gallager, E.R. Berlekamp, Lower bounds to error probability for coding in discrete memoryless channels II. Inf. Control **10**(2), 523–552 (1967)
349. P.C. Shields, *The Ergodic Theory of Discrete Sample Paths* (American Mathematical Society, 1991)
350. P.C. Shields, Two divergence-rate counterexamples. J. Theor. Probab. **6**, 521–545 (1993)
351. Y. Shkel, V.Y.F. Tan, S. Draper, Unequal message protection: asymptotic and non-asymptotic tradeoffs. IEEE Trans. Inf. Theory **61**(10), 5396–5416 (2015)
352. R. Sibson, Information radius. Z. Wahrscheinlichkeitstheorie Verw. Geb. **14**, 149–161 (1969)
353. C.A. Sims, Rational inattention and monetary economics, in *Handbook of Monetary Economics*, vol. 3 (2010), pp. 155–181
354. M. Skoglund, Soft decoding for vector quantization over noisy channels with memory. IEEE Trans. Inf. Theory **45**(4), 1293–1307 (1999)
355. M. Skoglund, On channel-constrained vector quantization and index assignment for discrete memoryless channels. IEEE Trans. Inf. Theory **45**(7), 2615–2622 (1999)
356. M. Skoglund, P. Hedelin, Hadamard-based soft decoding for vector quantization over noisy channels. IEEE Trans. Inf. Theory **45**(2), 515–532 (1999)
357. M. Skoglund, N. Phamdo, F. Alajaji, Design and performance of VQ-based hybrid digital-analog joint source-channel codes. IEEE Trans. Inf. Theory **48**(3), 708–720 (2002)
358. M. Skoglund, N. Phamdo, F. Alajaji, Hybrid digital-analog source-channel coding for bandwidth compression/expansion. IEEE Trans. Inf. Theory **52**(8), 3757–3763 (2006)
359. N.J.A. Sloane, A.D. Wyner (eds.), *Claude Elwood Shannon: Collected Papers* (IEEE Press, New York, 1993)
360. K. Song, Rényi information, loglikelihood and an intrinsic distribution measure. J. Stat. Plan. Inference **93**(1–2), 51–69 (2001)
361. L. Song, F. Alajaji, T. Linder, On the capacity of burst noise-erasure channels with and without feedback, in *Proceedings of IEEE International Symposium on Information Theory*, Aachen, Germany, June 2017, pp. 206–210
362. J. Soni, R. Goodman, *A Mind at Play: How Claude Shannon Invented the Information Age* (Simon & Schuster, 2017)
363. J.F. Sowa, *Conceptual Structures: Information Processing in Mind and Machine* (Addison-Wesley Pub, MA, 1983)
364. Y. Steinberg, S. Verdú, Simulation of random processes and rate-distortion theory. IEEE Trans. Inf. Theory **42**(1), 63–86 (1996)
365. Y. Steinberg, N. Merhav, On hierarchical joint source-channel coding with degraded side information. IEEE Trans. Inf. Theory **52**(3), 886–903 (2006)
366. K.P. Subbalakshmi, J. Vaisey, On the joint source-channel decoding of variable-length encoded sources: the additive-Markov case. IEEE Trans. Commun. **51**(9), 1420–1425 (2003)
367. M. Taherzadeh, A.K. Khandani, Single-sample robust joint source-channel coding: achieving asymptotically optimum scaling of SDR versus SNR. IEEE Trans. Inf. Theory **58**(3), 1565–1577 (2012)
368. G. Takahara, F. Alajaji, N.C. Beaulieu, H. Kuai, Constellation mappings for two-dimensional signaling of nonuniform sources. IEEE Trans. Commun. **51**(3), 400–408 (2003)
369. Y. Takashima, M. Wada, H. Murakami, Reversible variable length codes. IEEE Trans. Commun. **43**, 158–162 (1995)
370. C. Tan, N.C. Beaulieu, On first-order Markov modeling for the Rayleigh fading channel. IEEE Trans. Commun. **48**(12), 2032–2040 (2000)

371. I. Tal, A. Vardy, How to construct polar codes. IEEE Trans. Inf. Theory **59**(10), 6562–6582 (2013)
372. I. Tal, A. Vardy, List decoding of polar codes. IEEE Trans. Inf. Theory **61**(5), 2213–2226 (2015)
373. V.Y.F. Tan, S. Watanabe, M. Hayashi, Moderate deviations for joint source-channel coding of systems with Markovian memory, in *Proceedings IEEE Symposium on Information Theory*, Honolulu, HI, June 2014, pp. 1687–1691
374. A. Tang, D. Jackson, J. Hobbs, W. Chen, J.L. Smith, H. Patel, A. Prieto, D. Petrusca, M.I. Grivich, A. Sher, P. Hottowy, W. Davrowski, A.M. Litke, J.M. Beggs, A maximum entropy model applied to spatial and temporal correlations from cortical networks in vitro. J. Neurosci. **28**(2), 505–518 (2008)
375. N. Tanabe, N. Farvardin, Subband image coding using entropy-coded quantization over noisy channels. IEEE J. Sel. Areas Commun. **10**(5), 926–943 (1992)
376. S. Tatikonda, *Control Under Communication Constraints*, Ph.D. Dissertation, MIT, 2000
377. S. Tatikonda, S. Mitter, Control under communication constraints. IEEE Trans. Autom. Control **49**(7), 1056–1068 (2004)
378. S. Tatikonda, S. Mitter, The capacity of channels with feedback. IEEE Trans. Inf. Theory **55**, 323–349 (2009)
379. H. Theil, *Economics and Information Theory* (North-Holland, Amsterdam, 1967)
380. I.E. Telatar, Capacity of multi-antenna Gaussian channels. Eur. Trans. Telecommun. **10**(6), 585–596 (1999)
381. R. Thobaben, J. Kliewer, An efficient variable-length code construction for iterative source-channel decoding. IEEE Trans. Commun. **57**(7), 2005–2013 (2009)
382. C. Tian, S. Shamai, A unified coding scheme for hybrid transmission of Gaussian source over Gaussian channel, in *Proceedings International Symposium on Information Theory*, Toronto, Canada, July 2008, pp. 1548–1552
383. C. Tian, J. Chen, S.N. Diggavi, S. Shamai, Optimality and approximate optimality of source-channel separation in networks. IEEE Trans. Inf. Theory **60**(2), 904–918 (2014)
384. N. Tishby, F.C. Pereira, W. Bialek, The information bottleneck method, in *Proceedings of 37th Annual Allerton Conference on Communication, Control, and Computing* (1999), pp. 368–377
385. N. Tishby, N. Zaslavsky, Deep learning and the information bottleneck principle, in *Proceedings IEEE Information Theory Workshop*, Apr 2015, pp. 1–5
386. S. Tridenski, R. Zamir, A. Ingber, The Ziv-Zakai-Rényi bound for joint source-channel coding. IEEE Trans. Inf. Theory **61**(8), 4293–4315 (2015)
387. D.N.C. Tse, P. Viswanath, *Fundamentals of Wireless Communications* (Cambridge University Press, Cambridge, UK, 2005)
388. A. Tulino, S. Verdú, Monotonic decrease of the non-Gaussianness of the sum of independent random variables: a simple proof. IEEE Trans. Inf. Theory **52**(9), 4295–4297 (2006)
389. W. Turin, R. van Nobelen, Hidden Markov modeling of flat fading channels. IEEE J. Sel. Areas Commun. **16**, 1809–1817 (1998)
390. R.E. Ulanowicz, Information theory in ecology. Comput. Chem. **25**(4), 393–399 (2001)
391. V. Vaishampayan, S.I.R. Costa, Curves on a sphere, shift-map dynamics, and error control for continuous alphabet sources. IEEE Trans. Inf. Theory **49**(7), 1658–1672 (2003)
392. V.A. Vaishampayan, N. Farvardin, Joint design of block source codes and modulation signal sets. IEEE Trans. Inf. Theory **38**, 1230–1248 (1992)
393. I. Vajda, *Theory of Statistical Inference and Information* (Kluwer, Dordrecht, 1989)
394. S. Vembu, S. Verdú, Y. Steinberg, The source-channel separation theorem revisited. IEEE Trans. Inf. Theory **41**, 44–54 (1995)
395. S. Verdú, α-mutual information, in *Proceedings of Workshop Information Theory and Applications*, San Diego, 2015
396. S. Verdú, T.S. Han, A general formula for channel capacity. IEEE Trans. Inf. Theory **40**(4), 1147–1157 (1994)

397. S. Verdú, S. Shamai, Variable-rate channel capacity. IEEE Trans. Inf. Theory **56**(6), 2651–2667 (2010)
398. W.R. Wade, *An Introduction to Analysis* (Prentice Hall, Upper Saddle River, NJ, 1995)
399. D. Wang, A. Ingber, Y. Kochman, A strong converse for joint source-channel coding, in *Proceedings International Symposium on Information Theory*, Cambridge, MA, 2012, pp. 2117–2121
400. S.-W. Wang, P.-N. Chen, C.-H. Wang, Optimal power allocation for (N, K)-limited access channels. IEEE Trans. Inf. Theory **58**(6), 3725–3750 (2012)
401. Y. Wang, F. Alajaji, T. Linder, Hybrid digital-analog coding with bandwidth compression for Gaussian source-channel pairs. IEEE Trans. Commun. **57**(4), 997–1012 (2009)
402. T. Wang, W. Zhang, R.G. Maunder, L. Hanzo, Near-capacity joint source and channel coding of symbol values from an infinite source set using Elias Gamma error correction codes. IEEE Trans. Commun. **62**(1), 280–292 (2014)
403. W. Wang, L. Ying, J. Zhang, On the relation between identifiability, differential privacy, and mutual-information privacy. IEEE Trans. Inf. Theory **62**(9), 5018–5029 (2016)
404. T.A. Welch, A technique for high-performance data compression. Computer **17**(6), 8–19 (1984)
405. N. Wernersson, M. Skoglund, T. Ramstad, Polynomial based analog source channel codes. IEEE Trans. Commun. **57**(9), 2600–2606 (2009)
406. T. Weissman, E. Ordentlich, G. Seroussi, S. Verdú, M.J. Weinberger, Universal discrete denoising: known channel. IEEE Trans. Inf. Theory **51**, 5–28 (2005)
407. S. Wicker, *Error Control Systems for Digital Communication and Storage* (Prentice Hall, Upper Saddle RiverNJ, 1995)
408. M.M. Wilde, *Quantum Information Theory*, 2nd edn. (Cambridge University Press, Cambridge, 2017)
409. M.P. Wilson, K.R. Narayanan, G. Caire, Joint source-channel coding with side information using hybrid digital analog codes. IEEE Trans. Inf. Theory **56**(10), 4922–4940 (2010)
410. T.-Y. Wu, P.-N. Chen, F. Alajaji, Y.S. Han, On the design of variable-length error-correcting codes. IEEE Trans. Commun. **61**(9), 3553–3565 (2013)
411. A.D. Wyner, The capacity of the band-limited Gaussian channel. Bell Syst. Tech. J. **45**, 359–371 (1966)
412. A.D. Wyner, J. Ziv, Bounds on the rate-distortion function for stationary sources with memory. IEEE Trans. Inf. Theory **17**(5), 508–513 (1971)
413. A.D. Wyner, The wire-tap channel. Bell Syst. Tech. J. **54**, 1355–1387 (1975)
414. H. Yamamoto, A source coding problem for sources with additional outputs to keep secret from the receiver or wiretappers. IEEE Trans. Inf. Theory **29**(6), 918–923 (1983)
415. R.W. Yeung, *Information Theory and Network Coding* (Springer, New York, 2008)
416. S. Yong, Y. Yang, A.D. Liveris, V. Stankovic, Z. Xiong, Near-capacity dirty-paper code design: a source-channel coding approach. IEEE Trans. Inf. Theory **55**(7), 3013–3031 (2009)
417. X. Yu, H. Wang, E.-H. Yang, Design and analysis of optimal noisy channel quantization with random index assignment. IEEE Trans. Inf. Theory **56**(11), 5796–5804 (2010)
418. S. Yüksel, T. Başar, *Stochastic Networked Control Systems: Stabilization and Optimization under Information Constraints* (Springer, Berlin, 2013)
419. K.A. Zeger, A. Gersho, Pseudo-Gray coding. IEEE Trans. Commun. **38**(12), 2147–2158 (1990)
420. L. Zhong, F. Alajaji, G. Takahara, A binary communication channel with memory based on a finite queue. IEEE Trans. Inform. Theory **53**, 2815–2840 (2007)
421. L. Zhong, F. Alajaji, G. Takahara, A model for correlated Rician fading channels based on a finite queue. IEEE Trans. Veh. Technol. **57**(1), 79–89 (2008)
422. Y. Zhong, F. Alajaji, L.L. Campbell, On the joint source-channel coding error exponent for discrete memoryless systems. IEEE Trans. Inf. Theory **52**(4), 1450–1468 (2006)
423. Y. Zhong, F. Alajaji, L.L. Campbell, On the joint source-channel coding error exponent of discrete communication systems with Markovian memory. IEEE Trans. Inf. Theory **53**(12), 4457–4472 (2007)

424. Y. Zhong, F. Alajaji, L.L. Campbell, Joint source-channel coding excess distortion exponent for some memoryless continuous-alphabet systems. IEEE Trans. Inf. Theory **55**(3), 1296–1319 (2009)

425. Y. Zhong, F. Alajaji, L.L. Campbell, Error exponents for asymmetric two-user discrete memoryless source-channel coding systems. IEEE Trans. Inf. Theory **55**(4), 1487–1518 (2009)

426. G.-C. Zhu, F. Alajaji, Turbo codes for non-uniform memoryless sources over noisy channels. IEEE Commun. Lett. **6**(2), 64–66 (2002)

427. G.-C. Zhu, F. Alajaji, J. Bajcsy, P. Mitran, Transmission of non-uniform memoryless sources via non-systematic Turbo codes. IEEE Trans. Commun. **52**(8), 1344–1354 (2004)

428. G.-C. Zhu, F. Alajaji, Joint source-channel Turbo coding for binary Markov sources. IEEE Trans. Wireless Commun. **5**(5), 1065–1075 (2006)

429. J. Ziv, The behavior of analog communication systems. IEEE Trans. Inf. Theory **16**(5), 587–594 (1970)

430. J. Ziv, A. Lempel, A universal algorithm for sequential data compression. IEEE Trans. Inf. Theory **23**(3), 337–343 (1977)

431. J. Ziv, A. Lempel, Compression of individual sequences via variable-rate coding. IEEE Trans. Inf. Theory **24**(5), 530–536 (1978)

Index

© Springer Nature Singapore Pte Ltd. 2018
F. Alajaji and P.-N. Chen, *An Introduction to Single-User Information Theory*,
Springer Undergraduate Texts in Mathematics and Technology,
https://doi.org/10.1007/978-981-10-8001-2

Printed in the United States
By Bookmasters